SECURING THE FUTURE

Regional and National Programs to Support the Semiconductor Industry

Charles W. Wessner, Editor

Board on Science, Technology, and Economic Policy

Policy and Global Affairs

NATIONAL RESEARCH COUNCIL
OF THE NATIONAL ACADEMIES

THE NATIONAL ACADEMIES PRESS
Washington, D.C.
www.nap.edu

THE NATIONAL ACADEMIES PRESS 500 Fifth Street, N.W. Washington, DC 20001

NOTICE: The project that is the subject of this report was approved by the Governing Board of the National Research Council, whose members are drawn from the councils of the National Academy of Sciences, the National Academy of Engineering, and the Institute of Medicine. The members of the committee responsible for the report were chosen for their special competences and with regard for appropriate balance.

International Standard Book Number 0-309-08501-2

Limited copies are available from Board on Science, Technology, and Economic Policy, National Research Council, 500 Fifth Street, NW, W547, Washington, DC 20001; 202-334-2200.

Additional copies of this report are available from the National Academies Press, 500 Fifth Street, N.W., Lockbox 285, Washington, DC 20055; (800) 624-6242 or (202) 334-3313 (in the Washington metropolitan area); Internet, http://www.nap.edu

Printed in the United States of America

THE NATIONAL ACADEMIES

Advisers to the Nation on Science, Engineering, and Medicine

The **National Academy of Sciences** is a private, nonprofit, self-perpetuating society of distinguished scholars engaged in scientific and engineering research, dedicated to the furtherance of science and technology and to their use for the general welfare. Upon the authority of the charter granted to it by the Congress in 1863, the Academy has a mandate that requires it to advise the federal government on scientific and technical matters. Dr. Bruce M. Alberts is president of the National Academy of Sciences.

The **National Academy of Engineering** was established in 1964, under the charter of the National Academy of Sciences, as a parallel organization of outstanding engineers. It is autonomous in its administration and in the selection of its members, sharing with the National Academy of Sciences the responsibility for advising the federal government. The National Academy of Engineering also sponsors engineering programs aimed at meeting national needs, encourages education and research, and recognizes the superior achievements of engineers. Dr. Wm. A. Wulf is president of the National Academy of Engineering.

The **Institute of Medicine** was established in 1970 by the National Academy of Sciences to secure the services of eminent members of appropriate professions in the examination of policy matters pertaining to the health of the public. The Institute acts under the responsibility given to the National Academy of Sciences by its congressional charter to be an adviser to the federal government and, upon its own initiative, to identify issues of medical care, research, and education. Dr. Harvey V. Fineberg is president of the Institute of Medicine.

The **National Research Council** was organized by the National Academy of Sciences in 1916 to associate the broad community of science and technology with the Academy's purposes of furthering knowledge and advising the federal government. Functioning in accordance with general policies determined by the Academy, the Council has become the principal operating agency of both the National Academy of Sciences and the National Academy of Engineering in providing services to the government, the public, and the scientific and engineering communities. The Council is administered jointly by both Academies and the Institute of Medicine. Dr. Bruce M. Alberts and Dr. Wm. A. Wulf are chair and vice chair, respectively, of the National Research Council.

www.national-academies.org

Steering Committee for Government-Industry Partnerships for the Development of New Technologies *

Gordon Moore, *Chair*
Chairman Emeritus, *retired*
Intel Corporation

M. Kathy Behrens
Managing Director
of Medical Technology
Robertson Stephens Investment
Management
and STEP Board

Michael Borrus
Managing Director
The Petkevich Group, LLC

Iain M. Cockburn
Professor of Finance
and Economics
Boston University

Kenneth Flamm
Dean Rusk Chair
in International Affairs
LBJ School of Public Affairs
University of Texas at Austin

James F. Gibbons
Professor of Engineering
Stanford University

W. Clark McFadden
Partner
Dewey Ballantine

Burton J. McMurtry
General Partner
Technology Venture Investors

William J. Spencer, *Vice-Chair*
Chairman Emeritus
International SEMATECH, *retired*
and STEP Board

Mark B. Myers
Visiting Executive Professor of
Management
The Wharton School
University of Pennsylvania
and STEP Board

Richard Nelson
George Blumenthal Professor of
International and Public Affairs
Columbia University

Edward E. Penhoet
Chief Program Officer, Science and
Higher Education
Gordon and Betty Moore Foundation
and STEP Board

Charles Trimble
Chairman
U.S. GPS Industry Council

John P. Walker
Chairman
and Chief Executive Officer
Axys Pharmaceuticals, Inc.

Patrick Windham
President, Windham Consulting
and Lecturer, Stanford University

---

*As of October 2002.

Project Staff*

Charles W. Wessner
Study Director

Sujai J. Shivakumar
Program Officer

Adam Korobow
Program Officer

Alan Anderson
Consultant

David E. Dierksheide
Program Associate

Christopher S. Hayter
Program Associate

Tabitha M. Benney
Program Associate

McAlister T. Clabaugh
Program Associate

*As of October 2002.

For the National Research Council (NRC), this project was overseen by the Board on Science, Technology and Economic Policy (STEP), a standing board of the NRC established by the National Academies of Sciences and Engineering and the Institute of Medicine in 1991. The mandate of the STEP Board is to integrate understanding of scientific, technological, and economic elements in the formulation of national policies to promote the economic well-being of the United States. A distinctive characteristic of STEP's approach is its frequent interactions with public and private-sector decision makers. STEP bridges the disciplines of business management, engineering, economics, and the social sciences to bring diverse expertise to bear on pressing public policy questions. The members of the STEP Board* and the NRC staff are listed below.

*As of October 2002.

STEP Staff*

Stephen A. Merrill
Executive Director

Russell Moy
Senior Program Officer

Craig M. Schultz
Research Associate

Camille M. Collett
Program Associate

Christopher S. Hayter
Program Associate

David E. Dierksheide
Program Associate

Charles W. Wessner
Program Director

Sujai J. Shivakumar
Program Officer

Adam Korobow
Program Officer

McAlister T. Clabaugh
Program Associate

Tabitha M. Benney
Program Associate

*As of October 2002.

National Research Council
Board on Science, Technology, and Economic Policy

Sponsors

The National Research Council gratefully acknowledges the support of the following sponsors:

National Aeronautics and Space Administration

Office of the Director, Defense Research & Engineering

National Science Foundation

U.S. Department of Energy

Optoelectronics Industry Development Association

Office of Naval Research

National Institutes of Health

National Institute of Standards and Technology

Sandia National Laboratories

Electric Power Research Institute

International Business Machines

Kulicke and Soffa Industries

Merck and Company

Milliken Industries

Motorola

Nortel

Procter and Gamble

Silicon Valley Group, Incorporated

Advanced Micro Devices

Contents

Preface

This report was prepared by the Steering Committee for Government-Industry Partnerships for the Development of New Technologies under the auspices of the National Research Council's Board on Science, Technology, and Economic Policy (STEP). The STEP Board has undertaken a major study of the programs and issues associated with public-private collaboration for the development of new technologies. This report is one component of a multifaceted review of U.S. and foreign programs to develop new technologies, often described as public-private partnerships. The project's multidisciplinary Steering Committee, led by Gordon Moore, Chairman Emeritus of Intel, and Bill Spencer, Chairman Emeritus of SEMATECH, is charged with addressing such issues as the rationale for government-industry cooperation to develop new technologies, current practices, sectoral differences, means of evaluation, the experience of foreign-based partnerships, and the roles of government laboratories, universities and other non-profit research organizations. Overall, the study will have produced some 10 analyses of public-private partnerships.

This report focuses on public-private cooperation in the semiconductor industry.[1] The deliberations and analysis it contains are intended to improve policy makers' understanding of the diversity and scale of regional and national programs designed to support the semiconductor industry around the world. It draws together field research, empirical analysis, and the presentations and discussions of the leaders and industrial participants in the principal regional and national

[1]There have been few gatherings in a public forum of industry leaders of this caliber, from around the world, with an approved public record.

programs.[2] These same leading figures also discuss the research challenges facing the semiconductor industry. Also included are the Committee's specific recommendations concerning public support needed for research in the disciplines that underpin this enabling industry.

Semiconductors are pervasive and an importance source of productivity in the modern economy. Their rapid technological evolution—characterized by continuously increasing productivity and contemporaneously decreasing cost—has been a source of growth in emerging industries, while concurrently revitalizing more traditional industrial sectors.[3] The strong performance and development of the U.S. economy in recent years is rooted in the investment in and subsequent application of information technologies ultimately driven by modern semiconductor technology.[4] Semiconductors also play a crucial role in ensuring our national security by allowing advances in the capabilities of new devices, new technologies, and new applications for national defense. The pervasive impact of the microelectronics sector on economic growth—through improved communications, advances in health care, and better national security technologies—underscores the importance of the United States' position in semiconductor production and development.

The discussion and research in this report clarify the extent to which the SEMATECH model, developed in the United States in response to needs of the industry in the 1980s, has been emulated abroad.[5] Correspondingly, it notes the degree to which the principle of cooperative government-industry research activ-

---

[2]This report focuses on programs in Europe, Japan, Taiwan, and the United States. The scope of the analysis was expanded, through the analysis in the Introduction and the paper by Thomas Howell in this volume, to cover Korea and Singapore. Nonetheless, the Committee does not intend the report to be interpreted as a full account of all programs; the report is an overview of the some of the major programs in some of the main semiconductor-producing nations.

[3]The U.S. electronics industry, which includes semiconductors, is larger than the U.S. steel, automobile, and aerospace industries combined. As of August 2001, the semiconductor industry employed some 284,000 people in the United States alone. The industry, in turn, provides the enabling technologies for the $425 billion U.S. electronics industry. For an analysis of the role of new information technologies in the recent trends in high productivity growth, often described as the "New Economy," see Council of Economic Advisers, *Economic Report of the President,* H.Doc.107-2, Washington, D.C.: USGPO, January 2001. Also see National Research Council, *Measuring and Sustaining the New Economy, Report of a Workshop,* D. Jorgenson and C. Wessner, eds.,Washington, D.C.: National Academy Press, 2002.

[4]See National Research Council, *U.S. Industry in 2000: Studies in Competitive Performance,* Washington, D.C.: National Academy Press, 2000.

[5]As noted by Macher, Mowery, and Hodges, "The SEmiconductor MAnufacturing TECHnology (SEMATECH) consortium was created in 1987 to develop semiconductor manufacturing technology, using a combination of industry and federal government funding." See Jeffrey T. Macher, David C. Mowery, and David A. Hodges, "Semiconductors," *U.S. Industry in 2000: Studies in Competitive Performance,* David C. Mowery, ed., Washington, D.C.: National Academy Press, 1999, p. 247. Its initial membership included 14 firms constituting 80 percent of the U.S. semiconductor manufacturing industry. In their comprehensive review of the consortium, Browning and Shetler write that "at least three goals emerged in the early days of SEMATECH: (1) to improve manufacturing processes;

ity has been adopted and accelerated—often with success—in other semiconductor-producing countries and regions.

The considerable technical challenges that must be addressed by the industry, and the ambitious foreign programs designed to do this, are reminders that continued U.S. leadership cannot be taken for granted. In fact, the development of new production models, such as the foundry system, as well as increases in national support for domestic production facilities, present serious competitive challenges to the U.S. industry. Overcoming these and other challenges will require continued policy engagement and public investment through renewed attention to basic research and cooperative mechanisms such as public-private partnerships.

This type of cooperative activity to develop promising technologies is not new.[6] Indeed, beginning with the mid-1980s, the United States has undertaken a remarkably wide range of public-private partnerships in high-technology sectors.[7] There are public-private consortia of many types and multiple aims; some leverage the social benefits associated with federal R&D activity, while others seek to enhance the position of a national industry. Still other public-private consortia address the need to deploy R&D to meet other government missions.[8] The U.S. economy continues to be distinguished by the extent to which individual entrepreneurs and researchers take the lead in developing innovations and starting new businesses, yet, in doing so, they often harvest crops sown on fields made fertile by the government's long-term research investments.[9]

Americans have long held the conviction that new technologies offer the best means of meeting societal challenges, whether in the realms of defense, energy, or the environment.[10] The substantial federal investment in research and devel-

---

(2) to improve factory management; and (3) to improve the industry infrastructure, *especially the supply base of equipment and materials.*" See Larry D. Browning and Judy C. Shetler, *SEMATECH: Saving the U.S. Semiconductor Industry*, College Station: Texas A&M University Press, 2000, p. 205.

[6]For a brief summary of this tradition of partnerships, see National Research Council, *The Advanced Technology Program: Assessing Outcomes*, Charles W. Wessner, editor, Washington, D.C.: National Academy Press, 2001.

[7]See Chris Coburn and Dan Berglund, *Partnerships: A Compendium of State and Federal Cooperative Technology Programs*, Columbus, OH: Battele Press, 1995.

[8]See Albert Link, "Public/Private Partnerships as a Tool to Support Industrial R&D: Experiences in the United States." Paper prepared for the working group on Innovation Policy, Paris, 1998, p. 20. Partnerships can also be differentiated by the nature of public support. Some partnerships involve a direct transfer of funds to an industry consortium. Others focus on shared use of infrastructure, such as laboratory facilities.

[9]David B. Audretsch and Roy Thurik, *Innovation, Industry, Evolution, and Employment*, Cambridge, UK: Cambridge University Press, 1999.

[10]See Linda R. Cohen and Roger G. Noll, *The Technology Pork Barrel*, Washington, D.C.: The Brookings Institution, 1991. The authors observe that "the government's optimism about technology knows neither programmatic, partisan, nor ideological bounds" (p.1). They cite William Ophuls' observation that American public policy has a long history of technological optimism. See William Ophuls, *Ecology and the Politics of Scarcity: Prologue to a Political Theory of the Steady State*, San Francisco: Freeman, 1977.

opment reflects this conviction. Around the globe, policy makers now recognize that the breadth of potential applications of new technologies, their greater complexity, and the rising costs and technical risks of developing these new technologies require a supportive policy framework.[11] Against this background, various forms of public-private cooperation are increasingly seen as effective means to bring new, welfare-enhancing and wealth-generating technologies to the market.[12]

THE ROLE OF THE BOARD ON SCIENCE, TECHNOLOGY, AND ECONOMIC POLICY

Since 1991 the National Research Council's Board on Science, Technology, and Economic Policy (STEP) has undertaken a program of activities to improve policy makers' understanding of the interconnections of science, technology, and economic policy and their importance for the American economy and its international competitive position. The Board's activities have corresponded with increased recognition by policy makers of the importance of technology to economic growth. The new economic growth theory emphasizes the role of technology creation, which is believed to be characterized by significant growth externalities.[13] A consequence of the renewed appreciation of growth externalities is recognition of the economic geography of economic development. With growth externalities coming about in part from the exchanges of knowledge among innovators, certain regions become centers for particular types of high-growth activities.[14]

Some economic analysis suggests that high technology is often characterized

[11]For a review of the policies and programs to support the development of national industries, see the paper by Thomas Howell, "Competing Programs: Government Support for Microelectronics," in this volume.

[12]See David Vogel, *Kindred Strangers: The Uneasy Relationship Between Politics and Business in America,* Princeton: Princeton University Press, pp. 113-137, 1996. Vogel notes that arguments, both for and against government participation in the development of new technologies, largely overlook the prevailing tradition in U.S. industrial policy. He points out that, given the constraints of the American federal system and the strength of private capital markets, U.S. industrial policy focuses more on government-industry partnerships, in contrast to the direct subsidies or government ownership found in other countries.

[13]Paul Romer, "Endogenous Technological Change," *Journal of Political Economy,* 98(5):71-102, 1990. See also Gene Grossman and Elhanan Helpman, *Innovation and Growth in the Global Economy,* Cambridge, MA: MIT Press, 1993.

[14]Paul Krugman, *Geography and Trade*, Cambridge, MA: MIT Press, 1991, p. 23, points out that the British economist Alfred Marshall initially observed in his classic, *Principles of Economics,* how geographic clusters of specific economic activities arose from the exchange of "tacit" knowledge among businesses.

by increasing rather than decreasing returns, justifying to some the proposition that governments can capture permanent advantage in key industries by providing relatively small but potentially decisive support to bring national industries up the learning curve and down the cost curve.[15] In part, this is why the economic literature now recognizes the relationship between technology policy and trade policy.[16] Recognition of these linkages and the corresponding ability of governments to shift comparative advantage in favor of the national economy provide intellectual underpinning for government support for high-technology industry.[17] Another widely recognized rationale for government support for high technology exists in cases in which a technology generates benefits which cannot be fully captured by the innovating firms. These benefits to other firms in the economy are often referred to as spillovers.[18] There are also cases in which the cost of a given technology may be prohibitive for individual companies, even though potential benefits to society are substantial and widespread.[19]

EARLY PARTNERSHIPS

Recognition of the benefits of new technologies and the need to provide incentives to the private sector to develop them dates back to the origins of the Republic.[20] Driven by the exigencies of national defense and the requirements of

---

[15]Paul Krugman, *Rethinking International Trade*, Cambridge, MA: MIT Press, 1990.

[16]In addition to Krugman, *Ibid*, see J. A. Brander and B. J. Spencer, "International R&D Rivalry and Industrial Strategy," *Review of Economic Studies,* 50(4):707-722, 1983, and "Export Subsidies and International Market Share Rivalry," *Journal of International Economics*, 18(1-2):83-100, 1985. See also A. K. Dixit and A. S. Kyle, "The Use of Protection and Subsidies for Entry Promotion and Deterrence," *American Economic Review,* 75(1):139-152, 1985, and P. Krugman and M. Obstfeldt, *International Economics: Theory and Policy,* 3rd ed., New York: Addison-Wesley Publishing Company, 1994.

[17]For a review of governments' efforts to capture new technologies and the industries they spawn for their national economies, see National Research Council, *Conflict and Cooperation in National Competition for High-Technology Industry,* Washington, D.C.: National Academy Press, 1996, p. 28–40. For a critique of these efforts see Paul Krugman, *Peddling Prosperity,* New York: W. W. Norton Press, 1994.

[18]See, for example, Martin N. Baily and A. Chakrabarti, *Innovation and the Productivity Crisis,* Washington, D.C.: The Brookings Institution, 1998; and Zvi Griliches, *The Search for R&D Spillovers,* Cambridge, MA: Harvard University Press, 1990.

[19]See Ishaq Nadiri, *Innovations and Technological Spillovers,* NBER Working Paper No. 4423, 1993; and Edwin Mansfield, "Academic Research and Industrial Innovation," *Research Policy,* 20(1):1-12, 1991. See also, Council of Economic Advisers, *Supporting Research and Development to Promote Economic Growth: The Federal Government's Role,* Washington, D.C.: Executive Office of the President, 1995.

[20]The earliest articulation of the government's role with regard to the composition of the economy was Alexander Hamilton's 1791 *Report on Manufactures,* in which he urged that the federal government provide incentives to industry. Although controversial at the time and still a subject of debate today, U.S. policy has largely reflected Hamilton's belief in the benefits of an active federal role in the development of the American economy.

transportation and communication across the American continent, the federal government has played an instrumental role in developing new production techniques and technologies. To do so, government has often turned to individual entrepreneurs with innovative ideas. For example, in 1798 the federal government laid the foundation for the first machine tool industry with a contract to the inventor, Eli Whitney, for interchangeable musket parts.[21] A few decades later, in 1842, a hesitant Congress appropriated funds to demonstrate the feasibility of Samuel Morse's telegraph.[22] Both Whitney and Morse fostered significant innovations that led to whole new industries. Indeed, Morse's innovation was the first step on the road toward today's networked planet.

The support for Morse's new invention was not an isolated case. The federal government increasingly saw economic development as central to its responsibilities. Examples of federal contributions to U.S. economic development abound. The government played a key role in the development of the U.S. railway network, the growth of agriculture through the Morrill Act (1862) and the creation of the agricultural extension service, and support of industry through the creation of the National Bureau of Standards in 1901.[23]

Throughout the 20th century, the federal government had an enormous impact on the structure and composition of the economy through regulation, procurement, and a vast array of policies to support industrial and agricultural development. Between World War I and World War II, these policies included support for the development of key industries with commercial and military applications, such as radios and aircraft frames and engines. The requirements of World War

[21] Whitney missed his first delivery date for the arms and encountered substantial cost overruns, a set of events that is still familiar. However, his focus on the concept of interchangeable parts, and the machine tools to make them, was prescient. In David A. Hounshell's excellent analysis of the development of manufacturing technology in the United States, he suggests that Simeon North was ultimately the most successful in achieving interchangeability and the production of components by special-purpose machinery. See David A. Hounshell, *From the American System to Mass Production, 1800-1932*, Baltimore: Johns Hopkins University Press, 1985, p. 25-32. By the 1850s, the United States had begun to export specialized machine tools to the Enfield Arsenal in Great Britain. The British described the large-scale production of firearms, made with interchangeable parts, as "the American system of manufactures." See David C. Mowery and Nathan Rosenberg, *Paths of Innovation: Technological Change in 20th Century America*, New York: Cambridge University Press, 1998, p. 6.

[22] For a discussion of Samuel Morse's 1837 application for a grant and the congressional debate, see Irwin Lebow, *Information Highways and Byways,* New York: IEEE, 1995, pp. 9-12. For a more detailed account see Robert Luther Thompson, *Wiring a Continent: The History of the Telegraph Industry in the United States 1823-1836*, Princeton, NJ: Princeton University Press, 1947.

[23] See Richard Bingham, *Industrial Policy American Style: From Hamilton to HDTV*, New York: M.E. Sharpe, 1998, for a comprehensive review. In the case of the transcontinental railroad, Stephen Ambrose describes Abraham Lincoln as the "driving force" behind its development. Lincoln was intimately involved, helping to decide the project's route, financing, and even the gauge of the tracks: *Nothing Like It in the World: The Men Who Built the Transcontinental Railroad, 1863-1869.* New York: Simon and Schuster, 2000.

II generated a huge increase in government procurement and support for high-technology industries. At the industrial level, there were "major collaborative initiatives in pharmaceutical manufacturing, petrochemicals, synthetic rubber, and atomic weapons."[24] An impressive array of weapons based on new technologies was developed during the war, ranging from radar and improved aircraft to missiles and, not least, the atomic bomb. Many of these military technologies found civilian applications after the war.

Both during and after the war, the government made unprecedented investments in computer technology.[25] During the war it played a central role in creating the first electronic digital computers, the ENIAC and the Colossus.[26] Following the war, the federal government began to fund basic research at universities on a significant scale, first through the Office of Naval Research and later through the National Science Foundation (NSF) and the Public Health Service.[27] At the same time, the continued reluctance of commercial firms, such as IBM and NCR, to invest large sums in what they considered to be risky research and development projects with uncertain markets forced the government to continue sponsoring the development of the new technology now referred to as computers.[28] In this early phase, the National Bureau of Standards [the precursor of the National Institute of Standards and Technology (NIST)] made a significant contribution, through its SEAC machine, to the development of the modern computer.[29] Throughout the Cold War, the United States continued to emphasize

---

[24]David Mowery, "Collaborative R&D: How Effective Is It?," *Issues in Science and Technology*, Fall 1998, p. 37.

[25]Kenneth Flamm, *Creating the Computer*, Washington, D.C.: The Brookings Institution, 1988, Chapters 1-3.

[26]For a detailed account of ENIAC's creation, see Scott McCartney, *ENIAC: The Triumphs and Tragedies of the World's First Computer*, New York: Walker and Company, 1999; and Flamm, *op. cit.*, p. 39.

[27]The National Science Foundation was initially seen as the agency that would fund basic scientific research at universities after World War II. However, disagreements over the degree of Executive Branch control over the NSF delayed passage of its authorizing legislation until 1950, even though the concept for the agency was first put forth in 1945 in Vannevar Bush's report, *Science: The Endless Frontier*. The Office of Naval Research bridged the gap in basic research funding during these years. For an account of the politics of the NSF's creation, see G. Paschal Zachary, *Endless Frontier: Vannevar Bush, Engineer of the American Century*, New York: The Free Press, 1997, p. 231. See also Daniel Lee Kleinman, *Politics on the Endless Frontier: Postwar Research Policy in the United States*, Durham, NC: Duke University Press, 1995. Computer science did not, however, mature as a separate academic discipline until the 1960s. In the interim, the military supported the fledgling computer industry on national security grounds.

[28] See Flamm, *op. cit.* , p. 75.

[29]As Kenneth Flamm observes, besides being the first operational von-Neumann-type stored-program computer in the United States, the Bureau of Standards' SEAC, or Standards Eastern Automatic Computer, pioneered important technology concepts. All of the logic was implemented with newly

technological superiority as a means of ensuring U.S. security. Government funds and cost-plus contracts helped to support systems and enabling technologies such as semiconductors and new materials, radar, jet engines, advanced computer hardware and software, and missiles.

In the post-Cold War period, the evolution of the American economy continues to be profoundly marked by government-funded research in areas such as microelectronics, robotics, biotechnology, and the human genome, and through earlier investments in communications networks such as ARPANET—the forerunner of today's Internet.

PROJECT PARAMETERS

To advance our understanding of the operation and performance of partnerships, the STEP Board has undertaken a major study of programs relying on public-private collaboration for the development of new technologies. The project's multidisciplinary Steering Committee[30] includes members from academia, high-technology industries, venture capital firms, and the realm of public policy. The intent of the study is to focus on best practices rather than general questions of principle regarding the appropriateness of government involvement in partnerships. The Committee's charge is to take a pragmatic approach to address such issues as the rationale and organizing principles of government-industry cooperation to develop new technologies, current practices, sectoral differences, means of evaluation, the experience of foreign-based partnerships, and the roles of government laboratories, universities, and other non-profit research organizations.

As a program-based assessment of partnerships, the study has given particular attention to generic partnership programs such as the Small Business Innovation Research Program (SBIR) and Advanced Technology Program (ATP), and to the needs emerging from the growth in health-related funding and the relative decline in R&D support in areas such as information technologies. A series of 10 reports on these programs and topics contributes to the Committee's Summary report.

The Committee's analysis has included a significant but necessarily limited portion of the wide variety of cooperative activity that takes place between the government and the private sector.[31] The selection of specific programs to re-

---

developed germanium diodes (10,000 were used); the vacuum tubes within (750) were only for providing power and electrical pulse-shaping circuitry. The computer also used standardized, replaceable circuit modules, an innovation soon adopted throughout the industry. Thus the first computer to use solid-state logic was also the first modern computer to be completed in the United States. See Flamm, *op. cit.*, p. 74.

[30]For the Committee membership, see the front matter of this volume.

[31]For example, DARPA's programs and contributions have not been reviewed. For an indication of the scope of cooperative activity, see Coburn and Berglund, *op. cit.*; and the RaDiUS database, <HtmlResAnchor www.rand.org/services/radius/>.

view has been conditioned by the Committee's desire to carry out an analysis of current partnerships directly relevant to contemporary policy making. The Committee also recognizes the importance of placing each of the studies in the broader context of U.S. technology policy, which continues to employ a wide variety of ad hoc mechanisms developed through the government's decentralized decision-making and management process.

The Committee's desire to ensure that its deliberations and analysis are directly relevant to current policy making has allowed it to be responsive to requests from both the Executive Branch and Congress for examinations of various policies and programs of current policy relevance. These would include a White House and State Department request for an evaluation of opportunities for greater transatlantic cooperation as a result of the signature of the U.S.-E.U. Agreement on Science and Technology Cooperation, a request by the Defense Department's Under Secretary for Technology and Acquisitions to review the Fast Track initiative of the SBIR program at the Department of Defense, and a request by NIST for an assessment of the Advanced Technology Program, in compliance with Senate Report 105-235.[32] The Committee has also focused its attention on the emerging needs, synergies, and opportunities between the fields of biotechnology and computing, with special attention directed to the differences and similarities in government support for technology development in biotechnology and computing, the different uses of intellectual property in these sectors, and the need for balanced investments across disciplines. To meet its proposed objectives, the study has focused on the assessment of current and proposed programs, drawing on the experience of previous U.S. initiatives, foreign practice, and emerging areas (e.g., bioinformatics) resulting from U.S. investments in advanced technologies. A summary of the partnerships taken up by the study is included in Box A.

SUPPORT FOR ANALYSIS OF COOPERATIVE PROGRAMS

There is broad support for this type of objective analysis among federal agencies and the private sector. Government agencies supporting this analysis include the Department of Defense, the Department of Energy, the National Science Foundation, the National Institutes of Health, especially the National Cancer Institute and the National Institute of General Medical Sciences, the National Aeronautics and Space Administration, the Office of Naval Research, and the National Institute of Standards and Technology. Sandia National Laboratories and the Electric

[32]See Senate Report 105-235, Departments of Commerce, Justice, and State, the Judiciary, and Related Agencies Appropriation Bill, 1999. Report from the Committee on Appropriations to accompany bill S. 2260, which included the Commerce Department FY1999 Appropriations Bill.

Box A
Partnerships Reviewed by the *Government-Industry Partnerships* Study

The NRC study of *Government-Industry Partnerships for the Development of New Technologies* has reviewed a wide range of partnerships. The study can be divided into four primary areas: analysis of current U.S. partnership programs, potential partnerships, industry-national laboratory partnerships, and international collaboration and benchmarking. The analysis of current U.S. partnerships has focused on the Small Business Innovation Research Program, the Advanced Technology Program, and partnerships in biotechnology and computing. The review of potential partnerships for specific technologies, based on the project's extensive generic partnerships analysis, has focused on needs in biotechnology and computing and on opportunities in solid-state lighting. The industry-laboratory analysis has reviewed the potential of science and technology parks at Sandia National Laboratories and NASA Ames Research Center. International collaboration and benchmarking studies have included outlining new opportunities resulting from the U.S.-E.U. Science and Technology Agreement and, in this volume, a review of regional and national programs to support the semiconductor industry, focusing on Japan, Europe, Taiwan, and the United States.

Power Research Institute have also contributed. Private support is provided by a diverse group of private corporations. All sponsors are listed in the front matter.

ACKNOWLEDGMENTS

Highlights of the conference on *Regional and National Programs to Support the Semiconductor Industry* include presentations by leaders of semiconductor industry in Europe, Japan, Taiwan, and the United States. A complete list of participants is included in Annex B of this volume. The Proceedings section of this volume contains detailed summaries of their presentations and discussions. On behalf of the National Academies, we wish to express our appreciation and recognition for the insights, experiences, and perspectives made available by the participants. We are very much in debt to the senior executives, researchers, and Committee members who joined experts from the United States for this exceptional meeting.

Recognition is also due to Thomas Howell of Dewey Ballantine LLP, and Kenneth Flamm of the University of Texas. Both authors have contributed significant original research to this report. Howell's compendium of national and regional programs, which is largely based on in-country and language-of-origin research, is especially rich. No comparable review exists. Similarly, Flamm has prepared a careful analysis of SEMATECH's contribution to the industry and a review of the existing literature of this exceptional consortium. His empirical analysis and greater rigor cast new light on the contributions of the SEMATECH consortium.

Given the quality and the number of presentations, summarizing the papers and conference proceedings has been a challenge. We have made every effort to capture the main points made during the presentations and the ensuing discussions. We apologize for any inadvertent errors or omissions in our summary of the proceedings.

A number of individuals with the National Academies deserve recognition for their contributions to the preparation of this report. Among the STEP staff, Adam Korobow contributed a great deal to the preparation of the report and quality and originality of its research. He is joined by Alan Anderson, who prepared the proceedings summary, and Sujai Shivakumar, who also assisted in the preparation of the report. Christopher Hayter and McAlister Clabaugh each contributed a great deal to the preparation and quality of the report. David Dierksheide deserves particular recognition for his skill, persistence, and dedication during the review and preparation of this report. He and Chris Hayter put in many long hours to ensure a quality product. Without their sustained efforts among many other competing priorities, this report would not have been possible.

NRC REVIEW

This report has been reviewed in draft form by individuals chosen for their diverse perspectives and technical expertise, in accordance with procedures approved by the National Academies' Report Review Committee. The purpose of this independent review is to provide candid and critical comments that will assist the institution in making its published report as sound as possible and to ensure that the report meets institutional standards for objectivity, evidence, and responsiveness to the study charge. The review comments and draft manuscript remain confidential to protect the integrity of the deliberative process.

We wish to thank the following individuals for their review of this report: Avram Bar-Cohen, University of Maryland; John Chipman, University of Minnesota; David Hodges, University of California, Berkeley; Thomas Kalil, University of California, Berkeley; Martha Krebs, University of California, Los Angeles; Egbert Maynard, Microelectronics Advanced Research Corporation; Lawrence Thompson, Ultratech Stepper, Inc.; and Rosemarie Ham Ziedonis, University of Pennsylvania.

Although the reviewers listed above have provided many constructive comments and suggestions, they were not asked to endorse the conclusions or recommendations, nor did they see the final draft of the report before its release. The review of this report was overseen by Gerald Dinneen. Appointed by the National

Academies, he was responsible for making certain that an independent examination of this report was carried out in accordance with institutional procedures and that all review comments were carefully considered. Responsibility for the final content of this report rests entirely with the authoring committee and the institution.

STRUCTURE

Following the Executive Summary, Part I of this report presents an introduction of the current trends within industry and the policies to encourage its growth, followed by a summary of the conference presentations and the papers. Part II presents the Findings and Recommendations, which are the collective responsibility of the Steering Committee. Part III summarizes the Conference Proceedings. It is especially rich in that it sets out the views of the conference participants in some detail. Part IV presents two commissioned papers which, though subject to NRC editing, remain the responsibility of the authors.

The report's goal is to advance our understanding of the contributions of this unique industry, the exceptional technical challenges it faces, and the substantial programs that are under way around the world to address them. If the American economy is to continue to benefit from the leading position of this industry in the global marketplace, we must renew and strengthen our commitment to the institutions and scientific research that are essential to its continued progress.

Gordon Moore William Spencer Charles Wessner

Executive Summary

The substantial impact of information technologies on the growth and resilience of the U.S. economy represents a development now recognized at the highest level of U.S. policy making.[1] Maintaining this positive linkage between improvements in information technology and better economic performance is an appropriate goal for public policy.

Semiconductors are pervasive and an important source of productivity in the modern economy. Their rapid technological evolution—characterized by continuously increasing productivity and contemporaneously decreasing cost—is a source of growth throughout the economy, both in emerging industries and in more traditional industrial sectors.[2] A significant element of the strong performance of the U.S. economy in the last decades is rooted in investment in and subsequent application of information technologies, which are ultimately driven by advances in semiconductor technology.[3] Semiconductors also play a crucial role in ensuring our national security by allowing advances in the capabilities of

[1]Alan Greenspan, *Technological Innovation and the Economy*, Remarks Before the White House Conference on the New Economy, Washington, D.C. April 5, 2000, Federal Reserve Board.

[2]For an analysis of the role of new information technologies in the recent trends in high productivity growth, often described as the "New Economy," see Council of Economic Advisers, *Economic Report of the President,* H.Doc. 107-2, Washington, D.C.: USGPO, January 2001. Also see, National Research Council, *Measuring and Sustaining the New Economy, Report of a Workshop,* D. Jorgenson and C. Wessner, eds., Washington, D.C.: National Academy Press, 2002.

[3]See National Research Council, *U.S. Industry in 2000: Studies in Competitive Performance,* Washington, D.C.: National Academy Press, 2000.

new devices and new applications for national defense. The pervasive impact of the microelectronics sector on the nation's well-being—through improved communications, advances in health care, and better national security technologies—underscores the importance of the United States' role as the world's preeminent semiconductor producer.

This report focuses attention on the regional and national programs that have emerged around the world both to nurture local semiconductor industries and to help maintain the industry's exceptional growth rates. Specifically, the report highlights public-private partnerships in Europe, Japan, Taiwan, and the United States that seek to address the technical challenges faced by the global semiconductor industry. A unique feature of this report is that it provides the views of leaders in semiconductor research (from industry and academia) from Japan, Europe, Taiwan, and the United States. These experts came together to discuss common technical challenges facing the industry and the programs various nations and regions have undertaken to address them. In addition, the report contains original research, including an assessment of the major U.S. consortium, SEMATECH, and a summary of the programs of major producing countries and regions of the world. The diversity and scale of these programs underscore the sustained policy attention and support the industry receives in many parts of the world.

Most policy makers understand and accept that U.S. industry competes in a global marketplace. It is less widely appreciated that while the competition may be global in scope, the outcomes of this competition have important local and, ultimately, national consequences. Globalization therefore implies the need to learn about the policies and programs of all participants in this industry. Learning the scope, structure, and focus of other nations' programs is potentially valuable, both as a point of comparison and as a means of learning from the experiences of others in designing and managing cooperative programs.

KEY ISSUES OF THE REPORT

This report addresses three significant developments and the associated policy implications of these developments.

Productivity Growth

The first development, noted above, is the major contribution of the semiconductor industry to the productivity growth that has characterized the U.S. economy in the latter half of the 1990s as well as the early part of the new decade.[4] Given the industry's positive impact on economic growth, sustaining the

[4]The contribution of semiconductors to the economy is not reviewed here. Recent analysis by the Board on Science, Technology, and Economy Policy does document this impact of information tech-

technical advance that has made increases in semiconductor power possible is of major policy interest.

Growing Technical Challenges—Declines in Research Support

The second development concerns the technical challenges faced by the industry as it strives to sustain the remarkable technological progress predicted by Moore's Law.[5] Leading figures in the industry and academic experts are concerned that the federal government is not allocating adequate resources to the basic research required to maintain technical advance in what is now the largest manufacturing industry in the United States. At the very least, the analysis suggests that measures should be taken to reverse the disturbing *decline* in U.S. public support for the basic research on which this industry ultimately depends.

Significant Program Growth Abroad

The third development involves the significant growth in programs abroad that support national and regional semiconductor industries, how this support is fueling the structural changes, and its consequences in the global industry. The emergence of specialized design firms (referred to as "fabless" semiconductor companies because they do not engage in the production of the actual memory chip but rather only design them) and the rise of specialized manufacturing firms (the dedicated foundries), especially in Taiwan and mainland China, represent a structural shift in the industry that may present a challenge to U.S. firms over time. These structural changes may be accelerating in part as a result of programs to support national industries.

A further development involves the perception that SEMATECH contributed to the resurgence of the American industry. This perception has led to its emulation in many producing countries—often on a significantly larger scale and

---

nologies on productivity. See National Research Council, *Measuring and Sustaining the New Economy*. See also Dale W. Jorgenson and Kevin J. Stiroh, "Raising the Speed Limit: U.S. Economic Growth in the Information Age." *Brookings Papers on Economic Activity*. No 1: 125-211, 2000.

[5]In 1965, just seven years after the invention of the integrated circuit, Gordon Moore predicted that the number of transistors that would fit on an integrated circuit, or chip, would double every year. He tentatively extended this forecast for "at least 10 years." Dr. Moore's extrapolation proved to be highly accurate in describing the evolution of the transistor density of a chip. By 1975, some 65,000 transistors fit on a single chip. More remarkably, Moore's general prediction has held true to the present day, when microcircuits hold hundreds of millions of transistors per chip, connected by astonishingly complex patterns. Beyond its technical accuracy, the implications of "Moore's Law" have been far-reaching. Since the doubling in chip density was not accompanied by commensurate increases in cost, the expense of each transistor was halved with each doubling.

with greater underlying political support.[6] In light of the growing significance of R&D collaboration in both the equipment and device industries, providing policy and financial incentives to encourage such cooperation is increasingly important to support the transition to successive generations of new, enabling technologies.

Given the recognition of the contribution of semiconductors to the U.S. economy and the fundamental technical and structural challenges facing the industry, this report identifies measures that can be undertaken involving the industry, universities, and public policy to ensure continued U.S. leadership in this enabling technology.

COMMITTEE RECOMMENDATIONS

The Committee's recommendations outline a series of modest steps that nonetheless may prove important to the long-term welfare, economic growth, and security of the United States.

Resources for University-based Semiconductor Research

To better address the technical challenges faced by the semiconductor industry and to better ensure the foundation for continued progress, more resources for university-based research are required.

The Committee believes that universities have an important role in maintaining a balance between applied science and fundamental research. This balance is key in generating ideas for future research.

The Committee suggests consideration of the development of three-way partnerships among industry, academia, and government to catalyze progress in the high-cost area of future process and design. These partnerships would:

a. **Sponsor more initiatives that encourage collaboration between universities and industry**, especially through student training programs, in order to generate research interest in solutions to impending and current industry problems.
b. **Increase funding for current programs.**[7] Research programs that are already operational, such as the Focus Center Research Program devel-

---

[6]See Part IV in the Recommendations and Findings. See also Thomas Howell's discussion of foreign programs in "Competing Programs: Government Support for Microelectronics," in this volume. See also Kenneth Flamm and Qifei Wang's research in this volume, "Sematech Revisited: Assessing Consortium Impacts on Semiconductor Industry R&D."

[7]The president's FY 2003 budget makes important steps in this direction. It calls for a 3 percent increase, to $1.9 billion, in the *Networking and Information Technology Research and Development*

oped by the SRC, could usefully be augmented through substantially increased direct government funding. These centers also represent opportunities for collaborative research with other federal research programs, such as those supported by the National Science Foundation.

c. **Create Incentives for students**. A key role for universities is to ensure the flow of technical innovation and skills that originate with students. In order to address the undersupply of talented workers and graduate students in the industry, more incentive programs should be established. Since professors typically respond to appropriate research incentives, augmented federal support for programs that encourage research in semiconductors would attract professors and graduate students.[8] In addition, specific incentive programs could be established to attract and retain talented graduate students.

Program (NITRD). This particular program could play a key role in funding the basic research that confronts the technical challenges in the semiconductor industry. The NITRD coordinates key advanced information technology research across multiple agencies to make broad advances in computing and networking. These advances manifest themselves in the development of new technologies such as computing platforms and software, which can support advances research in physics, materials science and engineering as well as biomedical and earth and space sciences. The 2003 budget envisions emphasizing critical areas of research such as networks security issues; high-assurance software and systems; micro- and embedded-sensor technologies; and revolutionary architectures to reduce cost, size, and power requirements of high-end computing. The budget emphasizes research on the social and economic impacts of developments in the fields of information technology. For the text of the president's proposed initiatives, see *Fiscal Year 2003, Analytical Perspectives*, Budget of the United States Government, U.S. Government Printing Office, Washington, D.C., 2002, p. 164.

[8]See Paula Stephan and Grant Black, "Bioinformatics: Emerging Opportunities and Emerging Gaps," in National Research Council, *Capitalizing on New Needs and New Opportunities: Government-Industry Partnerships in Biotechnology and Information Technologies,* Washington, D.C.: National Academy Press, p. 244.

I

INTRODUCTION

Introduction

The semiconductor industry is one of the major contributors to modern economic growth.[1] As one recent National Academies' study notes[2]:

> "...often called the 'crude oil of the information age,' semiconductors are the basic building blocks of many electronics industries. Declines in the price/performance ratio of semiconductor components have propelled their adoption in an ever-expanding array of applications and have supported the rapid diffusion of products utilizing them. Semiconductors have accelerated the development and productivity of industries as diverse as telecommunications, automobiles, and military systems. Semiconductor technology has increased the variety of products offered in industries such as consumer electronics, personal communications, and home appliances."

This pervasiveness in use establishes semiconductors as the premier *general-purpose technology* of our post-industrial era.[3] In its impact, the semiconductor is in many ways analogous to the steam engine of the first industrial revolution.[4]

---

[1]Dale W. Jorgenson. "Information Technology and the U.S. Economy," *The American Economic Review*, 91(1): 1-32, 2001.

[2] This excerpt is taken from Jeffrey T. Macher, David C. Mowery, and David A. Hodges, "Semiconductors," *U.S. Industry in 2000: Studies in Competitive Performance*, David C. Mowery, ed., Washington, D.C.: National Academy Press, 1999, p. 245.

[3]For a full discussion and definition of general-purpose technologies and their impact on economic growth and development, see Helpman, E. and M. Trajtenberg "Diffusion of General Purpose Technologies," pp. 85-119 in *General Purpose Technologies and Economic Growth,* E. Helpman, ed. Cambridge and London: MIT Press, 1998.

[4]*Ibid.*

The invention of the first transistor in 1947 at Bell Telephone Laboratories heralded the beginning of the modern era in technological advancement. Four years later, in 1951, Bell Labs sponsored a conference in which the capabilities of the transistor were demonstrated to leading scientists and engineers for the first time. Although the attendees from outside Bell Labs did not yet possess the capability of producing a transistor, the conference conveyed the enormous potential of transistors, and many eager scientists returned in the spring of 1952 for the Bell sponsored *Transistor Technology Symposium*.[5] The foundation of the modern day high-tech revolution was established at this symposium as the attendees shared their knowledge and ideas about the capabilities and applications of the transistor. Bell Labs assembled the knowledge shared at the eight-day conference into two volumes, entitled *Transistor Technology*.[6]

As a matter of antitrust settlement and corporate policy, in 1955 Bell Labs established an important precedent in creating the merchant semiconductor industry through a decision to share its intellectual property on diffused-base transistor technology.[7] This decision allowed other researchers access to the knowledge describing methods for creating this new technology. Four years later, in 1959, the first integrated circuit (IC) was created, and the semiconductor industry began its rapid ascent from the cradle of the research lab to become the largest value-added manufacturing industry in the United States.[8]

SUSTAINED, PREDICTABLE GROWTH

The scale of this industry's growth—exceptional both because of its rapidity and its predictability over time—and its contributions to the economy are not always fully appreciated. The U.S. Semiconductor industry is a major generator of high-wage jobs, employing 283,875 in 2000. The industry's sales reached \$102 billion[9] in a global market estimated at \$204 billion. The value of U.S.

---

[5]For a more in-depth discussion of the events leading up to the Technology Transistor Symposium, go to: <http://www.pbs.org/transistor/index.html>. See also the Institute of Electrical and Electronics Engineers website, which also gives an excellent account of the transistor's history. < http://www.ieee.org/organizations/history_center/>.

[6]*Ibid.* The book also became known as "Mother Bell's Cookbook."

[7]For an excellent description of the early evolution of the semiconductor industry see Kenneth Flamm *Mismanaged Trade? Strategic Policy and the Semiconductor Industry*, Washington, D.C.: Brookings Institution Press, 1996, pp. 30-31. See the paper by Thomas Howell, "Competing Programs: Government Support for Microelectronics," in this volume.

[8]*Source: U.S. Census Bureau, Annual Survey of Manufactures, 1999, Statistics for Industry Groups and Industries,* Series M99(AS)-1, in *Statistical Abstract of the United States*; 2001, 121st edition. U.S. Census Bureau, U.S. Department of Commerce.

[9]Global market sales in 2000 were about \$204 billion according to the SIA (Semiconductor Industry Association). For more information on the semiconductor industry, see <http://www.semichips.org/ind_facts.cfm>.

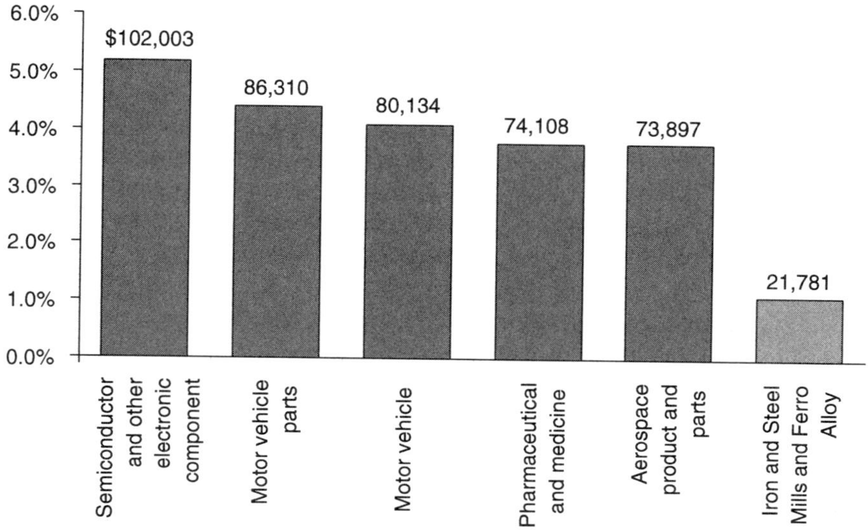

FIGURE 1 Semiconductor Value Added: Largest Five Value-Added Manufacturing Industries Compared with Other Major Sectors (Value Added as a Percentage of Value Added by Manufacturers—1999).
SOURCE: US Census Bureau, *Annual Survey of Manufacturers.*

semiconductor sales has averaged 50 percent of total worldwide sales in the past six years.

The semiconductor industry in 1999 was the largest value-added industry in manufacturing—almost five times the size of the *Iron and Steel* sector in that year (see Figure 1). It is, in fact, larger in terms of valued added than the *Iron and Steel* and *Motor Vehicle* industries (excluding *Motor Vehicle Parts*—a separate industry classification) combined. As noted below, the electronics industry, largely based on semiconductors, is the largest U.S. manufacturing industry.[10]

While the manufacturing sector's contribution to GDP has been shrinking (accounting for just under 16 percent of GDP in 2000), U.S. semiconductor industry sales, as a percentage of output in the manufacturing sector, have increased steadily in the past 15 years, climbing from 1.5 percent of manufacturing GDP in 1987 to reach 6.5 percent in 2000 (See Figure 2).[11]

---

[10]Bureau of Economic Analysis, *Statistical Abstract of the United States: 2001*, Department of Commerce, Table 641, Washington, D.C: U.S. Government Printing Office, 1999, p. 418.

[11]National Research Council calculations derived from sales data from the *Semiconductor Research Association* and GDP data from the *Bureau of Economic Analysis.*

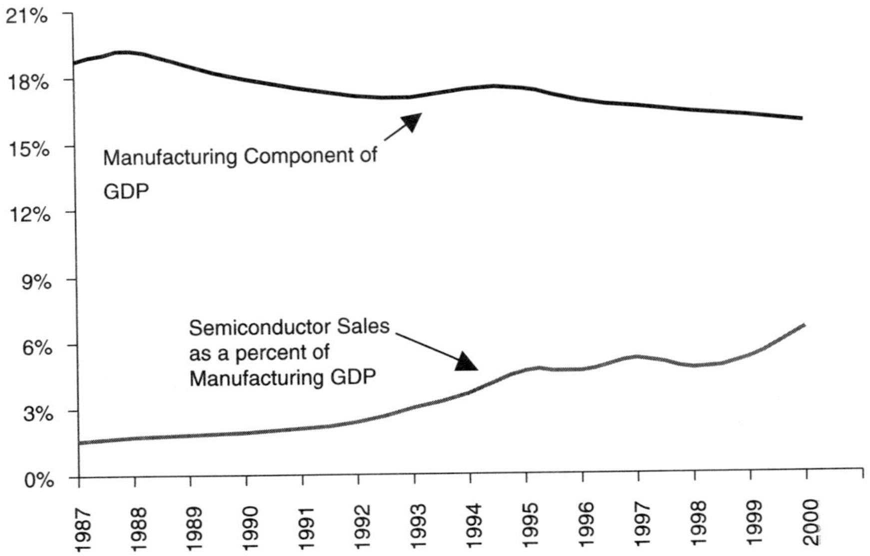

FIGURE 2 Semiconductor Sales as a Percent of Manufacturing GDP.
SOURCES: Semiconductor Sales; SIA GDP; Bureau of Economic Analysis.

These positive trends reflect the strong global economic position of the U.S. industry in a technology which is seen as fundamental to the economy. Given the industry's contribution to economic growth, other countries have taken a proactive approach to encouraging the development of their national semiconductor industries in order to ensure themselves a place in the technologies that underpin the knowledge-based economy.[12]

The growing impact of information technologies on economic growth, in large part the result of improvements in semiconductors, has attracted increased attention from leading economists. Yet, the underlying technological challenges facing the semiconductor industry pose a complex set of issues for both the industry and national policy. If the U.S. and global economy are to continue to benefit from the vast increases in semiconductor power characterized by Moore's Law, a series of impending technical challenges must be overcome. How these challenges are addressed will likely affect future national U.S. competitiveness and leadership in this enabling industry. The first firm, or geographically con-

[12]See the Proceedings and Howell, *op.cit.*, as well as earlier National Research Council Analysis.

centrated group of firms, that resolves the technical challenges facing the industry could develop a position of leadership in semiconductor design and production in the years ahead.

To help their companies meet these technical challenges, a number of countries are making substantial public investments in cooperative R&D. In addition, other firms are pursuing strategies that may ultimately challenge the current business models of U.S. firms.

EARLY PUBLIC SUPPORT FOR THE INDUSTRY

The birth and proliferation of the semiconductor was facilitated by substantial public support in transistor research. By 1952, the U.S. Army's Signal Corps Engineering Laboratory had funded 20 percent of total transistor-based research at Bell Labs.[13] The eagerness of the Defense Department to put to use this innovative and radical new technology encouraged the Signal Corps to fund half of the transistor work by 1953.[14]

Public support for the nascent semiconductor industry became more prevalent after 1955 when R&D funds were allotted to other companies after the U.S. Department of Justice's ongoing antitrust suit against Bell Labs pressured Bell into sharing its patents on transistor diffusion processes.[15] According to one estimate, the government directly or indirectly funded 40 to 45 percent of all industrial R&D in the semiconductor industry between the late 1950s and early 1970s.[16] On the demand side, federal consumption dominated the market for integrated circuits (ICs), which found their first major application in the Minuteman II guided missile. In the 1960s, military requirements were complemented by the needs of the Apollo Space Program.[17]

Public support played a critical and catalyzing role in the development and initial growth of the semiconductor industry. The groundbreaking inventions that launched the industry were made at Bell Labs, which was in part sustained by U.S. communications policy as well as by defense funding.[18] As the initial in-

---

[13]Flamm, *op.cit.*, pp. 30-34.

[14]*Ibid.*

[15]*Ibid.*

[16]*Ibid.* Government research contracts were not an unmixed blessing. Their heavy paperwork and rigidities acted as a significant constraint and could slow the redirection of research to more promising avenues. See Gordon Moore's comments and presentation in the Proceedings of this report.

[17]For an overview of the government's early role in the semiconductor industry, and its contributions over time, see Flamm, *op.cit.* pp. 1-38.

[18]See Michael Borrus, *Competing for Control: America's Stake in Microelectronics*, Cambridge, MA: Ballinger, 1988.

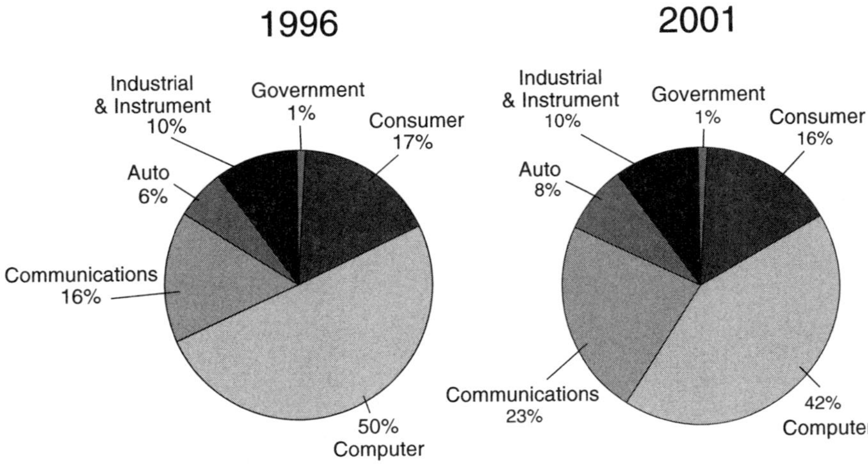

FIGURE 3 Worldwide Semiconductor End Use.
SOURCE: World Semiconductor Trade Statistics and SIA, September 2002.

vention revealed its potential, the government first encouraged the dissemination of the technology, then served as a source of sustained procurement for the most advanced products possible. This well-financed demand contributed directly to the early growth of the industry.[19] In 1963, federal contracts accounted for 35.5 percent of total U.S. semiconductor shipments.[20] Over the following decades, the semiconductor industry has grown enormously, and the government's share of semiconductor consumption is now only about 1 percent of a much larger industry (see Figure 3).

THE ECONOMIC CONSEQUENCES OF "FASTER AND CHEAPER"

As noted above, the history of the semiconductor industry has been characterized by rapid growth, concurrently decreasing costs, and growing economic importance. For example, the industry is characterized by high growth rates, averaging 17 percent per annum.[21] Semiconductors are also an enabling technol-

[19]See also Martin Kenney, *Understanding Silicon Valley: The Anatomy of an Entrepreneurial Region.* Stanford, CA: Stanford University Press, 2000. Government procurement continues to play a role, albeit a much smaller one. See Flamm, *Creating the Computer: Government, Industry, and High-technology,* Washington, D.C.: Brookings Institution, 1988

[20]See Table 1-8 in Flamm, *Mismanaged Trade? Strategic Policy and the Semiconductor Industry,* p. 37.

[21]Semiconductor Industry Association, "World Market Shares 1991-2001," Data for 1991-2000. San Jose, CA: Semiconductor Industry Association, 2002. See website: <http://www.semichips.org>.

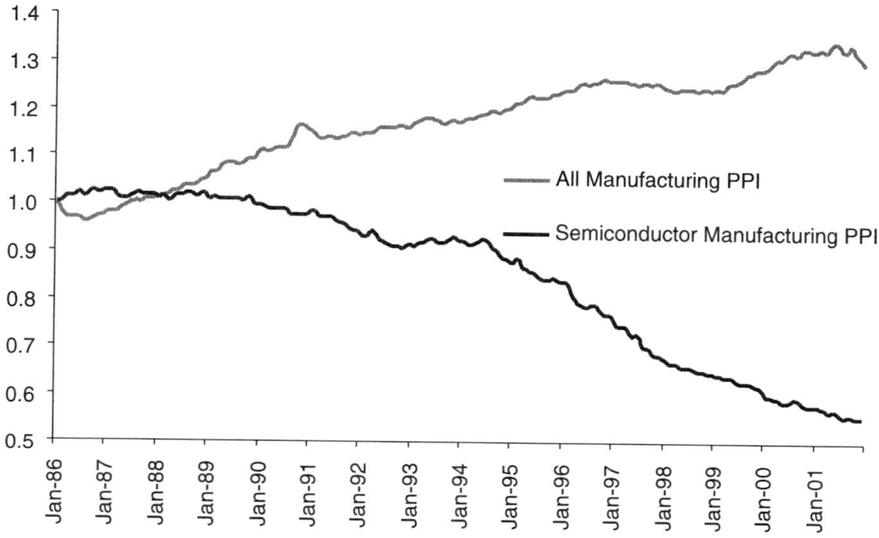

FIGURE 4 Semiconductor and All Manufacturing Producer Prices (Index 1986=1.00).
SOURCE: Producer Price Index, Bureau of Labor Statistics.

ogy with widespread and steadily growing applications (e.g., in medical technologies and research.)[22] As semiconductor prices have steadily declined, investment in information technologies has increased.[23] In the early 1950s, for example, a transistor was manufactured at a cost of between $5 and $45. Today, transistors on a microchip cost less than a hundred-thousandth of a cent apiece, which makes their marginal cost essentially zero.[24] While the manufacturing sector as a whole has experienced an increase in prices since the mid-1980s, the semiconductor industry has exhibited a deflationary trend (Figure 4), which accelerated in the middle 1990s. The significance of this deflationary trend in semiconductor prices has not only made powerful consumer electronics products more

[22]See National Research Council, *Capitalizing on New Needs and New Opportunities: Government-Industry Partnerships in Biotechnology and Information Technologies*, Washington, D.C.: National Academy Press, 2001.

[23]Semiconductor prices have declined at an annual rate of 30 percent in the past three decades. For an in-depth, technical discussion of semiconductor price evolution and its impact on information technology investment, see Jorgenson, *op.cit.*

[24]The exponential increase in power of the integrated circuit in the past several decades has been commensurately matched by a decrease in cost of each additional transistor on a chip. For a brief discussion of the decreasing cost of each new generation of integrated circuits, see National Academy of Engineering website; <http://www.greatachievements.org/greatachievements/ga_5_2.html>.

accessible, but has spurred increased business investment in information technology, which has in turn catalyzed improvements in productivity.

The ability to increase device power and decrease device cost underlies the semiconductor industry's growth. In 1965, just seven years after the invention of the integrated circuit, Gordon Moore predicted that the number of transistors that would fit on an integrated circuit, or chip, would double every year. He tentatively extended this forecast for "at least 10 years." [25] At that time, the world's most complex chip had 64 transistors. Dr. Moore's extrapolation proved to be highly accurate in describing the evolution of the transistor density of a chip. By 1975, some 65,000 transistors fit on a single chip. More remarkably, Moore's general prediction has held true to present day, when microcircuits hold hundreds of millions of transistors per chip, connected by astonishingly complex patterns.[26]

The implications of Moore's Law have been far-reaching. Since the doubling in chip density was not accompanied by commensurate increases in cost, the expense of each transistor was halved with each doubling. With twice as many transistors, a chip could store twice as much data. Higher levels of integration meant that greater numbers of functional units could be placed onto the chip, and more closely spaced devices—such as the transistors—could interact with less delay. Thus, these advances gave users increased computer processing power at a lower price, consequently spurring chip sales and a demand for yet more power.[27] Beginning in the late 1970s, the use of semiconductors became more pervasive, spreading from computers to air traffic control systems, microwave

---

[25]See Gordon E. Moore, "Cramming More Components onto Integrated Circuits," *Electronics* 38(8) April 19, 1965. Here, Dr. Moore notes that "[t]he complexity for minimum component costs has increased at a rate of roughly a factor of two per year. Certainly over the short term, this rate can be expected to continue, if not to increase. Over the longer term, the rate of increase is a bit more uncertain, although there is no reason to believe it will not remain nearly constant for at least 10 years. That means by 1975, the number of components per integrated circuit for minimum cost will be 65,000." See also, Gordon E. Moore, "The Continuing Silicon Technology Evolution Inside the PC Platform," *Intel Developer Update,* Issue 2, October 15, 1997, where he notes that he "first observed the 'doubling of transistor density on a manufactured die every year' in 1965, just four years after the first planar integrated circuit was discovered. The press called this 'Moore's Law' and the name has stuck. To be honest, I did not expect this law to still be true some 30 years later, but I am now confident that it will be true for another 20 years."

[26]*Ibid.* See also Michael Polcari's presentation in the Proceedings, which discusses the progression of Moore's Law.

[27]For a complete analysis of the impact of the increase in the power of the semiconductor accompanied by its subsequent decline in price, and its positive influence on economic growth, see Jorgenson, *op.cit.* See also G. Dan Hutcheson and Jerry D. Hutcheson, "Technology and Economics in the Semiconductor Industry," *Scientific American*, <http://www.sciam.com/specialissues/1097solidstate/1097hutch.html>.

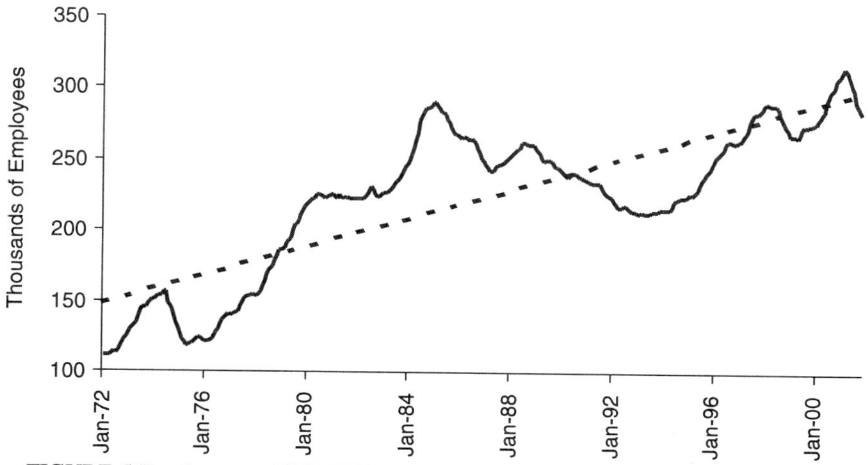

FIGURE 5 Employment 1972-2001: Semiconductor and Related Devices Industries.
SOURCE: Bureau of Labor Statistics, Form 790.

ovens, video cameras, watches, grocery checkout machines, automobiles, touch-tone phones, wireless communications, and satellite broadcasts.

A DRIVER OF MODERN INDUSTRY

The semiconductor has become the engine of growth for many fledging industries, as well as a source of revitalization and increased efficiency for more established industries (see Box A). Consequently, semiconductors, as well as related industries, have acquired significant global visibility and have become targets of national economic priority in many countries. As of August 2001, the semiconductor industry employed some 284,000 people in the United States alone.[28] The industry, in turn, provides enabling technologies for the $425 billion U.S. electronics industry.[29] Figure 5 exhibits the employment trends in *Semiconductor and Related Device* industries dating back to 1972. The cyclicality of the industry is evident, but employment has increased steadily by more than two-and-a-half times since 1972.

Importantly, the semiconductor industry is a substantial source of high-wage jobs. In addition to the increase in overall industry employment, real average

[28]According to the Semiconductor Industry Association, the semiconductor industry employs some 283,875 within the U.S. See <http://semichips.org/ind_facts.cfm>.

[29]This recent estimate is from Cahners Business Information at <http://www.cahners.com/2001>.

BOX A
WHY DO NATIONAL POLICIES FOCUS ON THE SEMICONDUCTOR INDUSTRY?

The semiconductor industry, characterized by yearly increases in performance and concurrent price decreases, has had a distinctive positive impact on the economy. It is—

- ***An Enabling Industry.*** Semiconductors serve as key inputs to a wide variety of intermediate and final products and services, ranging from construction to finance and banking. Semiconductors make positive contributions to the productivity of these sectors.
- ***A Key Contributor to Enhanced Economic Growth.*** Performance increases and price decreases in semiconductor-based products are a boon not only for consumers; they also make lower-priced, higher-powered investment goods available for all sectors of the economy—business large and small. Faster and cheaper information technologies, when brought on line by companies, can increase worker productivity. Semiconductors are a driver of the high-tech revolution.
- ***A Source of High-Wage Job Creation.*** The semiconductor industry is a knowledge-based manufacturing industry which creates high-wage jobs. By contrast, the manufacturing sector as a whole has witnessed a stagnation and slight decline of average pay over the past 30 years (See Figure 6).[a]
- ***A Source of Competitive Advantage.*** Increases in semiconductor productivity lead to more rapid advances in information technology. Possessing the latest technologies can often translate into a competitive advantage for firms investing in high-tech equipment.
- ***A Key Element in National Defense.*** Semiconductors have played and will continue to play an increasing role in promoting national security. From the original *Minuteman II* missile to future wireless-network battlefield capabilities such as BARS (*Battlefield Augmented Reality System*), semiconductor advances will have direct, positive consequences for improving our defense against old and new threats.[b]

---

[a]Source: Earnings—Bureau Labor Statistics, Form 790; Consumer Price Index, All Items—Bureau Labor Statistics *Statistical Abstract of the United States*; 2001, 121st edition. U.S. Census Bureau, U.S. Department of Commerce.

[b]The Office of Naval Research (ONR) is sponsoring research on *Battlefield Augmented Reality Systems*. The project examines how information can be relayed between a tactical command center and soldiers in an urban battlefield environment.

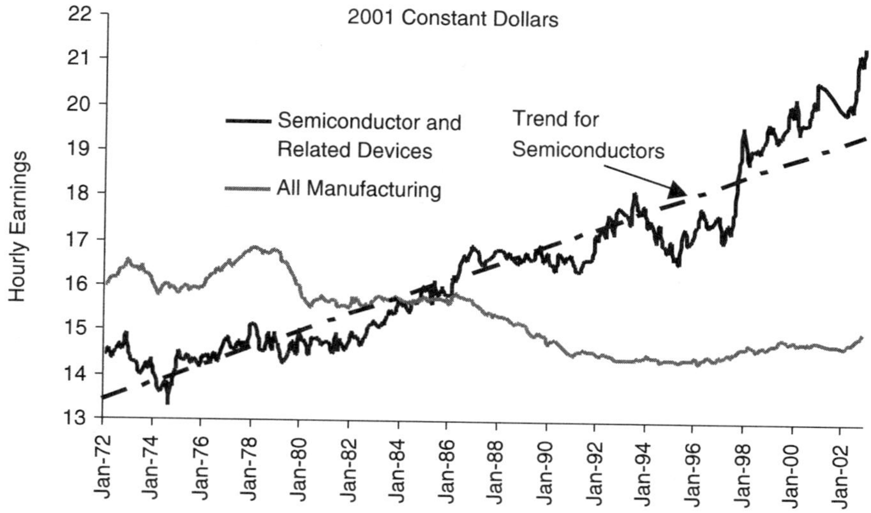

FIGURE 6 Average Hourly Earnings 1972-2001: Semiconductor and Related Device Industries and All Manufacturing (2001 Constant Dollars).
SOURCES: Earnings—Bureau of Labor Statistics, Form 790; Consumer Price Index, All Items—Bureau of Labor Statistics.

hourly earnings in the semiconductor and related device industries, shown in Figure 6, have risen a remarkable 50 percent in the past 30 years—from roughly $14.50 in 1972 to about $21.50 today in 2001 dollars. This sizeable increase in real average hourly earnings for the semiconductor industry stands in stark contrast to the stagnation and then decline in real wages in the manufacturing sector as a whole over the same 30-year period. Real wages for the overall manufacturing sector declined by about 6 percent over this period.[30]

AN ENABLING INDUSTRY

Some believe that the advances in semiconductors and computers are responsible for the increases in productivity throughout the economy. For example, the semiconductor lies at the heart of the computer (desktop computers, workstations, servers, etc.), which is the foundation of increases in firm productivity and provides the platform for the Internet. The Internet subsequently provides the platform for the World Wide Web, which then provides a foundation for e-com-

[30]This decline is likely underestimated since the figures used here for overall manufacturing real wages include the semiconductor and related device industries.

merce. In addition, the semiconductor drives the rapidly evolving world of wireless communication and a vast and growing universe of computer-enabled information digitization.

As was noted in the opening remarks of this symposium, the semiconductor industry carries an importance far beyond the specific trade, employment, and revenue figures of the industry itself. The industry is responsible for "much of the recent productivity gains in computers, communications, and software."[31]

The "digital economy" and the corresponding positive synergistic relation to the remarkable period of productivity growth in the 1990s has been acknowledged by the Federal Reserve in its public discourse, as well as reflected in its monetary policy. During the middle part of that decade, Alan Greenspan, Chairman of the Federal Reserve Board, suggested that the nation's expenditures on semiconductor-based products and the gains in productivity expected to accompany these expenditures characterize a far-reaching economic transformation:

> We are living through one of those rare, perhaps once-in-a-century events....The advent of the transistor and the integrated circuit and, as a consequence, the emergence of modern computer, telecommunication, and satellite technologies have fundamentally changed the structure of the American economy.[32]

Reflecting this investment, high technology firms also create positive spillovers, which affect society in many ways. Spillovers benefit other commercial sectors by generating new products and processes that can lead to productivity gains. A substantial literature in economics underscores the potential for high returns from technological innovation; it shows private innovators obtaining rates of return in the 20 to 30 percent range and spillover (or social return) averaging about 50 percent.[33]

High-technology products are a major source of growth in the major industrialized countries. Sectors such as aerospace, biotechnology, and information systems contribute to the growing global market for high-technology manufactured goods. While subject to pronounced cyclical swings, high-technology firms are

---

[31]See Bill Spencer's comments in the introduction of the proceedings in this volume. The view is supported by recent research. See Dale W. Jorgenson and Kevin J. Stiroh, "Raising the Speed Limit: U.S. Economic Growth in the Information Age," *Brookings Papers on Economic Activity 1*, 2000, p. 2. See also National Research Council, *Measuring and Sustaining the New Economy*, D. Jorgenson and C. Wessner, eds., Washington, D.C.: National Academy Press, 2002.

[32]Testimony of Alan Greenspan before the U.S. House of Representatives' Committee on Banking and Financial Services, July 23, 1996.

[33]For example, see M. Ishaq Nadiri, *Innovations and Technological Spillovers*, NBER Working Paper 4423, August 1993. See also, Council of Economic Advisers, *Supporting Research and Development to Promote Economic Growth: The Federal Government's Role,* Washington, D.C.: Government Printing Office, 1995.

also associated with high-value-added manufacturing and with the creation of high-wage employment.[34] Together, these contributions provide the productivity gains that underpin recent economic performance.[35]

SEMICONDUCTORS AND PRODUCTIVITY

In the late 1970s and early 1980s, there was a significant loss in global market share by many U.S. industries to Japanese producers. An extensive literature in the social sciences has focused on the decline in U.S. industry during this period. One influential study reported that U.S. manufacturers in general had lost the ability to compete internationally, especially with Japan, and that the U.S. industrial weakness was part of a longer period of decline.[36]

Part of this pessimism was connected to the multi-decade trend of low productivity growth, loss of market share for many U.S. industries, and rapid growth in the trade deficit. Despite large investments in purported labor-saving systems—especially information technology—productivity remained low through the first half of the 1990s, just as it had since 1973. As recently as 1997, many economists were convinced of the validity of the so-called productivity paradox—Robert Solow's casual but oft-repeated remark made in 1987: "We see the computer age everywhere except in the productivity statistics."[37]

In the middle 1990s, however, several significant trends became apparent. One was the relatively rapid and widespread adoption of the Internet and other information technology (IT), which allowed not only individuals but also businesses to benefit from previously unavailable low-cost communication. Another was a sudden acceleration in the decline of semiconductor and computer prices. In a recent paper, Dale W. Jorgenson and Kevin J. Stiroh describe this acceleration as a "point of inflection," where the price decline abruptly rose from 15 percent annually to 28 percent. In response to this rapid price decline, investment in computer technology exploded, and its contribution to growth rose more than

---

[34]Laura Tyson, *Who's Bashing Whom? Trade Conflict in High Technology Industries,* Washington D.C.: Institute for International Economics, 1992. For the impact of the telecommunication industry's downturn on the semiconductor industry, see Richard Gawel, "Semiconductor Equipment Shipments Plummet 35%" *Electronic Design,* Aug 20, 2001, Volume: 49, Issue: 17, p. 38; and Bolaji Ojo, "IC Equipment Makers Batten Down—Focusing On Next-Generation Technology As Demand Plummets," *EBN,* Jul 23, 2001, Special Volume/Issue: Issue: 1272, p.20.

[35]See Jorgenson and Stiroh, *op. cit.* For the most recent, as well as historical data on productivity, see Bureau of Labor Statistics website at <http://www.bls.gov/lpc/home.htm>.

[36]Robert M. Solow, Michael Dertouzos, and Richard Lester, *Made in America,* MIT Press, Cambridge, MA, 1989.

[37]Robert Solow, "We'd Better Watch Out," *New York Times Book Review,* July 12, 1987.

fivefold. Jorgenson and Stiroh find that computers contributed 0.46 percentage points to the 2.4 percent productivity growth in the period from 1995 to 1998.[38] Software and communications equipment contributed an additional 0.30 percentage points per year over the same period. Preliminary estimates through 1999 revealed further increases for all three categories (computers, communications equipment, and software).[39]

Jorgenson and Stiroh's analysis builds the case for "raising the speed limit," that is, for revising upward the intermediate-term projections of growth for the U.S. economy. They noted that after a 20-year slowdown, dating from the early 1970s, average labor productivity had grown by 2.4 percent per year during the period 1995-1998, exceeding the rate for 1990-1995 by a full percentage point. Even the most recent downturn in the U.S. economy seems to have left much of this increase in productivity intact.[40] In short, Jorgenson and Stiroh's research supports the notion that the economy is on a higher-productivity path, similar to that experienced from the early 1950s through the early 1970s, as suggested by Figure 7.

In related work, the National Research Council's Board on Science, Technology, and Economic Policy produced an analysis examining what some see as the resurgence of U.S. industry.[41] This analysis finds that over the previous 15 years, many industries in the United States had succeeded in regaining competitive positions relative to their counterparts abroad.[42] Importantly, of the industries reviewed, over half had been transformed by the use of information technology, which rests fundamentally on developments in semiconductor technology and software. One sector of focus in this study is the U.S. semiconductor industry, which the research found to have returned to international pre-eminence by the late 1990s.[43]

In sum, the American economy has benefited from the contributions of the information technology industry, not least through its contributions to productiv-

---

[38]An MGI study is more cautious but finds that the semiconductor industry alone contributed 0.20 percentage points to the 1.33 percent jump in productivity from the 1995-1999 period. For more of MGI's conclusions concerning the impact of the semiconductor industry on growth, see <http://www.mckinsey.com/knowledge/mgi/feature/index.asp>.

[39]Jorgenson and Stiroh, *op.cit.* See also National Research Council, *Measuring and Sustaining the New Economy.*

[40]"Productivity Growth May Be Here to Stay," *Wall Street Journal,* January 7, 2002. p. A1.

[41]See National Research Council, *U.S. Industry in 2000: Studies in Competitive Performance.*

[42]In some cases the perceived decline of U.S. industry was overstated. See Macher, Mowery, and Hodges, *op.cit.*

[43]*Ibid.* For a discussion of the factors leading to the resurgence of the U.S. semiconductor industry, see below.

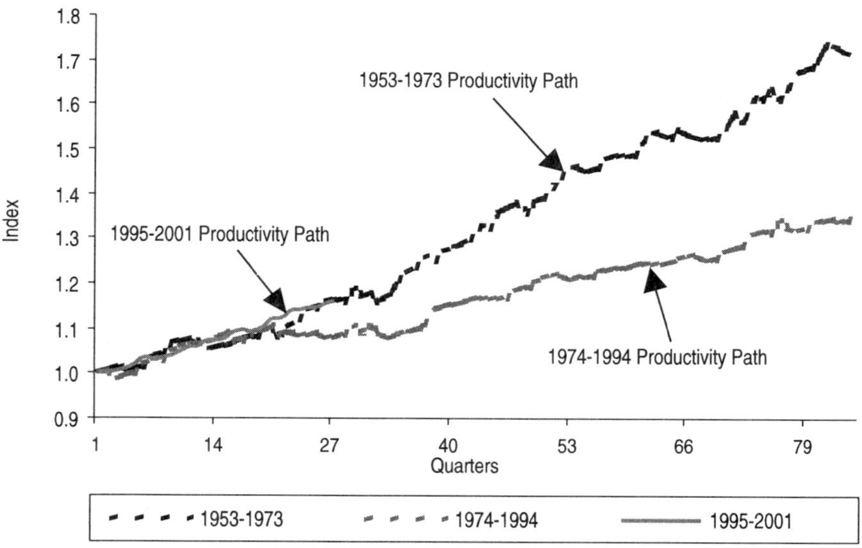

FIGURE 7 Productivity Growth for Three Periods: 1953–1973; 1974–1994; 1995–2001.
SOURCE: Labor Productivity, Bureau of Labor Statistics.

ity.[44] Leading researchers now believe that Solow's paradox has been resolved.[45] The information technology revolution is finally visible in productivity statistics.[46] As the Council of Economic Advisers noted, "even though economists

[44]Council of Economic Advisers, *Economic Report of the President*, Washington, D.C.: U.S. Government Printing Office, 2001, p. 34 and passim.

[45]*Ibid.* See also Dale W. Jorgenson, "Presidential Address to the American Economic Association, New Orleans, Louisiana, January 2001. See also Joseph H. Haimowitz, "Has the Surge in Computer Spending Fundamentally Changed the Economy?" *Economic Review*, Federal Reserve Bank of Kansas City, Second Quarter, pp. 27-42, 1998. For a historical perspective on the sources of U.S. economic growth, see Robert J. Gordon, "U.S. Economic Growth Since 1870: What We Know and Still Need to Know," *American Economic Review*, Papers and Proceedings, 89(2): 123-128, 1999. See also Martin Bailey and Robert Z. Lawrence, "Do we have an e-conomy?" NBER Working Paper 8243, April 2001, and Alan S. Blinder, "The Internet and the New Economy," Policy Brief No. 60, Brookings Institution, Washington, D.C., June 2000.

[46]*Ibid.* See also the discussion by Flamm, "Microprocessors and Computers," *Measuring and Sustaining the New Economy*. In recent years Flamm has marshaled economic evidence to demonstrate that the semiconductor industry has been the key force in the revival of industries related to information technology. See also Martin Neil Baily and Robert J. Gordon, "The Productivity Slow-

differ as to the correct way to adjust for responses to the business cycle, the finding that a structural acceleration has taken place is robust."[47] The council finds that a breakdown of the sources of this accelerated productivity suggests three lessons:

- "The information technology sector itself has provided a direct boost to productivity growth....;
- "The spread of information technology throughout the economy has been a major factor in the acceleration of productivity through capital deepening...;" and
- "Outside the information technology sector, organizational innovations and better ways of applying information technology are boosting the productivity of skilled workers."[48]

The sustainability of this growth resurgence, however, depends on the rate of current and future technological progress, which itself depends on the level and effectiveness of the nation's R&D investments, both private and public, as well as on the maintenance of supportive macroeconomic policy.[49]

TECHNICAL CHALLENGES AND SOARING CAPITAL COSTS

For more than 30 years the growth of the semiconductor industry has been largely associated with the ability to steadily and quickly shrink the transistor and increase its speed without increasing costs. If the increases in productivity observed since 1995 depend on the increases in semiconductor power characterized

---

down, Measurement Issues and the Explosion of Computer Power," *Brookings Papers on Economic Activity*, 2: 347-420, 1988; Alan S. Blinder, "The Speed Limit: Fact and Fantasy in the Growth Debate," *The American Prospect*, 34 (September/October): 57-62, 1997; Erik Brynjolfsson and Shinkyu Yang, "Information Technology and Productivity: A Review of the Literature," *Advances in Computers*, 43 (February): 179-214, 1996; Council of Economic Advisers, *The Annual Report of the Council of Economic Advisers*, Washington D.C.: U.S. Government Printing Office, 2000; Robert J. Gordon, "Has the New Economy Rendered the Productivity Slowdown Obsolete?" Manuscript, Northwestern University, June 12, 1999; Dale W. Jorgenson, "Information Technology and Growth," *AER*, 89(2): 109-115, 1999; Kevin J. Stiroh, "Computers, Productivity, and Input Substitution," *Economic Inquiry*, XXXLI(2): 175-191, 1998; Kevin J. Stiroh, "Is there a New Economy?" *Challenge*, 42(4):82-101, 1999; Jack E. Triplett, "Economic Statistics, the New Economy, and the Productivity Slowdown," *Business Economics*, XXXIV(2):13-17, 1999.

[47]Council of Economic Advisers, *Economic Report of the President*, 2001 p. 28. The report cautions, however, that it is uncertain whether the structural trend that emerged in 1995-2000 will continue or moderate again.

[48]*Ibid*, p. 33.

[49]See the analysis of these issues by Jorgenson and Stiroh, *op. cit.*

by Moore's Law, then a continuation of productivity increases will likely depend on the ongoing benefits associated with the process of "scaling" in microelectronics.[50]

Other challenges facing the industry include the need to substantially improve packaging technology in order to house and interconnect the next generation of chips emerging from the silicon foundries. Further, development and progress are also needed in the area of chip-level CAD (Computer Aided Design) tools as well. Absent dramatic innovation in these two areas, it may prove impossible to exploit the enhanced functionality, gate density, and speed of future semiconductor products, creating a disincentive for new-product adoption and leading to stagnation in semiconductor sales.[51]

The reduction in semiconductor size, however, may now be approaching important critical limits. In a recent paper, Paul Packan of Intel Corporation described some of these limits, including odd and undesirable quantum effects that appear under extreme miniaturization. For example, the *gates* that regulate the flow of electrons within semiconductor devices have become so short that electrons can *tunnel* through them even when they are closed. In addition, *dopants*, impurities mixed with silicon that increase its ability to hold localized charges, must be added in progressively higher concentrations as device size shrinks in order to enable them to hold the same charges. At a certain concentration, dopant atoms begin to interact with each other to form clusters that no longer hold a charge. Transistor dimensions have shrunk to such an extent that small changes in the exact number and precise distribution of individual dopant atoms can change the behavior of the device. Packan's conclusion is expressed in sobering words:

> "These fundamental issues have not previously limited the scaling of transistors and represent a considerable challenge for the semiconductor industry. There are currently no known solutions to these problems. To continue the performance trends of the past 20 years and maintain Moore's Law of improvement will be the most difficult challenge the semiconductor industry has ever faced."[52]

---

[50]See Bill Spencer's discussion of semiconductors in National Research Council, *Measuring and Sustaining the New Economy*.

[51] For a discussion of the issues involved with chip packaging, see James Malatesta and Ron Bauer, "A Chip-Scale Packaging Primer," *Printed Circuit Design*; San Francisco, 17(3): 10-18, 2000. More issues in advanced chip packing are highlighted in Peter Singer, "Consortiums Address Advanced Packaging Requirements," *Semiconductor International*; 25(6): 46, 2002.

[52]Paul A. Packan, "Pushing the Limits: Integrated Circuits Run Into Limits Due to Transistors," *Science*, September 24, 1999. While scaling has driven the progress of semiconductor power for decades, it is not the only source of competitive advantage among firms. Innovation in both hardware and software presents the high-value-added features of U.S. firms and highlights the importance of establishing market leadership through innovation in product design.

The industry also faces the additional challenge of the soaring cost of manufacturing chips. When Intel was founded in 1968, a single machine used to produce semiconductor chips cost roughly $12,000. Today a chip-fabricating plant costs billions of dollars, and the expense is expected to continue to rise as chips become ever more complex. Adding to this concern is the realization that capital costs are rising far faster than revenue.[53] In 2000, for example, average total expenditures for a six-inch-equivalent "wafer" were $3,110, an increase of 117 percent over the average total costs for a six-inch wafer in 1989, and a 390 percent increase since 1978.[54]

The consensus in the engineering community is that improvements, both large and small, will continue to uphold Moore's Law for another decade or so, even as scaling brings the industry very close to the theoretical minimum size of silicon-based circuits.[55] To the extent that physical constraints or cost pressures limit the continued growth of the industry, however, they will necessarily influence the role of the industry in stimulating productivity growth in the broader economy. As capital costs rise, fabrication capacity increases, and alternative business models (e.g., the foundry system) gain prominence, the competitive position of some U.S. device manufacturers (e.g., the merchant semiconductor producers) may be challenged,[56] while other U.S. firms may prosper in the new environment.

Greater Vertical Specialization: The Emergence of the Foundries and Fabless Firms[57]

Significant shifts are occurring in the semiconductor industry with a strong

[53]Charles C. Mann, "The End of Moore's Law?" *Technology Review*, May/June 2000, <http://www.technologyreview.com/magazine/may00/mann.asp>.

[54]These statistics originate from the Semiconductor Industry Association, *2001 Annual Databook: Review of Global and U.S. Semiconductor Competitive Trends, 1978-2000*. A wafer is a thinly sliced (less than 1 millimeter) circular piece of semiconductor material which is used to make semiconductor devices and integrated circuits.

[55]See discussion by Bob Doering of Texas Instruments on "Physical Limits of Silicon CMOS and Semiconductor Roadmap Predictions," at the National Academies Symposium, *Productivity and Cyclicality in the Semiconductor Industry*, held at Harvard University, September 24, 2001.

[56]See remarks by George Scalise, President of the Semiconductor Industry Association, at the National Academies Symposium; *Productivity and Cyclicality in the Semiconductor Industry*, held at Harvard University, September 24, 2001.

[57]Much of the information and description in this section is adapted from a presentation by D.A. Hodges and R.C. Leachman, "The New Geography of Innovation in the Semiconductor Industry." For the full presentation, see <http://web.mit.edu/ipc/www/hodges.pdf>. See also R. C. Leachman and D. A. Hodges, "Benchmarking Semiconductor Manufacturing," IEEE Transactions on Semiconductor Manufacturing, TSM-9, pp. 158-169 (May 1996). <http://radon.eecs. berkeley.edu/~hodges/BenchmarkingSM.pdf>

trend toward more vertically specialized firms.[58] This increased "vertical disintegration" means that more firms are either dedicating their resources to the manufacture of chips designed by others (the foundry model) or are choosing to specialize solely in the design of chips (the "fabless" firm).

Several factors are encouraging greater vertical specialization. For example, the opportunities now available in "system-on-a-chip" design, a focus on specialized markets, new market channels, and the different skill requirements for design and manufacture are all contributing to this trend.[59] For design firms, the different time scales and levels of investment necessary for manufacturing have helped to accelerate the trend toward vertical specialization. Further accelerating these structural changes is the reduced attractiveness of niche markets for many Integrated-Device Manufacturers (IDMs). These structural shifts mean that effectively competing in the design and manufacture of chips requires differently skilled firm workforces making it difficult for some IDMs to competitively engage in both design and manufacturing.

Another major driver enabling the separation of design and manufacturing is the maturation and standardization of both electronic design automation and process technologies. These include CMOS (complimentary metal-oxide semiconductor) technology—the microelectronic technology used in almost all microprocessors, memory products, and application-specific integrated circuits (ASICS). More accurate physical process developments, more accurate process characterization for design, better software design tools, and the rise of foundries with state-of-the-art manufacturing technology have all facilitated this structural transformation.

Foundries

The manufacturing segment of the market is increasingly characterized by the foundry, whose focus is on high productivity and rapid turnaround from design to product. Foundries also permit production of smaller batches of specialized chips at commodity-like costs. These fabrication facilities ("fabs"), where firms produce semiconductors under contract with other companies, have expanded rapidly, particularly in East Asia. Taiwan Semiconductor Corporation (TSMC) and United Semiconductor Corporation (UMC) hold about 65 percent of global market share of foundry-based production, with firms from other Asian nations and the United States (IBM) holding the remainder.[60] Tight quality control,

---

[58]The global supply chain in the semiconductor industry begins with a $10 billion (in year 2000) raw material segment, where Japan and Germany dominate in silicon refining. The next step in the chain is manufacturing equipment—an approximately $45 billion per year segment where Japan and the European Union lead in optical design, while the U.S. leads in other equipment with companies like Applied Materials and KLA-Tencor. The next segment of the market, Electronic Design Automation (EDA), is dominated by U.S. firms such as Synopsis, Cadence, and Mentor. *Ibid.*

[59]*Ibid.*

[60]*Ibid.*

rigorous manufacturing discipline, a fast, standardized production process, and short cycle times are typical characteristics of a successful foundry. Such foundries also provide top-of-the-line Internet-based customer service, while also protecting proprietary customer designs. The big gain is that foundries permit firms to bring products to market without raising and investing the capital required for an advanced manufacturing facility. This model of production can offer substantial cost savings in manufacturing new generations of chips. These cost savings may be accentuated by lower capital costs that reflect the impact of preferential tax treatment and more direct government subsidies for industries that are viewed as strategic by government policy makers in countries such as Taiwan.[61]

Fabless Design Firms

The emergence of design houses or "fabless" firms—firms which specialize in the design of semiconductors only and do not produce them—is yet another sign of vertical specialization in the industry and functions congruently with the foundry model of production. The modest amount of investment necessary for market entry, the short time to market, and the prospect of rapid growth have established design firms as high-risk, high-reward entrepreneurial vehicles. This degree of specialization has also accelerated the pace of product innovation. Success in the design segment of the semiconductor supply chain is determined by providing the right product features at the right market time. The integration of system and circuit designers and the achievement of flawless design discipline are key to a successful design firm. Fabless firms also need access to high-quality manufacturing—through foundries—and for new design firms, a path to market entry.

Over time, the development of new technologies, especially manufacturing technologies, may become more closely associated with the foundries themselves. Significant technical capability and know-how may be transferred, particularly as design houses, such as those in Hsinchu Park in Taiwan, are increasingly involved. The consequences of this phenomenon are not clear, nor are they unidirectional. As noted above, in the near term, the availability of low-cost, high-quality fabrication facilities can work to the benefit of U.S. design firms.[62] In periods of surplus manufacturing capacity, design firms can do well by benefiting

[61]Because the foundry concept was considered risky at TSMC's founding, 44 percent of TSMC's initial capitalization was provided by the Taiwanese Cabinet's Development Fund. See the discussion of foundries in Howell, *op. cit.*

[62]According to *IC Insights, Inc.* seven of the top fabless firms are U.S. based (e.g., Qualcomm, Nvidia, and Xilinx). See <www.siliconstrategies.com/story/OEG20020329S0036>.

from the foundries' rapid production of new products without bearing the burden of the long lead times and high fixed costs of modern fabs. Design firms may do less well in periods of high capacity utilization, particularly if they face competition in the same market from their suppliers.[63]

The push toward vertical specialization contrasts with much of the U.S. industry, notably the merchant device manufacturers, which typically house both design and production under one roof. This specialization poses interesting questions for U.S. producers, and ultimately for the trajectory of the U.S. economy. As described below, much of the resurgence in the U.S. industry as a whole is derived from its renewed capability in manufacturing, combined with its strength in design. Improved manufacturing was central to the recovery of the industry, and manufacturing continues to play an essential role in the development of new technology for the industry.[64] Manufacturing expertise and the construction of new fabrication facilities drive infrastructure development. The demands of manufacturing advance have conditioned the development of the nation's semiconductor technology infrastructure (i.e., suppliers of manufacturing equipment, test equipment, and materials). Should the locus of manufacturing continue to shift overseas, it will erode this process of technology and infrastructure development associated with manufacturing.[65]

In part, the trend toward expanded overseas manufacturing reflects the global scale of the industry and its rising capital costs for fabrication facilities. This trend also reflects the active industrial policies of leading East Asian economies.[66] The combination of greater vertical specialization and the impact of national policies to support local growth of the industry are changing the competitive environment. Increasingly, U.S. producers face challenges from the substantial capacity generated by government-supported fabrication facilities abroad.[67] The

[63]Leachman and Hodges identify inadequate access to manufacturing capability as one of the causes for failure of design firms. See Leachman and Hodges, *op.cit.* See also Howell, *op.cit.*

[64] See Macher, Mowery, and Hodges, *op. cit.*, passim.

[65]This would, in turn, compromise to some extent the contributions of the industry to higher wages and increased productivity. Over time, it could also mean a shift in the technological lead the country has enjoyed, with its attendant implications for national security. These concerns were the topic of a conference, *The Global Computer Industry Beyond Moore's Law: A Technical, Economic, and National Security Perspective*," a Joint Strategic Assessment Group (SAG) and Defense Advanced Research Projects Agency (DARPA) Conference, January 14-15, 2002, Tyson's Corner, VA.

[66]For example, Singapore and Malaysia have contributed significant public funds and have extended unprecedented tax incentives to companies constructing "fabs," while Taiwan has approximately 100 "design houses," also supported through various incentives by the government. See Howell, *op. cit.*

[67]See "The Great Chip Glut," *The Economist*, August 11, 2001.

recovery of the U.S. industry, described below, does not mean that its current competitive strength can be taken for granted.[68]

A related issue of significant concern is the impact of these trends on the R&D funding that drives the industry. To date, the foundries have tended to be fast followers—rapidly adopting the new manufacturing technologies that drive the industry but making relatively modest R&D investments of their own. As the foundries gain market share, it is not clear whether the R&D investments required to sustain the industry's exceptional growth will continue to be made.[69]

CHALLENGES OF MAINTAINING SUFFICIENT HUMAN CAPITAL

The unprecedented technical challenges faced by the industry underscore the need for talented individuals—the so-called "architects" of the future—to devise new solutions to these technical challenges.[70] This need is emerging at the same time as the pool of available skilled labor is shrinking. There is widespread concern about the supply of workers and researchers for the semiconductor industry. Almost without exception, top management and researchers from the leading consortia and companies expressed misgivings about the adequacy of the labor force to meet foreseeable demand as the industry begins to recover from its current steep cyclical downturn. The increasing technical challenges faced by the industry are compounding this need and may make competition for skilled labor an integral part of international competition within the industry.

Historically, the U.S. government has supported human resources through its system of funding basic research at universities, whereby the work and training of graduate students and postdoctoral scholars are supported by research grants to principal investigators. However, the rapid growth in demand for skilled engineers, scientists, and technicians is generating challenges for the industry and national policy on several fronts.

In recent years, federal funding for university research has declined steeply in sciences relevant to information technologies—such as mathematics, physics, and engineering (See Figure 8). This falloff in U.S. production of undergraduate

---

[68]As a recent National Research Council assessment of the resurgence of the industry concluded: "Some foreign producers, notably Taiwanese semiconductor firms, now are entering markets traditionally dominated by U.S. producers, a development that will intensify pressure on U.S. firms and increase the importance of manufacturing performance for competitive leadership." See Macher, Mowery, and Hodges, *op.cit.*, p. 283-284.

[69]A significant portion of the R&D burden is devolving to the equipment producers.

[70]David Tennenhouse, vice-president and Director of Research and Development at Intel, emphasized this point in his presentation at *The Global Computer Industry Beyond Moore's Law: A Technical, Economic, and National Security Perspective*, a Joint Strategic Assessments Group (SAG) and Defense Advanced Research Projects Agency (DARPA) Conference, January 14-15, 2002.

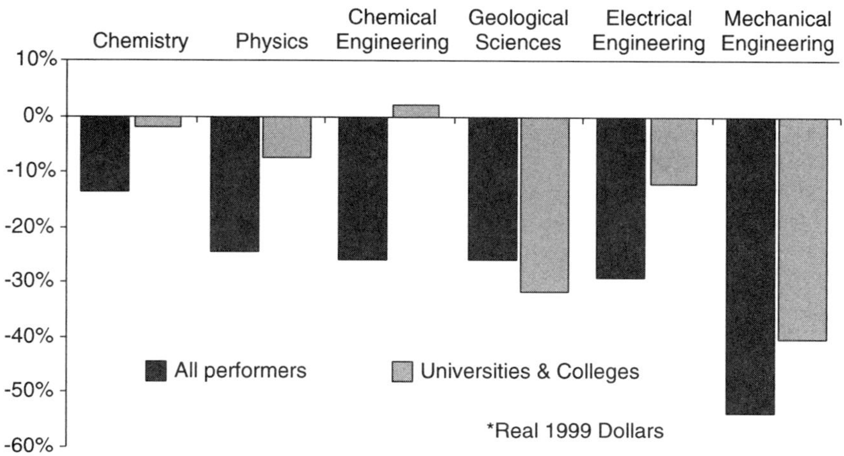

FIGURE 8 Real Changes in Federal Obligations for Research FY 1993–1999 (Real 1999 Dollars).[71]

engineers and graduate students in disciplines such as chemistry and physics is arguably linked to the decline in federal research support in these fields.

For more than a decade, U.S. graduate schools have depended on large numbers of foreign-born students and faculty to staff their laboratories and teach in their programs. The United States continues to attract foreign students, as well as scientists and engineers, who want to study, live, and work here. Increasingly, this group of highly skilled workers encounters significant inducements to return home.[72]

Most disconcerting from the U.S. perspective is the fact that the number of individuals graduating from U.S. universities with electrical engineering degrees has exhibited a declining trend since the mid-1980s (see Figure 9).[73] Some well-

[71]See M. McGeary and S.A. Merrill, "Recent Trends in Federal Spending on Scientific and Engineering Research: Impacts on Research Fields and Graduate Training," Appendix A in National Research Council, *Securing America's Industrial Strength.* Washington D.C.: National Academy Press, 1999.

[72]See Howell, *op. cit.*

[73]In 1988 approximately 24,000 people graduated from U.S. universities with bachelor's degrees in electrical and electronic engineering. By 1997 this total had fallen below 14,000, and it is not forecast to increase significantly in the foreseeable future, whereas current estimates put the production of engineers in China at about 150,000 per year. Engineering Workforce Commission statistics and SRC projections presented by Dr. Michael Polcari of IBM at the *Symposium on National Programs to Support the Semiconductor Industry*, October 2000.

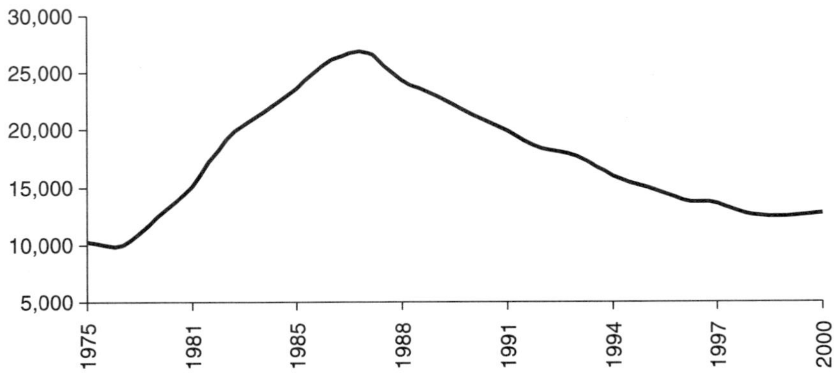

FIGURE 9 Electrical Engineering Graduates: Bachelor's Degrees Earned, 1975-2000.
SOURCE: National Science Foundation, *Science & Engineering Indicators 2000*, 1975-1987 Engineering Workforce Commission 1988-2000.

informed industry representatives see a growing problem. For example, John Kelley of Novellus observes that the problems facing the supplier industry are "fairly simple and straightforward." The first is the undersupply of talented graduate students. He said that the good news is that many of the students they have hired trained through the Semiconductor Research Corporation (SRC) and were prepared to "hit the ground running." The bad news is that there are not enough of them, and the situation seems to be worsening. Many graduates have moved away from the semiconductor industry into other areas, such as nanotechnology, he said, and the professors have been going "where the money is."[74]

In the sheer production of engineers, the United States lags its current and future competitors in the microelectronics industry (See Figure 10). Japan now produces about 63 percent more engineers per year than the U.S., while China produces more than twice as many—roughly 136 percent more.[75] While there may be issues of quality and industry-related experience, in sheer numbers Asia

[74]See Panel V of the proceedings in this volume. Other industry representatives and analysts echo this view. See the presentations of Michael Polcari of IBM, Kalman Kaufman of Applied Materials, and George Scalise of the Semiconductor Research Corporation in the Proceedings of this volume. For an analysis of the high demand in emerging areas such as bioinformatics, see Paula Stephan and Grant Black, "Bioinformatics: Emerging Opportunities and Emerging Gaps" in National Research Council, *Capitalizing on New Needs and New Opportunities: Government-Industry Partnerships in Biotechnology and Information Technologies.*

[75]National Research Council calculations derived from the National Science Foundation's, *Science and Engineering Indicators 2000*. It is important to note that material scientists are increasingly engaged by the semiconductor industry. A recent study (July 2002) by the OECD asserts that "there

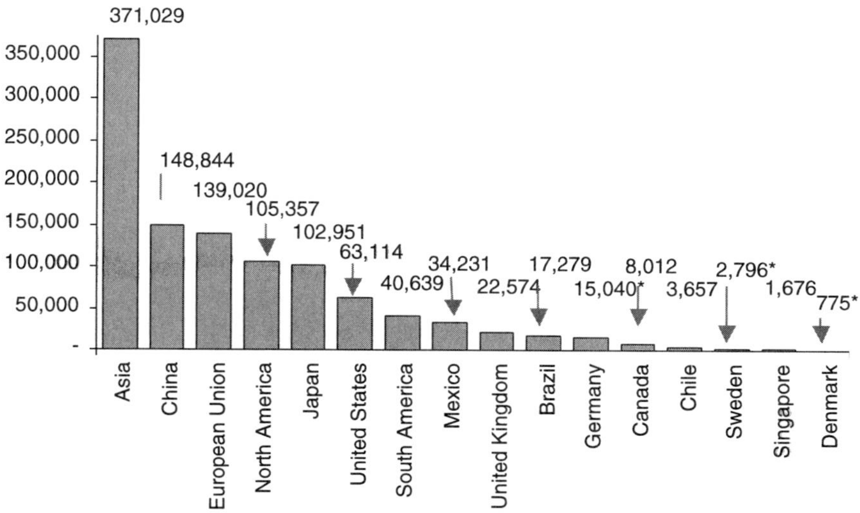

FIGURE 10 Number of First University Degrees in Engineering, 1997.
*Long Degree programs
SOURCE: National Science Foundation, *Science & Engineering Indicators 2000.*

as a region produces more engineers per year than the United States by almost a factor of six. The European Union produces more than double the U.S. output of engineers.[76]

In terms of the percentage of bachelor's degrees awarded each year in engineering (out of the total number of bachelor's degrees in the U.S.) compared to equivalent degrees and 4 to 6-year programs in other nations, the U.S. lags far behind (see Figure 11).[77] Almost half—roughly 46 percent—of all bachelor's

is indeed some evidence of *tightness in labour markets* for particular categories of IT workers," and further suggests that "the main issue of concern for policy makers and firms should be the gap between the skills of current and future IT workers and those sought by firms." For the details of this analysis see Vladimir Lopez-Bassols, "ICT Skills and Employment," STI Working Papers. Directorate for Science, Technology, and Industry, OECD, DSTI/DOC (2002) 10, July 17, 2002. See also National Science Foundation, Division of Science Resources Statistics, *Science and Engineering Degrees: 1966-2000.* NSF 02-327, Author, Susan T. Hill (Arlington, Va 2002) <http://www.nsf.gov/sbe/srs/nsf02327/pdf/nsf02327.pdf>.

[76]For further discussion of the implications of these statistics see comments by Mary L. Good, President of the American Association for the Advancement of Science, in "Scientist's Call to Action S.F. conference opens with plea for Cabinet position," *San Francisco Chronicle*, February 16, 2001

[77]Countries have different time periods for first-degree programs (i.e., first university degrees are not always academically equivalent). In European nations, for example, short degree programs are three years long, while long degree programs are 4 to 6 years long. In the analysis here, we use the long degree programs as our basis of comparison.

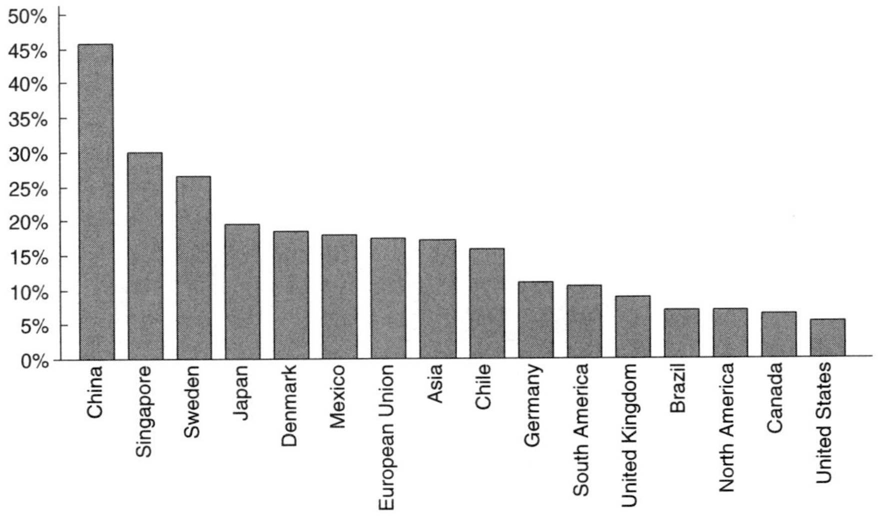

FIGURE 11 Percentage of Bachelor's Degrees in Engineering, 1997.
SOURCE: National Science Foundation, *Science & Engineering Indicators 2000.*

degrees in China each year are in the field of engineering, while the latest data show that out of all U.S. undergraduates each year, about 5 percent earn degrees in engineering. In terms of relative production of engineers (i.e., taking the ratio of engineering degrees to total degrees in each nation and then comparing it across countries) China outpaces the United States by more than 8.5 times, while countries such as Sweden and Japan outperform the United States by about 5 times and 3.7 times, respectively.[78]

These calculations are, of course, indicative of broad trends and make no qualitative assessment. They also reflect, at least in some cases, the national priorities of countries eager to master the technical requirements of the modern economy. Still, the disparities in the education of engineers are striking. Perhaps more of a cause for concern is that the declines in training of U.S. students in these fields are not based on estimates of national needs, but rather the result of unplanned reallocations of resources resulting from the post Cold War adjustments to the U.S. innovation system.[79] Over time, the results of these reductions

[78]National Research Council calculations derived from the National Science Foundation's, *Science and Engineering Indicators 2000.*

[79]As a recent report by this Committee observed, "for the most part, the shifts in federal research spending...have not been the result of a conscious national debate on priorities." One well-informed observer described some of the shifts in research funding as "random disinvestments," unintended but nonetheless injurious to national progress in R&D. See National Research Council, *Capitalizing on New Needs and New Opportunities: Government-Industry Partnerships in Biotechnology and Infor-*

may affect the ability of the United States to maintain its leading position in semiconductors, computers, and related industries, with potentially significant consequences for the nation's level of economic growth and national security.

NATIONAL PROGRAMS TO SUPPORT THE SEMICONDUCTOR INDUSTRY

The conviction that high-technology industries are fundamental to technological competency, national autonomy, economic growth, and high-wage, high-value-added employment is widespread in the global community and not least among the major trading partners of the United States.[80] Reflecting this conviction, many countries devote substantial resources to support and sustain high-technology industries within national and regional economies. "Governments believe that the future of their countries depends on the composition of their economies, and for the most part they see their success as nations defined by their relative success in these specific efforts."[81] This belief has stimulated increasingly vigorous international competition, especially in sectors that countries deem to be economically strategic.[82] Consequently, many governments have adopted policies to support nationally based firms in the hope of capturing the benefits of this industry, such as higher-wage jobs, increased competitiveness, and future government revenue. Information technology industries are often a target of these national policies. For example, as Laura Tyson noted in her 1992 study:

> The semiconductor industry has never been free of the visible hand of government intervention. Competitive advantage in production and trade has been heavily influenced by policy choices, particularly in the United States and Japan. Some of these choices, such as the provision of public support for basic science, R&D, and education in the United States, have had general, not industry-specific objectives. But other choices, such as the provision of secured demand for industry output through military procurement in the United States and through preferential procurement of computers and telecommunications equipment in Japan, have been industry specific in intent and implementation.[83]

---

mation Technologies, p. 61. See also Michael McGeary, "Recent Trends in Federal Funding of Research and Development Related to Health and Information Technology," in National Research Council, *Capitalizing on New Needs and New Opportunities: Government-Industry Partnerships in Biotechnology and Information Technologies,* Washington, D.C., National Academy Press, 2001.

[80]For a discussion of the importance of high-technology industries to national economies and the measures some countries adopt to capture these benefits, see National Research Council, *Conflict and Cooperation in National Competition for High-Technology Industry*, National Academy Press, Washington, D.C., 1996, especially box on pp. 33-35.

[81]*Ibid.*

[82]Alan Wm. Wolff, Thomas R. Howell, Brent L. Bartlett, and R. Michael Gadbaw, eds., *Conflict Among Nations: Trade Policies in the 1990s*, Westview Press, San Francisco, 1992, p. 528.

[83]Tyson, *op. cit.*, p. 85. For a review of government programs designed to develop and support the technologies underpinning the semiconductor industry, see Howell, *op. cit.*

As Tyson notes, the U.S. government provided early procurement-based funding to promote the development of semiconductors for both military and space exploration programs.[84] The U.S. government's subsequent role in assisting the commercial semiconductor sector was more controversial and more restrained.[85] However, the fall-off in R&D support identified above is not confined to semiconductors. The United States has also reduced the scale of its R&D investment in computers and computer architecture, in both absolute and relative terms.[86] The explanation for these reductions is complex, but these U.S. reductions run contrary to global trends. The lag effects of what have been described as "random disinvestments" may compromise the U.S. government's ability to achieve other societal goals over the long term.

In contrast, governments abroad are active in supporting their respective industries, notably semiconductors, as Box C indicates. It is important to recall that this policy interest and support is not a new development. The Japanese government, for example, recognizing the country's position as a late entrant in semiconductors, adopted a series of policies to jump-start its industry in the 1970s. Under the guidance of the Ministry of International Trade and Industry (MITI), the country made a sustained effort to promote a vibrant domestic semiconductor industry, notably through the successful VLSI Program.[87] In addition, the verti-

[84]Government procurement enabled U.S. firms to improve yield and efficiency through volume production and encouraged wider application of integrated circuit technology, first in military and then in commercial technologies. National Bureau of Standards, *The Influence of Defense Procurement and Sponsorship of Research and Development on the Development of the Civilian Electronics Industry*, June 30, 1977.

[85] Howell, *op. cit.*

[86] In a recent report for this study, Kenneth Flamm documents this downturn. He notes that it "would not be a source of concern if we were convinced that computing technology had matured" (i.e., that it was no longer an area with a high social payoff for the U.S. economy). Yet the contrary is the case. Given the potential for high-performance computing as a complement to technical advance in other high-payoff areas like biotechnology, Flamm suggests that it would be prudent for the United States "to plant more seedcorn in this particular field." Kenneth Flamm, "The Federal Partnership with U.S. Industry in U.S. Computer Research: History and Recent Concerns" in National Research Council, *Capitalizing on New Needs and New Opportunities: Government-Industry Partnerships in Biotechnology and Information Technologies,* p. 220.

[87]See the partial list of joint research and development projects in microelectronics sponsored by MITI in Tyson, *op. cit.*, p. 96. Tyson considers the VLSI (Very Large-scale Integration) Program to be the most successful of the Japanese programs. VLSI focused on cooperative R&D designed to help Japanese firms reach leading-edge capabilities in the production of both memory devices and logic circuits. *Ibid*, 97. The table drawn from Howell, *op.cit.,* gives an updated summary of the main national and regional programs. For a review of Japanese consortia, see Lee G. Branstetter and Mariko Sakakibara, "When Do Research Consortia Work Well and Why? Evidence from Japanese Panel Data," *American Economic Review,* 92(1): 143-59, 2002.

cally integrated structure of Japanese industry with its lower-cost capital proved to be a major advantage with respect to the capital-intensive investments required for manufacturing facilities, especially for DRAMs (Dynamic Random Access Memory), the technology driver at the time.[88] Competition in semiconductors in the early 1980s was thus characterized—to a considerable extent—by DRAM "capacity races."[89] Aided by lower capital costs and less constrained by capital markets, Japanese firms undertook a massive capacity build-up in the early 1980s, accelerating their gains through highly aggressive price-cutting. The worldwide DRAM market share of U.S. industry sank from roughly 90 percent in the late 1970s to less than 10 percent by 1984-85, with many U.S. firms exiting the DRAM market entirely.[90] The drastic effects of the Japanese competition led many informed U.S. observers to question the future viability of the U.S. semiconductor industry.[91]

U.S. Policy Initiatives

As competition from the Japanese producers intensified in the 1980s, the industry launched a series of initiatives, some in cooperation with the government, to strengthen its domestic capabilities (e.g., the Semiconductor Research Corporation) and later to stop what it considered to be unfair trade practices by Japanese producers through a series of bilateral trade agreements.[92] The range of these initiatives, as shown in Box B, was extensive.

---

[88] See Macher, Mowery, and Hodges, *op. cit.*, p. 264.

[89] *Ibid.*, p. 276.

[90] See Tyson, *op. cit.* Intel, Advanced Micro Devices, and National Semiconductor all withdrew from the DRAM market. Intel, now among the most profitable semiconductor manufacturers in the world, nearly collapsed in the 1984-1985 recession. Macher, Mowery, and Hodges, *op. cit.*, p. 246. As the MIT Commission on Industrial Productivity noted in 1989, "The technological edge that once enabled innovative American companies to excel despite their lack of financial and market clout has disappeared, and the Japanese have gained the lead." *Ibid.* Despite these difficulties, three important American semiconductor companies did remain in the DRAM race: Motorola, Texas Instruments, and then start-up Micron Technology; the last is now the world's second largest producer of DRAMs.

[91] Laura Tyson provides an excellent analysis of the competition for dominance in the semiconductor industry. See Tyson, *op. cit.* Chapter 4, "Managing Trade and Competition in the Semiconductor Industry," pp. 85-113. For a more recent and more comprehensive discussion of the Semiconductor Trade Agreement and its impact, see also Flamm, *Mismanaged Trade?: Strategic Policy in the Semiconductor Industry*.

[92] For a first-hand discussion of the U.S. concerns and the trade negotiations during this period, see Clyde Prestowitz, *Trading Places*, New York: Basic Books, 1988. See also Wolff, Howell, Bartlett, and Gadbaw, *op. cit.*

Box B
Industry-Government Cooperation on Semiconductors[a]

1982 The **Semiconductor Research Corporation (SRC)** is founded to provide funding for basic semiconductor research at American universities. The SRC becomes an independent affiliate of the Semiconductor Industry Association (SIA).

1983 The **U.S. – Japan Working Group on High Technology**, a bilateral government effort to address semiconductor trade conflicts, is created.

1984 The **National Cooperative Research Act** is signed into law by President Reagan, encouraging joint R&D consortia by reforming U.S. antitrust law.

The **Trade and Tariff Act** of 1984 becomes law, authorizing negotiation of high-tech trade issues and tariff elimination.

The **Semiconductor Chip Protection Act** becomes law, providing a new form of intellectual property protection.

1985 With the support of the industry, the United States and Japanese governments completely eliminate tariffs on imported semiconductors.

SIA files a Section 301 Petition with the U.S. government citing unfair Japanese market barriers. The industry argues that U.S. share of the Japanese market at the time is 8.5 percent versus a U.S. worldwide market share outside Japan of more than 70 percent.

1986 The U.S. Department of Commerce finds that Japanese semiconductor firms are selling (or dumping) memory chips in the U.S. market substantially below the cost of production.

The U.S. and Japan sign a bilateral **Semiconductor Trade Agreement** to eliminate dumping and open the Japanese market to foreign semiconductors.[b]

The Defense Science Board (DSB) taskforce report on U.S. semiconductor dependency becomes public; it calls for a semiconductor manufacturing technology institute involving government-industry collaboration.

1987 The industry consortium **SEMATECH** is founded by fourteen U.S. semiconductor manufacturers. Its mission is to sponsor and conduct research in semiconductor manufacturing technology for the U.S. industry.[c]

Citing Japan's failure to comply with the terms of the 1986 trade agreement, President Reagan imposes 100 percent duties on $300 million worth of Japanese goods.

1988 Congress approves formation of **the National Advisory Committee on Semiconductors (NACS)**, made up of top-level officials from government and industry, to report on proposals for a "national semiconductor strategy."[d]

1990 The semiconductor industry joins with the computer systems industry to support a new semiconductor trade agreement with Japan to ensure that the commitments made under the 1986 trade accord are fulfilled.

Semiconductor Trade Agreement renewed. The United States and Japan sign a new semiconductor trade agreement committing Japan to open its market to foreign semiconductors and providing a strong deterrent to dumping. Two significant improvements over the 1986 trade agreement are Japan's public recognition of a 20 percent foreign market share commitment and the inclusion of a "fast-track" approach to resolving dumping allegations.

1992 **The Semiconductor Roadmap** process begins. More than 150 technologists from industry, government, and academia gather in Dallas for a semiconductor industry technology workshop designed to produce a roadmap for the nation's semiconductor research needs for the next 15 years.

1993 SEMATECH announces achievement of one of its primary technical goals: demonstrating 0.35-micron manufacturing capability on all American-made equipment.

As President Clinton comes into office, Japan and the United States announce that foreign semiconductor manufacturers achieved a 20.2 percent share of Japan's chip market in the fourth quarter of 1992, in accordance with the U.S.-Japan semiconductor trade pact.

---

[a] Adapted from Andrew A. Procassini, *Competitors in Alliance: Industry Associations, Global Rivalries, and Business-government Relations*, Westport, CT: Quorum Books, 1995, pp. 195-196.

[b] Described below.

[c] SEMATECH incorporated in 1987 with thirteen members. An additional firm joined in 1988. In 1990, three member firms withdrew from the consortium. In January 1992, government funding was renewed for five additional years.

[d] The semiconductor industry's difficulties were also of serious concern to Federal officials interested in maintaining the highest level of U.S. national security. This concern is highlighted in the creation of the NACS. One of the NACS missions was to focus on the dependency of modern weapons systems on state-of-the-art semiconductor devices. Specifically, under the legislation to create NACS, Congress notes in its findings that "modern weapons systems are highly dependent on leading-edge semiconductor devices, and it is counter to the national security interest to be heavily dependent upon foreign sources for this technology." The charter further states that this Committee shall "identify new or emerging semiconductor technologies that will impact the national defense or United States competitiveness or both." For the objectives set forth for NACS, see <http://www4.law.cornell.edu/uscode/15/4632.html>.

Reducing Legal Constraints on Cooperative Research

In addition to concerns about foreign trade practices, there was a widespread perception in the 1980s that U.S. technological leadership was slipping and that greater cooperation among companies would be required. Existing antitrust laws and penalties were seen as too restrictive and as possibly impeding the ability of U.S. companies to compete in global markets. This perception resulted in the passage of the National Cooperative Research Act (NCRA) in 1984. This act encouraged U.S. firms to collaborate on generic, pre-competitive research. To gain protection from antitrust litigation, NCRA required firms engaging in research joint ventures to register them with the U.S. Department of Justice. By the end of 1996, more than 665 research joint ventures had been registered. Among these, SEMATECH has perhaps been one of the most significant private R&D consortia.[93] In 1993, Congress again relaxed restrictions—this time on cooperative production activities—by passing the National Cooperative Research and Production Act, which enables participants to work together to apply technologies developed by their research joint ventures.[94]

Trade Agreements

Efforts to address issues in U.S. manufacturing quality (see below) proceeded in parallel with efforts to resolve questions about Japan's trading practices. A series of trade accords between Japan and the United States did not resolve trade frictions between the two countries, nor did the agreements redress the steadily declining U.S. market share. As a result, a near-crisis sentiment spread through the U.S. industry during the mid-1980s.[95] At the urging of the industry, the federal government took several policy initiatives designed to support the U.S. industry.

---

[93]See Macher, Mowery, and Hodges, Chapter 10, *op. cit.*, p 277.

[94]See Kenneth Flamm and Qifei Wang, "SEMATECH Revisited: Assessing Consortium Impacts on Semiconductor Industry R&D"; A. L. Link, "Research Joint Ventures: Patterns From *Federal Register* Filings", *Review of Industrial Organization* 11, No. 5 (October): 617-28, 1996, and N.S. Vonortas, *Cooperation in Research and Development,* Norwell, MA: Lower Academic Publishers, 1997. See also, National Science Board, *Science and Engineering Indicators, 1998,* Arlington, VA: National Science Foundation, 1998.

[95]For an industry perspective, see the account by Charles E. Sporck (with Richard L. Molay), *Spinoff: A Personal History of the Industry that Changed the World*, Saranac Lake, New York: Saranac Lake Publishing, 2001. Sporck recounts that, in this period, when memory products (DRAMs) represented a major percentage of the industry, "the core strategy of the Japanese industry was to add manufacturing capacity at a pace unrelated to market share" and to price products below U.S. producers. p. 244.

After several unsatisfactory trade accords, there was a significant shift in U.S. policy on trade in semiconductors, notably through the conclusion of the 1986 Semiconductor Trade Agreement (STA) with Japan.[96] The agreement sought to improve access to the Japanese market for U.S. producers and to end dumping (selling products below cost) in U.S. and other markets.[97] After President Reagan's decision to impose trade sanctions, the STA brought an end to the dumping in the U.S. and other markets and succeeded in obtaining limited access to the Japanese market for foreign producers, in particular, Korean and, later, Taiwanese DRAM producers.[98] In fact, one of the most significant impacts of the accord was that it established a price floor for DRAMs, thus encouraging new entrants and, thereby, making the global DRAM market competitive once again.

As Laura Tyson points out, the trade agreement was a first in many respects. It was the first major U.S. trade agreement focused on a high-technology, strategic industry, and the first one motivated by concerns about the loss of competitiveness rather than the loss of employment.[99] It was unusual in that the agreement concentrated on improving market access abroad rather than restricting access to the U.S. market. And unlike other bilateral trade deals, it sought to regulate trade (i.e., end dumping) not only in trade between the United States and Japan but in other global markets as well. It also included, for the first time, the threat of trade sanctions should the agreement not be respected. As such, it signaled a significant shift in U.S. trade policy.[100]

The Creation of SEMATECH

A second major step in this regard was the industry's decision to seek a partnership between the government and a coalition of like-minded private firms to form the SEMATECH consortium, whose purpose was to revive a seriously weakened U.S. industry through collaborative research and pooling of manufac-

---

[96]See Prestowitz, *op. cit.*, and Wolff, Howell, Bartlett, and Gadbaw, *op.cit.* For additional discussion of the Semiconductor Trade Agreement, see National Research Council, *Conflict and Cooperation in National Competition for High-Technology Industry,* pp. 132-41. For a discussion of dumping/anti-dumping trade-policy debate, see pp. 82-87.

[97]As part of the agreement, dumping suits in the U.S. and the Section 301 case were suspended in return for agreement to improve market access and terminate dumping. A side letter called for a 20 percent market share for foreign firms within five years. Tyson, *op. cit.*, p. 109.

[98]For a discussion of the impact of the agreement, see Flamm, *Mismanaged Trade? Strategic Policy and the Semiconductor Industry*. For an earlier assessment see Tyson, *op. cit.*, pp. 136-143.

[99]*Ibid*, p. 109.

[100]*Ibid*.

turing knowledge.[101] A central element of the challenges facing the U.S. semiconductor industry was manufacturing quality. By the mid-1980s, the leading U.S. semiconductor firms had recognized the strategic importance of quality and begun to initiate quality improvement programs. A key element in this effort was the formation of the consortium, which in part reflected the belief that the Japanese cooperative programs had been instrumental in the success of Japanese producers.[102]

The decision to form the SEMATECH consortium represented a significant new experiment in government-industry cooperation in technology development. The Silicon Valley CEOs of the U.S. semiconductor companies hesitated about cooperating with each other, and they were even more hesitant about cooperating with the government—an attitude mirrored in some quarters in Washington.[103] Despite these hesitations, the consortium was conceived and funded under the Reagan administration. The formation of the consortium represented an unusual collaborative effort, both for the U.S. government and for the fiercely competitive U.S. semiconductor industry.[104]

---

[101]As noted, the Semiconductor manufacturers—secretive, often adversarial competitors—faced an early critical challenge to effective cooperation within SEMATECH in the fear that cooperation would reveal proprietary secrets to competitors. As Larry Browning and Judy Shetler note in their comprehensive study of the consortium, this initial reluctance centered around three questions: "What technology could they use for the mission of improving performance?" "What firm would they want to contribute a cutting-edge proprietary process?" and "What would be the use in working on anything else?" See Larry D. Browning and Judy C. Shetler, *SEMATECH: Saving the U.S. Semiconductor Industry,* College Station: Texas A&M University Press, 2000, p. 22.

[102]*Ibid.,* Chapter 1.

[103]*Ibid.* pp. 21-23. Browning and Shetler record that the Treasury and Council of Economic Advisers was adamantly opposed to government funding of a consortium; the Departments of Defense and Commerce were supportive. *Ibid*, p. 24.

[104]As Hedrick Smith noted "the mere formation of SEMATECH required a radically new mind-set at some of America's leading high-tech corporations." See Hedrick Smith, *Rethinking America*, New York: Random House, 1995, p. 385. In particular, Charlie Sporck, then CEO of National Semiconductor, and Bob Noyce, Intel co-founder, played a decisive role in garnering the political and industrial support for the formation of the consortium. There are corporate critics of SEMATECH. T.J. Rodgers of Cypress Semiconductors is a frequent critic. For a comprehensive statement of his views, see T. J. Rodgers, "Silicon Valley Versus Corporate Welfare," *CATO Institute Briefing Papers*, Briefing Paper No. 37, April 27, 1998. Rodgers notes that "My battles with SEMATECH started when our engineers were denied access to an advanced piece of wafer-making equipment, a chemical-mechanical polisher (CMP) machine manufactured by an Arizona company [that]...Sematech [had] contracted...to develop....Cypress was denied access to that critical piece of wafer-making equipment, which could have differentiated between winners and losers in the next-generation technology. At that point I became a vocal critic of SEMATECH...." (p. 9). Rodgers also objected to the SEMATECH dues structure, finding the $1 million minimum to be onerous for a relatively small semiconductor-producing firm. He adds "I believe that if Sematech had been formed as a private consortium with a smaller budget, it would have come to its current, more efficient model of operation much more quickly." (p. 10).

This unprecedented level of cooperation, and the important corresponding collaborative activity among the semiconductor materials and equipment suppliers, appear to have contributed to a resurgence in the quality of U.S. products and, indirectly, to the resurgence of the industry.[105] The collective accomplishments and impact of this cooperative activity appear to have been an essential element contributing to the recovery of the U.S. industry. Still, it should be underscored that the consortium's contribution and other public policy initiatives were by no means sufficient to ensure the industry's recovery. Essentially, these public policy initiatives can be understood as having *collectively* provided positive framework conditions for private action by U.S. semiconductor producers.[106]

From the government's perspective, its support for the consortium enabled it to achieve a substantial number of strategic goals. From the industry's perspective, the consortium contributed substantially to improvements in product quality and strengthened the U.S. equipment and supply industry. In combination with the Semiconductor Trade Agreements, the U.S. industry was able to increase its market share in Japan to over 20 percent by December 1992.[107] Perhaps the most appropriate measure of SEMATECH's contribution is the reaction of the market itself—that is, the willingness of industry participants to continue to provide matching funds over a sustained period; and then for these same firms to continue to fund the consortium with private resources and expand it with new members. The emulation of the consortium model by other nations represents an important development.

[105]As a research consortium, SEMATECH's contributions were necessarily indirect. As Browning and Shetler observe, "any effects caused by SEMATECH would, of course, be indirect because, as member-firm executives are disposed to point out, it was ultimately the member companies' factory-production that led to the increased U.S. semiconductor market share. SEMATECH's role has been to develop new manufacturing technologies and methods and transfer them to its member companies, which in turn manufacture and sell improved chips. SEMATECH's precise contribution to the market recovery is therefore difficult to directly assess." See Browning and Shetler, *op. cit.*, p. 208.

[106]Many factors contributed to the recovery of the U.S. industry. It is unlikely that any one factor would have proved sufficient independently. Trade policy, no matter how innovative, could not have met the requirement to improve U.S. product quality. On the other hand, by their long-term nature, even effective industry-government partnerships can be rendered useless in a market unprotected against dumping by foreign rivals. Most important, neither trade nor technology policy can succeed in the absence of adaptable, adequately capitalized, effectively managed, technologically innovative companies. See below.

[107]At the time, Japan was both the largest producer and consumer of semiconductors, hence the importance of access to the Japanese market. See Andrew A. Procassini, *Competitors in Alliance: Industry Associations, Global Rivalries, and Business-government Relations*, Westport, CT: Quorum Books, 1995, p. 194.

Expanding National Programs

In spite of the recent, pronounced downturn in the global semiconductor market, many governments remain active in their support of initiatives to promote the development of advanced microelectronics technology. Others steadily provide substantial incentives to add national industry manufacturing capacity (See Box C). Some nations are also providing substantial incentives to attract native-born and foreign talent to their national industry, in order to meet what some see as an emerging zero-sum competition for skilled labor.[108] In doing so, some national programs are altering the terms of global economic competition with policies that differ in important ways from those of the traditional leaders.[109]

The levels of investment and promotional activity across many countries attest to the importance governments attach to this industry. An important development is the emergence of China, which for reasons of scale and skill is likely to pose a major competitive challenge. This is especially true as cooperation increases with the highly competent Taiwanese industry, a trend which has accelerated dramatically as a result of an influx of foreign investment and skilled manpower. At present China represents a small part of the world semiconductor market, but much new capacity is scheduled to come onstream.[110] This expansion reflects a new Chinese government promotional effort designed to replicate Taiwan's success in microelectronics on a much larger scale in China, drawing heavily on Taiwanese and other foreign capital, management, and technology. China's new policy measures closely resemble those utilized by Taiwan, including the establishment of science-based industrial parks, tax-free treatment of semiconductor enterprises, passive government equity investments in majority pri-

---

[108]Howell, *op. cit.*

[109]As Thomas Howell documents through his extensive field research, there is now a broad area of well-funded programs to support national and regional semiconductor industries, as well as the international cooperation increasingly required in this global industry. See Howell, *op. cit.* For example, state-supported producers in Korea, Taiwan, Malaysia, and now China present special challenges in the competition for global markets in high-technology products. The 1996 STEP report identified this trend and predicted that it would accelerate. It has. See National Research Council, *Conflict and Cooperation in National Competition for High-Technology Industry*, p. 21.

[110]In September 2002, Shanghai-based Semiconductor Manufacturing Corp. (SMIC) had two fabs operational and planned at least two more, and Grace Semiconductor Manufacturing International, also based in Shanghai, had two fabs under construction and two more planned (interviews with senior executives at SMIC and Grace, Shanghai Zhangjiung Science & Technology Park, September 2002). In Suzhou, He Jian Technology Corporation, widely reported to be affiliated with Taiwan's UMC, had one fab under construction and five more planned (interview with officials of the Suzhou Industrial Park, Suzhou, September 2002). Taiwan's TSMC had announced plans to build at least one fab in Songjiang. This confirms the earlier view exemplified in "Is China's Semiconductor Industry Market Worth the Risk for Multinationals? Definitely!" *Cahners In-Stat Group* (March 29, 1999).

vately held semiconductor companies, and preferential financing by government banks.[111]

The sudden emergence of China as a significant site for semiconductor manufacturing is significant, but China is by no means the only player. Malaysia has opened a $1.7 billion wafer fab and has planned to construct two more.[112] In Taiwan, planners in mid-2000 envisioned that a total of 21 new 300-mm fabs and 9 new 200-mm fabs would be built by the year 2010.[113] These plans have been significantly reduced as Taiwan's government planners seek to adjust to the growing migration of the island's semiconductor manufacturing operations to China. Specifically, the government hopes to sustain a high concentration of 12-inch wafer fabs on the island, to enhance Taiwan's capabilities with respect to systems-on-a-chip, and to improve Taiwan's position in upstream (materials and semiconductor manufacturing equipment) and downstream (assembly, test, packaging) functions. The government of Singapore has publicly set a goal of 20 fabs by the year 2005. In South Korea, the government pressured commercial banks to finance the move into chip making by the country's family-controlled conglomerates.[114] In the proceedings for this report, speakers from Japan describe that nation's vigorous attempt to bring about a "national revival" in microelectronics. In addition to the programs described above, Japan has launched a number of government-supported industry-government R&D projects in 2001-2002. For example, the Millennium Research for Advanced Information Technology (MIRAI) was initiated by METI in 2001 to develop next-generation semiconductor materials and process technologies, such as measuring and mask technology

---

[111]The principal Chinese policies are spelled out in the *Tenth Five Year Plan (2001-2005)—Information Industry*, <http://www.trp.hku.hk/infofile/china/2002/10-5-yr-plan.pdf>, and Circular 18 of June 24, 2000, *Several Policies for Encouraging the Development of Software Industry and Integrated Circuit Industry*, published in Beijing Xinhua Domestic Service, 04:49 GMT, July 1, 2000. The municipal governments of Shanghai and Beijing have issued their own circulars articulating promotional policies to be implemented within their jurisdictions to augment the national-level measures. These are, respectively, Shanghai Circular 54 of December 1, 2000, *Some Policy Guidelines of This Municipality for Encouraging the Development of the Software Industry and the Integrated Circuit Industry*, Shanghai Gazette, January 2001; and Beijing Circular 2001-4, *Measures for Implementing 'Policies for Encouraging the Development of Software and Integrated Circuit Industries' Issued by the State Council*, Jing Zhen Fa No. 2001-4 (February 6, 2001).

[112]"The Great Chip Glut," *The Economist*, August 11, 2001. <http://www.economistgroup.com>.

[113]According to the World Fab Watch (WFW) database, which is prepared by Strategic Marketing Associates and contains information on over 1,000 fabs worldwide, these estimates are subject to significant and sudden shifts, which has proved to be the case. Strategic Marketing Associates, *World Fab Watch*, Santa Cruz, CA, 2002.

[114]See Howell, *op. cit.* Howell's figures and many of his conclusions are based primarily on personal interviews with industry officials in Asia. This type of field research on national policies for an industry is exceedingly rare in the U.S.

Box C
National Programs to Support the Semiconductor Industry

Many nations are actively and substantially supporting initiatives in their respective national semiconductor industries. Some of these programs are listed below: A more complete list can be found in Thomas Howell's analysis in this report.[a]

Country	Project	Period of Project	Level of Funding	Purpose
Japan	Next Generation Semiconductor R&D Center (Super clean room)	2001-08	$300 million ($60 million in 2001)[b]	Process and device technology for 70-mm generation
Japan	Future Information Society Creation Laboratory	2001-06	$300 million	Create small-scale, very short-term semiconductor production line
Japan	ASET	1995-	$500 million	Lithography, semiconductor manufacturing technology
Japan	Nanotechnology Programs	1985-	$350 million in FY 2001; METI labs conducting R&D	Basic R&D nanotechnology, includes microelectronics themes
Japan	Selete[c]	1996-	[d]	Manufacturing technology for 300-mm wafers

Box C Continued

Taiwan	ASTRO	2000-	Government will fund half	Technology induction, upgrading of local industry
European Union	MEDEA	1997-2000	$720 million (est.)	Process technology, design, applications
European Union	MEDEA Plus	2001-09	$1,350 million (est.)	Systems-on-a-chip, UV lithography
Germany	Semiconductor 300	1996-2000	$680 million	300-mm wafer technology
France	Crolles I and II	1998-	$136 million (est.)[e]	Pilot 300-mm fab
United States	MARCO	1997-	$75 million over 6 years	Basic microelectronics R&D
United States	National Nanotechnology Initiative	2000-	$270 million in 2000	Basic R&D on nanotechnology; includes same microelectronics themes
United States	DARPA	Permanent	$192 million in 2000 for "advanced electronics technology"	Advanced lithography; nanomechnisms; electronic modules

Box C Continued

United States**	SEMATECH	1989-1996	$850 million	Cooperative research facility to benchmark next-generation development of processes, products, and tools; forum for information exchange and coordination of research projects
United States	EUVL (Extreme Ultraviolet Lithography) CRADA[f]	1997-	$250 million	Advanced lithography

---

** International SEMATECH, as its name suggests, involves companies from many countries and does not receive direct U.S.-government support.

[a]See Table, *Examples of Government Supported Microelectronics R&D Initiatives*, in the appendix of Howell, *op.cit.*

[b]METI requested $60 million in FY2001 budget for first year of a seven-year project.

[c]Samsung is also a member of Selete.

[d]Privately funded but received NEDO contract to develop technology to cut PFC use.

[e]*Crolles I* reportedly received support of FF 900 million to FF 1 billion. Additional funds have been requested for *Crolles II.*

[f]The EUVL CRADA is in fact an international effort.

for 50-nm generation devices.[115] Many of these programs, at least in part, emulate the consortium model as well as other U.S. programs.[116]

This summary of national programs should not be interpreted as a criticism of them. The collective impact of these programs should help the semiconductor industry as a whole meet its increasingly complex technical challenges. At the same time, underlying these programs are genuine differences in national attitudes concerning a nation's knowledge and technology base. Some nations believe the development of a nation's manufacturing capacity in leading industries to be an appropriate national goal worthy of sustained support, a perspective apparent in policies of growing East Asian economies.[117] Both European and Japanese industry leaders are identifying what they see as the main semiconductor growth markets of the twenty-first century—wireless, wired telecommunications, and digital home appliances. As one senior European participant observed, Europe in recent years has taken a leading position in several areas: communications, automotive electronics, smart cards, and multimedia. These applications are driven by system innovations on silicon and require embedded technologies. The next big cooperative challenge will be to develop systems-on-a-chip—achieving the same functionality in one-fiftieth the space. To meet this challenge, he

[115]Government funding for this seven-year project was set at 3.8 billion yen for the first year. The project is being operated jointly by ASET and MITI's new semiconductor R&D organization, Advanced Semiconductor Research Center ("ASRC") in the Tsukuba Super Clean Room. MIRAI website, <http://unit.asit.go.jp/asrc/mirai/index.htm>; *Handotai Kojo Handobukku* (December 5, 2001), pp. 4-5.

[116]In 2002 METI launched a five-year industry-government R&D project to develop extreme-ultraviolet (EUV) lithography for 50-nm device manufacturing in conjunction with an association of 10 Japanese device and lithography equipment purchasers. The producers have formed the Extreme Ultraviolet Lithography System Development Association ("EUVA") to undertake the project. First-year government funding was set at 1.09 billion yen. Japan Patent Office General Affairs Department Technology Research Division, *Handotai Rokogijutsu Ni Kansaru Shutsugan Gijutsu Douko Chosa* (May 10, 2001), p. 17; METI, *Heisei Yonnendo Jisshi Hoshin* (March 8, 2002), p. 1; *Handotai Sangyo Shimbun* (January 16, 2002), p. 3. In July 2000, 11 Japanese semiconductor manufacturers established a new R&D company, Advanced SoC Platform Corporation ("ASPLA"), to standardize design and process technologies for systems-on-a-chip utilizing 90-nm design rules. METI reportedly will provide 31.5 billion yen for this effort, which will feature partnership with STARC and Selete. See also the presentations of Masataka Hirose, Toshiaki Masuhara, and Hideo Setoya in the proceedings of this volume.

[117]Some nations pursue consumer welfare as an implicit, if vaguely defined, goal, while other nations adopt explicit national economic strategies, designed to pursue national economic strength through the acquisition of the capability to manufacture high-technology products. See National Research Council, *Conflict and Cooperation in National Competition for High-technology Industry*, pp. 12-27 and pp. 51-54. See also Richard Samuel's *Rich Nation, Strong Army: National Security and Technological Transformation of Japan*, Ithaca, NY: Cornell University Press, 1994.

said, Philips will cooperate in both MEDEA Plus and ITEA, the Information Technology for European Applications.[118] U.S. companies have dominated computer applications of semiconductors, in particular personal computers. Growth prospects in these applications may prove to be more limited in the future.[119]

CHALLENGES TO U.S. PUBLIC POLICY

While federal funding for SEMATECH ended after 1996 at the industry's request, the debate has continued within Congress and the Executive branch as to whether and to what extent the U.S. government should continue to invest federal funds in supporting R&D in microelectronics.[120] Some observers argue that the role of the government support for R&D should be curtailed, asserting that federal programs in microelectronics represent "corporate welfare."[121] Advocates of R&D cooperation among universities, industry, and government to advance knowledge and the nation's capacity to produce microelectronics argue that such support is justified, not only by this technology's relevance to many national missions (not least, defense), but also by its benefits to the national economy and society as a whole.[122]

In fact, no consensus exists on the appropriate mechanisms or levels of support for research. Discussions of the need for such programs have often been dogged by doctrinaire views as to the appropriateness of government support for industry R&D and by domestic politics (e.g., balancing the federal budget) that have generated uncertainty about this form of cooperation, especially at the federal level.[123] This irresolution has resulted in a passive federal role in addressing

---

[118]See the presentation by Philips Semiconductor's Peter Draheim in this volume.

[119]"From Stagnation to Growth, The Push to Strengthen Design," *Nikkei Microdevices* (January 2001); "Three Major European LSI Makers Show Stable Growth Through Large Investments," *Nikkei Microdevices* (January 2001). See also Howell, *op. cit.*

[120]At a meeting in 1994, the SEMATECH board of directors reasoned that the U.S. semiconductor industry had regained strength in both the device-making and supplier markets, and thus voted to seek an end to matching federal funding after 1996. For a brief timeline and history of SEMATECH, see <http://www.sematech.org/public/corporate/history/history.htm>.

[121]See Rodgers, *op. cit.*

[122]Policy debates on public-private partnerships have often suffered from sloganeering, with no clear resolution. One side claims that the market is efficient and will therefore sort itself out without the involvement of government. The other side counters that markets are imperfect and that, in any event, government missions cannot depend on markets alone, nor can they wait for the appropriate price signals to emerge. Therefore public policy has a role—and always has. One contribution of this analysis, and of others in the series, is to document current cooperative activity and redirect attention away from this abstract rhetoric and demonstrate that carefully crafted partnerships can help accelerate innovation.

[123]See David M. Hart, *Forged Consensus: Science, Technology, and Economic Policy in the United States*, 1921-1953, Princeton: Princeton University Press, 1998, p. 230. For a broader review of these differing perspectives, see Richard Bingham, *Industrial Policy American Style: From Hamilton to*

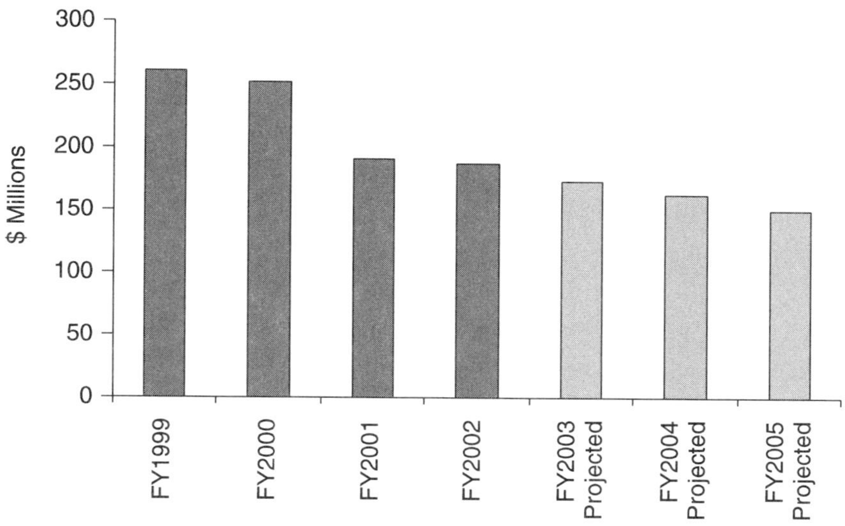

FIGURE 12 Defense Advanced Researech Projects Agency's Annual Funding of Microelectornics R&D.
SOURCE: DARPA

the technical uncertainties central to the continued rapid evolution of information technologies. Annual funding of microelectronics R&D through DARPA—the principal channel of direct federal financial support—has declined and is projected to decline further (see Figure 12).[124] As noted above, this trend runs counter to those in Europe and East Asia, where governments are providing substantial

---

HDTV, New York: M.E. Sharpe, 1998. See also I. Lebow, *Information Highways and Byways: From the Telegraph to the 21st Century*, New York: Institute of Electrical and Electronics Engineers, 1995. For a global perspective, see J. Fallows, *Looking into the Sun: The Rise of the New East Asian Economic and Political System,* New York: Pantheon Books, 1994; and J. A. Brander and B. J. Spencer, "International R&D Rivalry and Industrial Strategy," *Review of Economic Studies,* 50(4):707-722, 1983. There is much less ambivalence at the state level. See Christopher Coburn and Dan Berglund. *Partnerships: A Compendium of State and Federal Cooperative Technology Programs.* Columbus, OH: Battelle Press, 1995.

[124]This presentation understates the declines. Support for lithography, for example, fell from $54.4 million in FY 2001 to $32.6 in FY 2002 and is projected to stabilize at $25 million in FY 2003. Some reports suggest that overall support for microelectronics research actually fell from about $350 million in the early 1990s to about $55 million in 2000. See Scott Nance, "Broad Federal Research Required to Keep Semiconductors on Track," *New Technology Week,* October 30, 2000, and Sonny Maynard, Semiconductor Research Corporation, cited in presentation by Dr. Michael Polcari, "Current Challenges; A U.S. and Global Perspective," National Research Council, *Symposium on National Programs to Support the Semiconductor Industry* (October 2000).

direct and indirect funding in this sector. The declines in federal funding for research are of particular concern to U.S. industry.

Cooperative Research Programs

Continuing to advance microelectronics technology is becoming increasingly difficult. As semiconductors become denser, faster, and cheaper, they approach physical limits that will prevent further progress based on current chip-making processes. Significant research breakthroughs will be required to allow historic trends to continue; yet if these occur, in 15 years semiconductor memory costs could be one one-hundredth of today's costs and microprocessors 15 times faster.

Reflecting this concern, the industry has initiated several new programs aimed at strengthening the research capability of U.S. universities. The largest of these, carried out under MARCO (the Microelectronics Advanced Research Corporation), is the Focus Center Research Program (FCRP). In this program, the U.S. semiconductor industry, the federal government, and universities collaborate on cutting-edge research deemed critical to the continued growth of the industry (see Box D). As an industry-government partnership supporting university research in microelectronics, the FCRP research is long range (typically eight or more years out) and essential for the timely development of a replacement technology for the current chip-making process.[125]

There are currently four focus centers, addressing design and test; interconnect; materials, structures and devices; and circuits, systems, and software. The four focus centers now involve 21 universities. A brief description of each of the centers is provided in Appendix A. The FCRP plans to eventually include six national focus centers channeling $60 million per year into new research activities. However, the sharp downturn in the industry may jeopardize this commitment, and the federal government's commitment is also in doubt. The industry funds 75 percent of the program, and the government has funded the remaining 25 percent. The government's share has been supported through the Government-Industry Co-sponsorship of University Research (GICUR) program within the Office of the Secretary of Defense. When the industry and government embarked on the FCRP, the plan outlined a ramp-up which would now require $10 million in funds for semiconductors in 2003, $12 million in 2004, $13 million in 2005,

[125]The FCRP is part of MARCO, the Microelectronics Advanced Research Corporation, within the SRC. See MARCO website, <http://marco.fcrp.org>. MARCO has its own management personnel but uses the infrastructure and resources of the SRC. MARCO's supporters include the following: Advanced Micro Devices, Inc., Agere Systems, Agilent Technologies, Analog Devices, Inc., Conexant Systems, Inc., Cypress Semiconductor, IBM Corporation, Intel Corporation, LSI Logic Corporation, MICRON Technology, Inc., Motorola Incorporated, National Semiconductor Corporation, Texas Instruments Incorporated, Xilinx, Inc., Air Products & Chemicals, Inc., Applied Materials, Inc., KLA-Tencor Corporation, Novellus Systems, Inc., SCP Global Technologies, SpeedFam-IPEC, Teradyne, Inc., DARPA, and the Deputy Undersecretary of Defense for Science & Technology.

Box D
Characteristics of the MARCO Program

Characteristics of the MARCO program are as follows:

University-driven Management Philosophy. The goal is to identify the best professors in outstanding universities, outline broad areas of interest to the industry, and then delegate decisions about the research agenda to the university researchers.

Substantial Funding. The MARCO program awards are significant, often in the $10 million range, which is substantially larger than many normal academic grants. This enables researchers to focus on a substantial program of work without the need to continually seek supplemental funding.

New Technical Approaches. The substantial autonomy provided to the researchers is designed to encourage "out-of-the box" or non-traditional approaches to the technical problems the industry must address. New technical solutions and new manufacturing methods may be required to sustain the industry's current high rate of growth.

Student Training. An important component of the program is its ability to attract top students through the engagement of leading professors in major universities with first-class facilities.

Sustained Industry Commitment. The semiconductor industry's long-term commitment to the Semiconductor Research Corporation, and more recently its sustained investment in the MARCO program, reflect the widespread recognition that research and training are the key to its long-term success.

and $15 million in 2006. These funding targets have not been met, yet the centers seem to be providing valuable research results for the Department of Defense.

International SEMATECH continues to promote collaboration among major firms, which now include non-U.S. members. Among its activities, it is funding the development of new 300-mm tools and has taken a leadership position in pursuing the technology roadmaps in cooperation with industry. It has supported initiatives on mask-making tools, lithography using very-short-wavelength ultraviolet light (157 nm) from a special laser, next-generation lithography consensus, low dielectric-constant materials, and other innovations. It has also continued to benchmark the industry and to help improve manufacturing methods, among other contributions. The most recent budgets for NSF and DoD include increases in some important semiconductor areas that had been reduced during the 1990s.[126]

The responsibility of the government to ensure the availability of trained, educated manpower is widely accepted. Further, while immigration policy is admittedly complex—especially now, in light of September 11—it can be administered in a manner that facilitates the attraction of foreign talent to the United States. Top-quality research talent, whether foreign or domestic, will be required to address the technological "brick wall" confronting the industry. In the past, large, complex technical challenges have been surmounted through prolonged federal support of basic science and pre-competitive R&D. The nation that is able to produce, attract, and retain this talent may lead technical progress in the research and development clusters that will condition commercial success in the decades ahead.

The United States was able to muster an appropriate policy mix in the 1980s that helped U.S. firms succeed. Today's challenges in research and manpower will require similar innovative efforts. The events of September 11, 2001 and the dispersal of semiconductor technologies and expertise make these issues more pressing.

SYMPOSIUM SUMMARY

The presentations, discourse, and commissioned papers offered in this symposium may help inform the policy community of the challenges faced by the industry. Taken together, they offer an assessment of the industry's contributions as well as provide information on the technical challenges, research needs, and the range of foreign efforts currently characterizing the microelectronics sector. The analysis presented by such insight may raise questions about the scope and scale of public programs to support this unique industry in the United States.

The next section reviews the main points made by the speakers at the symposium. They present expert perspective on challenges in the research and development of new semiconductor technologies. Considerable effort has been made to accurately capture key points from the discourse of the symposium; however, the presentations themselves should be consulted for a fuller, more measured record of the participants' remarks.

---

[126]In light of the information and analysis presented in this report, the Committee believes these efforts should be strengthened. The Committee's recommendations are elaborated in the *Findings and Recommendations* section of this report.

PANEL I: THE U.S. EXPERIENCE: SEMATECH

Moderator **Clark McFadden** of Dewey Ballantine reviewed the SEMATECH program—the pioneering government-industry partnership initiated in 1987 to revitalize the U.S. semiconductor industry. He observed that it has been successful in meeting a variety of goals, such as attracting investment from its industrial participants, developing industry roadmaps, and fostering an industry-wide perspective on technology development. At the same time, he recognized that assessing its impact on the semiconductor industry as a whole, and on U.S. technology development, is a more difficult exercise. It is nonetheless true that the informed observations of the participants below suggest that the consortium is effective. Perhaps the most compelling evidence of the value of the consortium is that it still exists. Supported now only by private funds, it has retained most of its membership for over a decade while adding new member companies from other countries.

Gordon Moore, cofounder of Intel Corp., said that the primary early challenge for SEMATECH was to raise the quality and productivity of American industry. Its mission was "unusual" in promoting collective action by industry and cooperation between industry and government, but it succeeded in focusing attention on the fragmented tool-manufacturing industry. He concluded that a government-industry partnership can have "a positive impact on the U.S. industry." He added that the focus of industrial R&D on short-term, predictable results makes it extremely important for the government to support long-term research "across a very broad base."

Kenneth Flamm of the University of Texas at Austin offered an economist's view of SEMATECH. He said that it is generally perceived as a success by industry, but that only a few economic studies have been done.[127] His own review of the economic literature revealed that cooperation can have either positive or negative impacts on R&D. From a public policy perspective, he saw three motives for cooperation: information sharing; cooperation on projects that promise such large spillovers that a company would not do the projects at all in the absence of partners; and the creation of an institutional structure that can increase spillovers. In their empirical contribution to this report, Flamm and his co-author Qifei Wang reexamine the impact of SEMATECH on semiconductor industry R&D by updating and improving on the published work on this issue. Their results suggest that SEMATECH reduced the R&D expenditures of its membership

[127]See the paper presented in this report by Flamm and Wang, *op. cit.* Flamm and Wang note that there has been a limited amount of empirical attention focusing on the impact of R&D cooperation on industrial R&D outcomes. Further, the authors note that SEMATECH has been the subject of only three more rigorously oriented studies.

somewhat, in part by eliminating duplication. They conclude that the underlying models of R&D cooperation that ultimately must be the basis of a scientific effort to untangle the chains of causality are too simplified to capture the complexity of the real world of R&D consortia. Moreover, they note that "the only absolutely certain thing about SEMATECH is that a substantial portion of its member companies must have found it to be of net value—having actually run the experiment of ending public subsidy and finding that its consumers continued to buy its output."

Current Challenges: A U.S. and Global Perspective

Michael Polcari of IBM called the technical challenges facing the computer industry "unprecedented." It will be very difficult to maintain the industry's rapid increases in productivity, which, following Moore's Law, has approximately doubled every 18 to 24 months. To date, these increases have come primarily from scaling—the progressive shrinking of component size. The challenge for the near future is to find new solutions when scaling ends. Dr. Polcari listed many areas of anticipated improvement such as the importance of building the nation's capacity in the basic sciences, continuing to adequately fund high-risk research, and training more engineers.

David Mowery of the University of California at Berkeley listed five observations about SEMATECH:

(1) It proved to be dynamic and adaptable.
(2) The rigid requirements of the Government Performance and Results Act might reduce such flexibility in future collaborations.
(3) Its contribution was important for its "extension" role, in the sense of agricultural extension programs, and its collaborative agenda.
(4) It is difficult to evaluate economically because it is impossible to know what would have happened in its absence.
(5) More needs to be known about the importance of the government's catalytic role in providing funding for eight years. He suggested additional study on how "this unusual instrument of R&D collaboration" can evolve in response to changes occurring in the structure of this industry.

PANEL II: CURRENT JAPANESE PARTNERSHIPS: SELETE AND ASET

Akihiko Morino described Selete (SEmiconductor Leading Edge TEchnologies) as a joint venture company that performs R&D on behalf of the semiconductor industry. The mission of Selete is to develop semiconductor device and process technologies to the point where they can be produced at reasonable cost. Selete fosters collaboration among academia, industry, equipment and materials suppliers, research institutes, and overseas consortia, including International SEMATECH.

Hideo Setoya said that ASET (the Association of Super-Advanced Electronics Technologies) is a consortium of the electronics device industry that also includes equipment and materials suppliers. Of 14 members, six are non-Japanese companies or subsidiaries. Its mission is to perform pre-competitive research between the basic and applied levels. All research is performed by the staffs of member companies. It is 100 percent financed by the national government and open to the public.

Japanese Consortia for Semiconductor R&D

Yoichi Unno described SIRIJ, the Semiconductor Industry Research Institute of Japan, as a think tank founded in 1994 by 10 Japanese semiconductor companies to promote joint R&D. The objectives of SIRIJ are to promote development of next-generation technologies, study the future of the industry, and implement projects for international cooperation. It has recently added an educational program in LSI (Large Scale Integrated) design for small companies, a roadmap committee, and a team to study the needs of the industry.

University Research Centers for Silicon Technology

Masataka Hirose of Hiroshima University described the structure and missions of three university research centers for silicon technology, all sponsored by Monbusho, the Ministry of Education, Science, and Culture:

(1) The VLSI Design and Education Center, established in May 1996 at the University of Tokyo;
(2) The New Industry Creation Hatchery ("Incubation") Center, located at the Department of Electrical Engineering of Tohoku University;
(3) The Research Center for Nanodevices and Systems at Hiroshima University, of which he is director.

PANEL III: EUROPEAN PARTNERSHIPS

Michael Borrus of Petkevich and Partners told the audience that the panel's presentation might prove to be "a bit of a surprise to some of you." He said that European semiconductor activities have strengthened rapidly in recent years.

The MEDEA Program

Jürgen Knorr of Micro-Electronics Development for European Applications (MEDEA) confessed that it feels "very strange" to lead a program in support of a multinational semiconductor industry. The tradition, he said, has been to support one's national industry. MEDEA, however, is proof that the semiconductor industry is global, and each company's objectives must be shaped accordingly. Competition today is not as much between nations as between companies. MEDEA is an industry-initiated, industry-driven program supported by the national governments of 12 participating countries to stimulate trans-border R&D cooperation. A four-year program that ended in December 2000 has now been extended as MEDEA Plus under the guideline "system innovation on silicon."

Government-Industry Partnerships in Europe (1): Embedded Technologies and Systems-on-a-Chip

Peter Draheim of Philips Semiconductor said that Europe in recent years has taken a leading position in several areas: communications, automotive electronics, smart cards, and multimedia. These applications are driven by system innovations on silicon and require embedded technologies. Philips expects major breakthroughs in "portable infotainment," third-generation mobile communication, home networks, and enhanced digital TV. The next big cooperative challenge will be to develop systems-on-a-chip—achieving the same functionality in one-fiftieth the space. To meet this challenge, he said, Philips will cooperate in both MEDEA Plus and ITEA, the Information Technology for European Applications.

Government-Industry Partnerships in Europe (2): International Cooperation: SEMATECH and IMEC

Wilhelm Beinvogl of Infineon noted that the three major information technology (IT) players in Europe are all members of International SEMATECH. They are not only financial contributors but also significant technical contributors, especially to 300-mm technology. Another example of collaboration, he said, is that the IMEC institute in Belgium, a world leader in cooperative research, which is closely cooperating with International SEMATECH on a major project. He also described one "full-blown success story," a joint venture between Infineon and Motorola to move to the leading edge in transition to the next wafer size. He echoed the manpower needs cited by other speakers, calling the decrease in engineers "dramatic."

PANEL IV: THE TAIWANESE APPROACH

Patrick Windham of Windham Consulting said that the Taiwanese approach constitutes "a rare success story," and that Taiwan's journey to become the fourth-largest semiconductor producer in the world is a "remarkable" one.

Government-Industry Partnerships in Taiwan

Genda J. Hu of Taiwan Semiconductor Manufacturing Co. said that Taiwan's rise to success had depended on the government's decision in 1974 to focus on semiconductors as a key industry. The government set up a special agency to develop the industry, and helped establish several companies and secure rights to key technologies. It sought steadily to shift more responsibility to the private sector: In 1990, the government provided some 44 percent of total R&D spending to benefit the private sector; by 1999 the government's share had fallen to 6.5 percent. Another factor in success was the decision to concentrate on "fabless" designs and the manufacture of custom devices for other companies. TSMC is a member of International SEMATECH; UMC, another leading company, has an alliance with IBM and Infineon.

The Science Park Approach in Taiwan

Chien-Yuan Lin of National Taiwan University said that the government in Taiwan, unlike the U.S. government, has actively promoted economic development. In 1980, the government began Hsinchu Science Park as a government-industry partnership, providing major venture capital, some tax deductions or exemptions, the infrastructure (including the park itself), special public services (such as the "one-stop business service"), and other services, such as R&D and education. At the time of the symposium, the park held 291 units and was considered by some analysts to be a model S&T park. Two other parks have been initiated in Taiwan.

Discussion

Michael Luger of the University of North Carolina at Chapel Hill offered a "continuum" of government-industry parks. At one end are large, national consortia that have abundant basic research, high spillover, and few direct local applications. Next are programs supported directly by federal funds, which also emphasize basic research, including university R&D centers. Beyond them are state-funded R&D centers, which feature more applied research, fewer spillovers, and more concentrated spatial effects. Localization economies lead to clusters of firms and industries that are related through input-output linkages and other growth-stimulating relationships. Dr. Luger, building on the discussion of high-technology clusters initiated by Chien-Yuan Lin, shared insights on technology parks by highlighting a brief history of the Research Triangle Park in North Carolina.[128]

PANEL V: CHALLENGES FACING THE EQUIPMENT INDUSTRY

Kalman Kaufman of Applied Materials said that the semiconductor and electronics industry represents an increasingly significant force in the economy, and that equipment suppliers play an increasingly important role in the industry. He listed several "imperatives" for sustained success:

(1) Equipment suppliers must continue to invest heavily in technology development and commercialization.
(2) Governments in every country must ensure fair access to markets and technologies.
(3) Universities must teach and motivate more researchers and engineers.
(4) Semiconductor producers must reduce risk and improve the efficiency of the industry.
(5) National labs must bridge the widening gap between academic research and the "next-generation" industry requirements for generic, pre-competitive research.

The most important functions of a consortium, he said, are to foster cooperation and provide valuable information to the industry so it can change its roadmaps and learn how better to serve customers. He recommended that a national lab with close ties to universities be dedicated to pre-competitive generic research problems.

John Kelly of Novellus Systems said that problems facing the supplier industry are "fairly simple and straightforward." The first is the undersupply of talented graduate students, a "situation that seemed to be worsening." Many graduate students have moved away from semiconductors to other areas, such as nanotechnology, and professors have been going "where the money is." Another, more complex issue, he said, is the problem of shrinking resources for long-term research. The technological "brick wall," he said, could be very real "if we don't work on the right problems fast enough." A current challenge to the equipment industry is that it is no longer acceptable simply to deliver a tool to the customer. It must be delivered as part of a process, and the process has to be perfect. This requires far more work on long-term "fundamentals, materials, the real basics."

[128]For a review of science and technology parks, see Michael I. Luger, "Science and Technology Parks at the Millennium: Concept, History, and Metrics," in National Research Council, *A Review of the New Initiatives at the NASA Ames Research Center,* C. Wessner, ed.,Washington, D.C.: National Academy Press, 2001.

Papken Der Torossian of Silicon Valley Group (SVG) said that the technical challenges of moving from a three-year cycle to a two-year cycle require huge investments by the industry. Research spending will have to increase by almost 30 percent to accelerate the equipment cycle. Because these investments are huge and often long term, they cannot be borne by a single company. A consortium is one simple way of working with competitors, which is "the only way we're going to advance the science in the next few years." He praised SEMATECH for having created an environment for buyers and sellers to work together.

PANEL VI: THE INTERNATIONALIZATION OF COOPERATION

A U.S. Perspective

George Scalise of the Semiconductor Industry Association said that the SRC is a "structure that works well." It was founded in 1982 to address a lack of engineers coming out of college and a shortage of engineers trained in the then-new solid-state technology. It has created an "integrated, virtual semiconductor research laboratory" by funding projects at about 65 universities across the country. Through two other programs, it supports research in semiconductor design and testing, and in layout. SEMATECH, he suggested, could help promote international research programs on materials structures and devices, circuits systems, and software that will begin to fill a part of the research gap. A "consortium of consortia," he said, is needed to make more efficient use of the R&D dollar.

A Japanese Perspective

Toshiaki Masuhara of Hitachi said that university-industry consortia in both process and design R&D will be very important in the future and will require a great deal of funding. He offered five criteria for organizing a successful R&D consortium:

(1) Business merit: Is the technology applicable to industry and can the market accept the new technology?
(2) Technical merit: Are there pitfalls in application, technology matching, suitability, or reliability?
(3) Participants' merit: Does the consortium provide good opportunities for participants and a good career path?
(4) Academic merit: Can the work lead to research papers, advanced degrees, and faculty success?
(5) Industry manager's merit: Are managers willing to send the best R&D people from industry?

He added that inter-consortium collaboration will be needed if the industry is to avoid hitting the "red brick wall" of technical challenges.

A Taiwanese Perspective

Genda Hu discussed a planned Taiwanese consortium called ASTRO, which had been placed "on hold" due to issues beyond the control of the industry. The attempt to form that organization, he said, had been a clear demonstration that Taiwan intends to participate in R&D consortia. One objective of ASTRO is to facilitate participation in international R&D activities. Absent ASTRO, the best strategy for individual companies is to join international collaborations on their own, which almost every semiconductor company in Taiwan has done.

A European Perspective

Erik Kamerbeek of the European Semiconductor Equipment and Materials Association said that collaboration is common in Europe, which has a greater need for joint efforts than a single, large country like the United States. Among international consortia, the Information Society Technologies Programme is planned and organized by the European Commission with the support of industry. Programs are approved by the national representatives of the 15 EU countries. Another major IT program is MEDEA, in which each project is accepted by the ministers' conference of participating countries. All projects are initiated and guided by industry.

The views summarized above reflect the diversity in the national and regional approaches to meeting the needs of the semiconductor industry. They also affirm the common perception of the technical challenges the industry must overcome if it is to continue its extraordinary rate of growth.

II

FINDINGS AND RECOMMENDATIONS

Findings

The federal government has played a significant role in supporting the growth of the semiconductor industry since its inception.[1] The industry has benefited from close cooperation with government, both through generous procurement contracts such as those related to defense and space exploration, and through research consortia. This support for industry research and development is fully justified. The semiconductor industry's technological progress has enabled major advances in technologies directly relevant to core government missions including those in national security, communications, health, weather, the environment, and education. In addition, there is growing recognition of the importance of the industry's contributions to the productivity growth of the U.S. economy.[2]

The contribution of new technologies to growth, especially information technologies, is now recognized at the highest levels of U.S. policy making. Notably,

[1]As Laura Tyson observed in 1992: "The semiconductor industry has never been free of the visible hand of government intervention. Competitive advantage in production and trade has been heavily influenced by policy choices, particularly in the United States and Japan. Some of these choices, such as the provision of public support for basic science, R&D, and education in the United States, have had general, not industry-specific objectives. But other choices, such as the provision of secured demand for industry output through military procurement in the United States and through preferential procurement of computers and telecommunications equipment in Japan, have been industry specific in intent and implementation." Laura D'Andrea Tyson, *Who's Bashing Whom?: Trade Conflict in High Technology Industries*, Washington, D.C.: Institute for International Economics, 1992, p. 85.

[2]Dale Jorgenson and Kevin Stiroh, "Raising the Speed Limit: U.S. Economic Growth in the Information Age," in *Brookings Papers on Economic Activity*, (2), 2000, p. 125-212.

Federal Reserve Chairman Alan Greenspan has affirmed the contribution of new technologies to the low inflation, low unemployment, and the continued high growth rates that characterized the U.S. economy in the latter half of the 1990s.[3] Much of the technological advance that has made these productivity gains possible is dependent on the unprecedented decrease in cost of increasingly more powerful semiconductors.[4]

I. THE SEMICONDUCTOR INDUSTRY: A HISTORY OF COMPETITION AND COOPERATION

Firms in the U.S. semiconductor industry have a deserved reputation as fierce competitors in both American and foreign markets. Yet, at key points in the history of the American semiconductor industry, particularly in the decade of the 1980s, the industry launched cooperative efforts through organizations such as the Semiconductor Research Corporation (SRC—1982) and SEMATECH (1987).[5] This cooperative research has pooled expertise, lowered costs, and encouraged the dissemination of knowledge across the industry.[6] After two decades of relative declines, the decade of the 1990s witnessed a major resurgence in the competitive position of American industry in many sectors.[7] As

---

[3]Alan Greenspan, *Technological Innovation and the Economy*, Remarks Before the White House Conference on the New Economy, Washington, D.C. April 5, 2000, Federal Reserve Board.

[4]*Ibid.* See also National Research Council, *Measuring and Sustaining the New Economy,* Washington, D.C.: National Academy Press, 2002.

[5]The Semiconductor Research Corporation, founded in 1982, is based in Research Triangle Park, North Carolina, and has an office in San Jose, California. Its stated goal is to operate globally in order to provide competitive advantage to its member companies as the world's premier university research management consortium delivering relevantly educated technical talent and early research results. In the SRC's words; "The goal in 1982, as it is today, was to define common industry needs, invest in and manage the research that would expand the industry knowledge base and attract premier students to study semiconductor technology." In addition, the SRC also trains and produces graduates who are highly and relevantly skilled to perform at the frontier of semiconductor research.

The SEMATECH (SEmiconductor MAnufacturing TECHnology) consortium was a public-private industry partnership formed in 1987 in order to reinvigorate the semiconductor industry in the United States, which had lost significant market share to Japanese firms. The consortium eventually focused on encouraging cooperation among firms to establish standards and helped to develop roadmaps for the evolution of the industry. The consortium stopped receiving federal support in 1996 and has further evolved to include foreign firms. It is now known as International SEMATECH. Currently, partnerships are under way with members, equipment and materials suppliers, national laboratories, and other consortia.

[6]Kenneth Flamm and Qifei Wang, "Sematech Revisited: Assessing Consortium Impacts on Semiconductor Industry R&D," in this report.

[7]See National Research Council, *U.S. Industry in 2000: Studies in Competitive Performance*, Washington, D.C.: National Academy Press, 2000.

previous analysis by the National Research Council suggests, an important part of the improvement in the competitive position of American industry can be attributed to the growth in the application of information technologies, particularly after 1995.[8] A key challenge of the new century is to sustain the high rate of technological advance that has characterized the semiconductor industry, which underpins the information technologies which have in turn contributed to the growth of the American economy as a whole.[9]

A. A Steady Increase in U.S. R&D Investments

In aggregate terms, the outlook for R&D investments in the United States appears favorable. On December 20, 2001, Congress approved a record federal R&D budget for FY 2002 of $103.7 billion—a 13.5 percent increase over FY 2001. Total R&D funding reached a preliminary $264.6 billion in 2000, or 2.68 percent of total GDP. This amount reflects an increased R&D share of GDP from 2.63 percent in 1999.[10] Total U.S. R&D expenditures show a steady increase. For example, between 1995 and 2000, R&D expenditures increased at an average rate of 7.74 percent.[11]

B. Increases Mask Substantial Shifts

Overall, increases in R&D investments are widely recognized as a good thing in that the social returns on such investments (that is, the gains for society as a whole) are very high, on the order of 40 to 50 percent.[12] However, the composition of R&D investments also matters and, in this regard, current trends are a cause for concern. Federal support for R&D has not kept pace with private-sector investments, which have risen dramatically (see Figure 1). In 1980, the federal share of R&D was roughly 48 percent. In 1999, it had fallen to 28 percent of the

[8]*Ibid.* See also: Jorgenson and Stiroh, *op. cit.*; and National Research Council, *Measuring and Sustaining the New Economy.*

[9]*Ibid.*

[10]These figures use a preliminary estimate for 2000 from the National Science Foundation. The data are derived from the National Science Foundation's, *National Patterns of R&D Resources: 2000 Data Update,* "National expenditures for R&D, from funding sectors to performing sectors: 1993-2000."

[11]*Ibid.*

[12]There is a substantial literature on the social benefits or "spillovers" attributed to R&D investments, from Edwin Mansfield's early work in 1977 to more recent analysis by Zvi Griliches. See, for example, Martin N. Baily and A. Chakrabarti, *Innovation and the Productivity Crisis,* Washington, D.C.: The Brookings Institution, 1998; and Zvi Griliches, *The Search for R&D Spillovers,* Cambridge, MA: Harvard University Press, 1990.

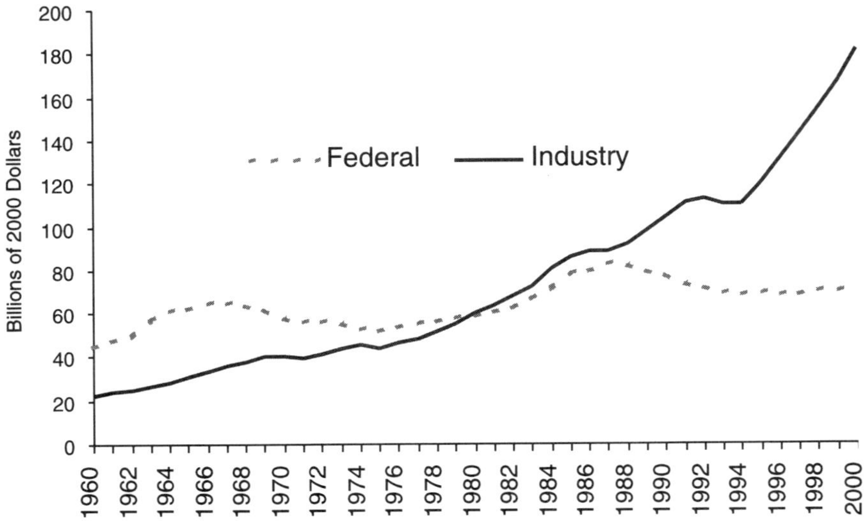

FIGURE 1. Total Real R&D Expenditures By Source of Funds 1960–2000.
SOURCE: National Science Foundation, *National Patterns of R&D Resources*.

total. In part, this reflects U.S. industry's commitment to developing new products and processes, and, in part, the declines reflect the budgetary constraints and uncertainties of the mid-1990s. Whatever the cause, the current differential trends are a source of concern because, as noted below, the government and industry focus on different phases of the innovation system.

The Nature of Industry R&D

The contribution of industry to the R&D budget has focused more on product development than on basic research. Of the $43.5 billion of federal R&D devoted solely to research in FY2001, $22 billion, or 50.7 percent, was channeled into basic research.[13] The semiconductor industry devotes $14 billion to R&D, or 14 percent of sales per year (as of 2000).[14] The industry also devotes a significant portion of its R&D effort to university-based research through the Semiconductor

[13]See American Association for the Advancement of Science; *Congressional Action on Research and Development in the FY 2001 Budget.*

[14]See <http://semichips.org/ind_facts.cfm>.

BOX A
THE MICROELECTRONICS ADVANCED RESEARCH CORPORATION (MARCO)[a]

MARCO, a cooperative program organized under the auspices of SRC, funds and operates a number of university-based research centers in microelectronics as part of its Focus Center Research Program (FCRP).

- The Focus Centers concentrate on those areas of microelectronics research that must be addressed to maintain the historic productivity growth curve of the industry.
- Focus Programs involve multiple universities and place strong emphasis on cross-fertilization of ideas during the basic research stage.
- Focus Program research is typically longer term—normally eight years away from commercialization.
- Advances made under the Focus Program can become proposals to the SRC to address long-term needs identified in the Industry Roadmap (ITRS).[b]

[a] MARCO was established in 1998. For additional information on MARCO, see the SRC website. <http://src.org/member/about/fcrp.asp>.

[b] See *International Technology Roadmap for Semiconductors,* <http://public.itrs.net/Files/2001ITRS/Home.htm>.

Research Corporation, the jointly funded Focus Center Research Programs (see Box A), and industry R&D collaboration through SEMATECH.[15]

In general, however, industry has been less inclined to fund the basic research on which the future growth of the economy ultimately depends.[16] Yet, much of the current technological progress the United States and, indeed, the rest

[15]The *Focus Center Research Program* is a national research network formed in 1998. It is jointly funded by the U.S. semiconductor industry and the federal government. Its purpose is to address core issues in technology development for the semiconductor industry. The program supports long-range, broad-based research that seeks to establish new perspectives and approaches to technological challenges facing the industry. In March 2001, the Semiconductor Industry Association announced that it would double the size of the FCRP. Subsequent economic conditions have made the timing of this decision uncertain.

[16]National Research Council, *Allocating Federal Funds for R&D*, Washington, D.C.: National Academy Press, 1995. See also National Research Council, *Capitalizing on New Needs and New Opportunities: Government-Industry Partnerships in Biotechnology and Information Technologies: New Needs and New Opportunities*, C. Wessner, ed., Washington, D.C.: National Academy Press, 2002.

of world enjoys today rests on inventions and investments made 30 and 40 years ago.[17] In addition, many of the large industry laboratories that once supported major technological advances, such as the transistor, no longer exist or have seen their research strategies substantially modified.[18]

C. The Expansion of Foreign National R&D Programs

As noted above, governments around the world have played an active role in the development of the semiconductor industry. In its early years, the U.S. industry received substantial support for research and development from the federal government, particularly to achieve national missions in defense (e.g., the Minuteman program) and in space exploration (e.g., the Apollo program). Governments around the world have also taken an active approach in supporting the entry of their national firms into the global semiconductor market.[19]

Japan's early VSLI program, for example, helped bring its producers to the forefront of the industry in only a few years.[20] In 1987, SEMATECH was founded to aid a beleaguered U.S. industry. Korea followed in the late 1980s and 1990s with generous state-supported financing for DRAM production by its *chaebols*. Taiwan's innovative policy mix of equity finance, technical support, favorable tax treatment, and the development of the Hsinchu Science and Technology Park Complex helped propel its industry forward in the 1990s. Enhanced R&D support and other programs are not confined to new entrants. Several European countries, operating in conjunction with the European Union, have put in

---

[17]David Tennenhouse, vice-president and Director of Research and Development at Intel, emphasized this point in his presentation at *The Global Computer Industry Beyond Moore's Law: A Technical, Economic, and National Security Perspective*, a Joint Strategic Assessments Group (SAG) and Defense Advanced Research Projects Agency (DARPA) Conference, January 14-15, 2002.

[18]Richard Rosenbloom and Bill Spencer, *Engines of Innovation: U.S. Industrial Research at the End of an Era*, Boston: Harvard Business School Press, 1996. For a condensed version, see Rosenbloom and Spencer, "The Transformation of Industrial Research" in *Issues in Science and Technology*, 8(3):68-74.

[19]See the proceedings below for a discussion of the wide range of programs under way in Japan, Taiwan, and Europe (the latter at the EU, regional, and national levels). The paper by Thomas Howell in this report, "Competing Programs: Government Support for Microelectronics," provides original documentation concerning the focus and funding of many of these programs. The steady growth of these programs and the levels of public support reflect both the perceived importance of the industry and the perceived success of the American model. See for example the chart summarizing program goals and funding in Thomas Howell, *op cit.*, in this report. Howell identifies about 17 programs currently under way in the microelectronics industry outside the United States.

[20]For a description of these programs and a prescient prediction of the recovery of European firms, see Thomas, Howell, Brent Bartlett, and Warren Davis, *Creating Advantage: Semiconductors and Government Industrial Policy in the 1990s,* Santa Clara, CA: SIA, 1992.

place programs that have contributed to a strengthened competitive position for European producers.

A wide variety of policy instruments—ranging from substantial government funding for national R&D programs to favorable tax treatment (e.g., short depreciation allowances) and, in the past, trade measures such as tariffs and private restraints of trade (i.e., restrictive internal market arrangements)—have been used to promote domestic semiconductor firms. Given the perceived contributions of SEMATECH, countries and regions interested in supporting the semiconductor industry have adopted the consortium model as a means of encouraging cooperation among firms within a national industry and as a vehicle for providing government support.

The combination of technical challenges facing the semiconductor industry and the perceived success of cooperative programs in the United States have led policy makers in several countries to increase government funding in support of their national semiconductor industries.[21] Box B describes current trends in national programs to support national semiconductor industries.

II. THE SEMICONDUCTOR INDUSTRY FACES SIGNIFICANT CHALLENGES

The substantial increases in semiconductor power—predicted by Moore's Law—are becoming more challenging to continue. To do so, the industry must overcome a series of technical hurdles, including the need for both new materials and designs. It must also address the need for skilled labor required to overcome these hurdles amidst emerging changes in the structure of the industry.

Given the economic importance of the industry, there is very limited research on the impact of SEMATECH on R&D in the semiconductor industry, its role in the resurgence of the U.S. industry, and its potential lessons for other U.S. consortia.[22] With regard to the industry as a whole, there is limited economic research as to the sources of the industry's pronounced cyclical swings, its contributions to productivity, and its subsequent impact on the economy at large. Scant public policy attention has been focused, as well, on the research requirements needed to keep this industry on its positive course, and on the skilled labor and advanced training needed to sustain this trajectory.

---

[21]*Ibid.*

[22]See Flamm and Wang, *op. cit.,* in this report.

Box B
Significant Global Trends in the Semiconductor Industry as a Result of National Programs

- **Substantial Support for Microelectronics in Japan.** The Japanese government has initiated a series of R&D initiatives and has provided substantial support in cooperation with the Japanese industry. These initiatives are intended to contribute to a "national revival" in the competitive position of the Japanese semiconductor industry.

- **Strong Support for Microelectronics in the European Union**. European Union and national government supported R&D projects—such as JESSI (the Joint European Submicron Silicon Initiative), MEDEA+ (the Micro-Electronics Development for European Applications), and IMEC (Inter-university Micro-Electronics Centre)—have helped to reverse declines in the European microelectronics industry, considerably improving its global competitive standing.

- **The comparative advantage of other nations in leading-edge mobile communications and digital technologies will be leveraged**. Japan and the EU have designed comprehensive strategies to challenge the U.S. leadership in microelectronics by leveraging their present and expected future advantages in mobile communications and digital home appliances.

- **The Challenge of the foundry model**. Taiwan's pioneering of a new business model—the dedicated foundry—has the potential to revolutionize the industry. Taiwan has emerged as a major production base, in part through government capital to launch the industry and through a supportive environment such as the Hsinchu Park Complex.

- **China: A future competitor**. China is making a concerted effort to become a significant competitor in microelectronics. Its government has made substantial efforts to attract foreign investment and promote the diffusion of more advanced foreign technology to the Chinese mainland, while promoting indigenous producers. The evolution of the Chinese microelectronics industry will continue to be augmented by the ongoing movement of Taiwanese manufacturing information technology to mainland China.

A. Need for Highly Skilled Human Capital

1. Continued Progress Depends on the Supply of Talented and Skilled Labor.

In order for the semiconductor industry to maintain high growth rates and respond to the growing challenges within the industry, the United States faces a long-term need to bolster support for highly skilled workers. While, at this writing, the industry is showing the effects of a historically sharp downturn, this cyclical feature should not mask the growing long-run demand for the skilled workers needed to keep the industry on its current growth path. Sustaining the industry's remarkable rate of technological advance requires persistent creativity and ingenuity. Such an innovative environment is sustained by a trained workforce well grounded in the disciplines—such as physics, mathematics, and engineering—that underpin the semiconductor industry. This long-term growth in the demand for skilled labor has emerged against the decline in U.S. federal funding for these disciplines (see Figure 1). The United States is also competing globally to generate and attract the human capital necessary to the long-term health and development of the semiconductor industry.[23]

2. Generating a Skilled and Qualified Workforce in the Microelectronics Industries.

Despite ongoing initiatives to address the skilled manpower needs of the industry by organizations such as the Semiconductor Research Cooperation (SRC), there is concern within the industry as to whether there will be *enough* skilled graduate students to meet future demand—a problem some believe is worsening. The SRC has documented a significant drop in the graduation rate of electrical engineering students in the United States from 1988 to the present and projects no recovery from these low levels in coming years.[24]

To compound this challenge, competition for the limited pool of talented workers in engineering fields is global, and U.S. industry will face increased difficulty in attracting the young, skilled workers it needs to continue growing. The United States exhibits one of the lowest yields, 5.3 percent, in producing engineers when comparing the number of bachelor's degrees in engineering to all bachelor's degrees. In the sheer production of engineers on a yearly basis, the U.S. is surpassed by Asian nations as a group by almost six times.[25] Other na-

---

[23]See Howell, *op. cit.*

[24]For a discussion of the decrease in engineering students and its implications for the microelectronics industry, see Michael Polcari's statements in the Proceedings of this report, "Current Challenges: A U.S. and Global Perspective," pp. 115-116. For a recent review of these challenges, see National Research Council, *Building a Workforce for the Information Economy,* Washington, D.C.: National Academy Press, 2001.

[25]This group of Asian nations consists of China, India, Japan, South Korea, and Taiwan.

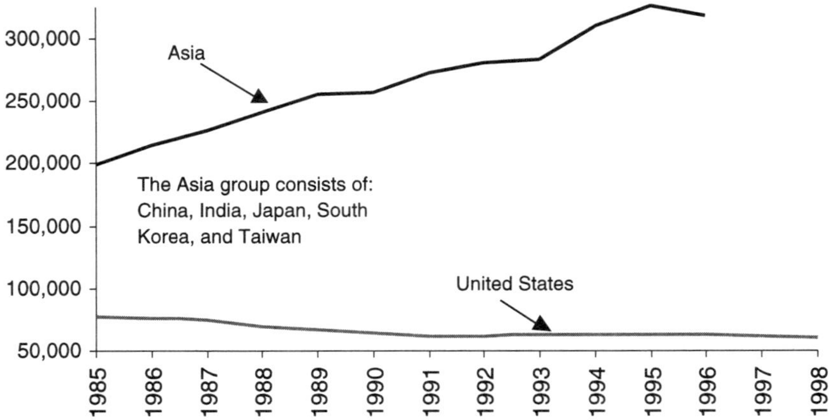

FIGURE 2 Bachelor's Degrees in Engineering: Asian and U.S. Universities.
SOURCE: National Science Foundation.

tions in Europe and Asia have recognized this growing demand for skilled labor in microelectronics and appear to be making a concerted effort to address this global challenge (see Figure 2).[26]

Recommendations for a resolution to this general trend are beyond the purview of this report. The declines in the supply of engineers and the reductions in the federal spending on these disciplines are likely to augment the challenge faced by the industry as it seeks new solutions to pressing technical problems.

---

[26]Germany, for example, has witnessed a major decrease in the supply of engineering graduates since the early 1990s, which has subsequently spurred a worldwide search to attract qualified labor to the German semiconductor industry. See Wilhelm Beinvogl's comments on the German industry's global search for skilled workers to satisfy skilled labor demand in the semiconductor industry, Proceedings, Panel III, "European Partnerships, Government-Industry Partnerships in Europe II," this report, p.148. Japan has not noticed the same falloff in the number of technical graduates on a per-annum basis. However, industry experts believe that there is an impending increase in demand for qualified and skilled workers, which, to the detriment of industry growth rates, may not be satisfied given current trends. Toshiaki Masuhara notes that, in the field of designing VLSI chips in Japan, a study completed several years ago found that there is a company demand four times larger than the number of students. See the Proceedings of this report, p. 136. This shortage exists as well in Taiwan, where it may become difficult to sustain the growth rates of Taiwanese firms without an increase in the supply of engineers and electronics personnel. Genda J. Hu from Taiwan's TSMC (Taiwan Semiconductor Manufacturing Company) noted that the company would have a difficult time maintaining its projected growth rate without an increase in the supply of skilled workers. When asked if this skilled-worker shortage was a problem in Taiwan, Mr. Hu responded; "on a large scale, there is a problem." See the Proceedings of this report, "Panel IV: The Taiwanese Approach," pp. 150-156 and 163.

B. Structural Changes in the Industry Present New Challenges and Opportunities

1. Vertical Specialization and the Emergence of the Foundry Model.

The semiconductor industry is becoming more vertically specialized. For example, the foundry model—low-cost, state-of-the-art fabrication facilities ("fabs") found in Taiwan and other Asian nations, where firms produce semiconductors under contract with other companies—is revolutionizing the industry. The foundry model of production can offer substantial cost savings in manufacturing new generations of chips. In some cases, these cost savings may be the result of lower costs of capital reflecting preferential tax treatment or more direct government subsidies for industries that are viewed as strategic by government policy makers.

The emergence of design houses—firms which specialize in the design of semiconductors only and do not produce them—is yet another sign of vertical specialization in the industry and functions congruently with the foundry model of production. Some analysts are concerned that, over time, the development of new technologies, especially manufacturing technologies, may become more closely associated with the foundries themselves. To the extent that offshore design houses are involved in this process, significant technical capability and know-how may be transferred. The consequences of this phenomenon are not clear, nor are they unidirectional. For example, in the near-term, the availability of these fabrication facilities can work to the benefit of some U.S. firms—especially those that focus on design. This specialization stands in contrast to the U.S. industry, which has typically housed both of these functions under one roof. In part, these trends reflect the global scale of the industry; they reflect as well the active industrial policies of leading East Asian economies. For example, Singapore and Malaysia have contributed significant public funds and have extended tax incentives to companies constructing "fabs," while Taiwan has approximately 100 "design houses," also supported through various government incentives.[27]

While far from certain, one foreseeable outcome of the growth in this model could entail a steady increase in U.S. manufacturers' commissioning "design houses" in Taiwan or Europe, and then contracting a foundry-firm to manufacture that particular design.[28] In any case, U.S. merchant device producers may face increasing challenges from the substantial capacity generated by government-supported fabrication facilities abroad.[29] At the same time, as noted above

[27] See Howell, *op. cit.*

[28] *Ibid.*

[29] See "The Great Chip Glut," *The Economist*, August 11, 2001.

and in the Introduction, fabless producers in the United States benefit from greater opportunities to obtain lower-cost access to state-of-the-art fabrication facilities. If the rate of growth and ultimate impact of this vertical disintegration within the industry are unclear, there is growing concern about the impact of these trends on the R&D funding that drives the industry. To date, the foundries have tended to be fast followers—rapidly adopting the new manufacturing technologies that drive the industry. As foundries gain market share, it is not clear that the necessary levels of R&D investment required to sustain the industry's growth will be made.

2. The Risk of Knowledge and Technology Transfers.

A further possible challenge to consider from a U.S. perspective concerns the relationship between the foundries and the design manufacturers. In order to take advantage of the foundry model as an outsourcing alternative to in-house production, designs must be shared with the foundries. As particular aspects of semiconductor design are necessarily shared with foreign-owned and controlled fabrication facilities, over time some elements of the U.S. industry may see their comparative advantage in design erode. At the same time, reverse engineering is immensely complex, time-consuming, and, in principle, limited by intellectual property protection. Nonetheless, a great deal of production knowledge occurs on a "learning by doing" basis. Again, the long-term consequences of this process are not clear, but the potential consequences of these developments for U.S. industry, economic growth, and national security suggest that better data collection and further analysis would be appropriate.[30]

C. Significant Technical Challenges

Discovering a method that allows for the continued improvement in semiconductor productivity presents another significant challenge to the semiconductor industry. Much of the progress to date has been the result of scaling—progressively shrinking component size. This process has been the basis of the growth in the semiconductor industry for the past 30 years. Once scaling reaches its limits, the productivity growth of the industry may also diminish. The long-term challenge will be to find new methods to continue the progress in semiconductor productivity, which entails development of new materials and design structures. (Some of these major material challenges are listed and described in Box C.)

[30]Reflecting the growing interest in this topic, *The Advisory Group on Electron Devices Forum* operating under the aegis of the Department of Defense organized a conference to review global economic and technological trends affecting U.S. leadership in microelectronics on September 24, 2002.

Another prominent challenge involves the evolution of lithography toward the use of smaller wavelengths. In the past, the industry has produced several generations of chips by using a specific lithography system with a particular wavelength.[31] This system of lithography has now evolved to one where new wavelengths are introduced and the system is changed in fewer generations. This development has placed a large burden on the equipment industry to keep pace with these rapid changes. For instance, in some cases where the limits of prevailing systems have been approached, companies have resorted to using other, less efficient methods to circumvent the true material barriers. These methods of circumvention, however, do not present real, long-term solutions and in many cases represent a "huge tax on the industry in terms of time and resources."[32]

The future of shorter wavelengths needed for the increase in semiconductor productivity appears to be approaching, and thus there is a need to prepare the shift to new lithographic systems such as EUV (Extreme Ultra-Violet).[33] However, these changes require significant research and development. Some experts have suggested that with current levels of investment, it is not reasonable to expect to find solutions relatively soon.[34]

III. INTERNATIONAL TRENDS

A. International Convergence in Semiconductor Technology

The gap between the United States and countries such as Taiwan and South Korea in semiconductor industry technology is narrowing. Korea and Taiwan exhibit global strength and significant progress in memory technologies and foundry-based fabrication facilities. Over time, these trends point towards a shared global leadership. Complacency about the strength of the U.S. industry vis-à-vis its global competitors would be misplaced. As one recent Academy analysis observed:

> The revival of the U.S. semiconductor industry is an impressive feat, for which government policymakers and industry managers, engineers, and researchers should share in the credit. But the unexpected nature of this revival, its rather complex causes, the contributions to it of cyclical factors, and the fragility of its foundation all suggest that competitive strength in this industry cannot be taken for granted....In other words, U.S. semiconductor firms must maintain their stra-

---

[31]For a more detailed discussion of the impending challenges in lithography, see Michael Polcari's discussion in the Proceedings of this volume, "Current Challenges: A U.S. and Global Perspective," pp. 111-117.

[32]*Ibid,* p. 112.

[33]*Ibid.*

[34]*Ibid.*

BOX C
Long-Term Challenges in the Semiconductor Industry[a]

Over the long term[b], some of the major *Performance Enhancement* challenges in the semiconductor industry entail:

- **Moving beyond current methods and materials.** This includes implementation of non-traditional (non-CMOS)[c] device structures and architectures, including new methods of chip interconnections and memory.
- **The need for Next-Generation Lithography (NGL) technologies.** Since optical lithography will not be viable due to the limits of scaling, the development of NGL technologies such as Extreme Ultraviolet (EUV) are needed.
- **Interconnect Issues.**[d] Scaling or miniaturization of established interconnect structures will no longer allow for the achievement of performance targets. There is a need for material innovation and design in this area.

Some of the major challenges in the long run for *Cost-Effective Manufacturing* in semiconductors are:

- **New Designs to Address Noise Management.**[e] Noise sensitivity is becoming a major issue in semiconductor devices in an age where advances in technology require semiconductors to run at high speeds with low noise, i.e., few disturbances or unwanted signals. In order for

tegic agility and strength in product innovation while avoiding significant erosion in their manufacturing capabilities in order to maintain their strength. The task will require imagination and collaboration among government, industry, and academia.[35]

B. An Increase in Global Partnerships

Global partnerships have become very common in the semiconductor industry. Both Europe and Asian countries have established consortia in order to cre-

[35] Jeffrey T. Macher, David C. Mowery, and David A. Hodges, "Semiconductors," *U.S. Industry in 2000, Studies in Competitive Performance*, David C. Mowery, ed., Washington, D.C.: National Academy Press, 1999, pp. 283-284.

semiconductors to continue along the path of their yearly power increases, more attention needs to be devoted to modeling, analysis, and estimation at all design levels.

- **Error-Tolerant Designs.** The cost of manufacturing, verification, and testing may be reduced by relaxing some of the 100 per cent correctness standards for devices, as the scaling of the design complexity and the increasing transistor count will greatly reduce the potential for failures.
- **Need for New, Larger-Area Starting Materials.** In order to maintain the productivity gains of semiconductors, there is a pressing need for research and engineering to find new, large-area starting substrate material.[f]

---

[a] For a more thorough and in depth technical description of the long term challenges facing the semiconductor industry, see *International Technology Roadmap for Semiconductors: Executive Summary*, 2001 Edition, <http://public.itrs.net/Files/2001ITRS/Home.htm>.

[b] *Ibid.* The Roadmap defines long-term as 2008-2016.

[c] Complimentary Metal Oxide Semiconductor, or CMOS, is the silicon-based material used in most current semiconductors.

[d] The interconnect can be thought of as a line which allows devices to communicate with each other. More specifically, an *interconnect* is a metal conductor line (copper in advanced semiconductors) connecting elements of an integrated circuit.

[e] Noise is a term used to describe any unwanted signal or disturbance that detracts from the performance of the semiconductor device. A more complete treatment of noise in semiconductor devices is given by; *Microwave Noise in Semiconductor Devices*, Hans Ludwig Hartnagel, Ramunas Katilius, Arvydas Matulionis, January 2001.

[f] *International Technology Roadmap for Semiconductors: Executive Summary*, 2001 Edition, *op. cit.*, pp. 14-15.

ate industry standards, map out future issues and challenges, and conduct collaborative research. In both Asia and Europe, policy makers recognize the potential contributions of consortia to overcoming the myriad technical challenges facing the semiconductor industry. Many of the leading figures in the industry, though not all, believe the effort to overcome the multiple technological challenges faced by the industry should be international in scope if these challenges are to be met in the required timeframes.[36] National and international consortia are likely to be a key element in encouraging the research that will aid in meeting these challenges.

---

[36]See the presentations of Genda Hu of the Taiwan Semiconductor Manufacturing Company and Masataka Hirose of Hiroshima University in this volume.

1. Private partnerships and government-industry partnerships represent an integrated national approach to develop semiconductor technology.

a. Japan's Pre-competitive R&D Programs. In Japan, Selete (SEmiconductor Leading Edge Technology Corporation), a joint-venture company established in 1996, conducts R&D on behalf of the Japanese semiconductor industry. Selete, which is not directly funded by government, has been successful in the promotion and evaluation of technologies, developing advanced technologies, and carrying out special projects. By comparison, ASET (Association of Super-Advanced Electronics Technologies) is completely funded by the government and focuses on equipment and chip R&D for 70- to 100-nm technology.

b. Europe's Multinational, Multi-firm Partnerships. MEDEA (Microelectronics Development for European Applications) is a multinational, multi-firm partnership. It is similar to SEMATECH in that it was jointly financed by government and industry. MEDEA has helped develop a better understanding between semiconductor suppliers and system houses, helped develop a better idea of where to focus R&D resources, and fostered closer cooperation among companies in different European countries—both vertically and horizontally. MEDEA officials report that these efforts demonstrate that collaboration in the semiconductor industry can have positive effects for society (employment in the industry increased) and that it can be a productive use of public funds. As a result of MEDEA's success, MEDEA-Plus was initiated in 2001 to address the challenges facing the semiconductor industry noted above.

c. Government-industry Partnering in Taiwan. The semiconductor industry in Taiwan was born out of government support and partnerships in the mid-1970s. Today, the major semiconductor companies in Taiwan are world leaders in their specialties. One of the most successful joint ventures between industry and government is Taiwan Semiconductor Corporation (TSMC). TSMC is a positive example of a government-industry equity partnership in terms of return to society on public investments.[37] The dynamic effects for the Taiwanese economy associated with the establishment of a rapidly growing, highly competitive industry are substantial.[38]

---

[37]The most evident return is in the investment itself. Originally, the government had invested about $100 million in TSMC; the stock was later sold for $400 million.

[38]The highest return is the long-run economic impact of having a leading semiconductor producer in Taiwan. Taiwan's science-park approach (notably Hsinchu Park) to creating partnerships between government and industry has shown substantial success. Though in the beginning there were many foreign companies in the park, by the late 1990s roughly 80 percent of the companies were either local or domestic. Direct equity investment by the government has proved effective. Most of the financial capital provided to the park originated from the government. Currently only about 4 percent of financial capital comes from the government.

IV. THE IMPACT OF SEMATECH: A GOVERNMENT-INDUSTRY PARTNERSHIP

SEMATECH is widely perceived as effective in accomplishing its goals. The consortium's members believe that participation in the consortium has been worthwhile, as evidenced by their continued participation and contributions. This positive assessment is further reflected in the industry's willingness to discontinue public funding while continuing to support the consortium.

The foreign competitors of the U.S. industry share the perception that SEMATECH contributed to the resurgence of the American semiconductor industry and have established a variety of similar programs. These programs are often on a significantly larger scale and have greater underlying political support. Furthermore, a significant number of foreign producers have affirmed their belief in the program's effectiveness by joining SEMATECH since it became an international consortium in 1999.[39]

These trends underscore the importance of public-private cooperation to support research and technology development in the semiconductor industry. In light of the growing significance of R&D collaboration in both the equipment and device industries, providing policy and financial incentives to encourage such cooperation is an increasingly important way to sustain the investments needed to transition to successive generations of new technologies.[40]

---

[39]This "market-based" judgment of the consortium's utility is, in the end, the most compelling. Economic analysis of SEMATECH's impact is extremely challenging. As Kenneth Flamm and Qifei Wang observe, "Finally, the underlying models of R&D cooperation which ultimately must be the basis of a scientific effort to untangle the chains of causality are simply too simplified at this point to capture the complexity of the real world of SEMATECH: a real world in which companies committed to R&D carried out within a joint venture while at the same time competing through internal R&D efforts which also may have spilled over to competitors, a real world in which the menu of consortium activities changes over time with experimentation and learning. At the end of the day, the only absolutely certain thing about SEMATECH is that a substantial portion of its member companies must have found it to be of net value—having actually run the experiment of ending public subsidy, and finding that its consumers continued to buy its output." See Flamm and Wang, *op. cit.*

[40]The operation of an effective consortium entails an agreement on achievable goals in accordance with a sense of shared interests. In addition, a consortium needs an effective management structure tightly linked to member interests, as well as a long-term commitment from its participants to contribute highly trained and qualified personnel and to provide financial support. For a further discussion of characteristics of successful national and international consortia, see National Research Council, *Conflict and Cooperation in National Competition for High-Technology Industry*, Washington, D.C.: National Academy Press, 1996, pp. 48-51.

A. Sources of SEMATECH Contributions

1. Flexible Objectives and Industry Leadership

By definition, an R&D consortium's contributions (Box D) are due in part to its ability to adapt its goals to the conditions of a rapidly evolving industry.[41] The cost sharing arrangement with the government and industry management of the research agenda has contributed to this flexibility. Indeed, while it benefited from strong leadership, no single entity dominated the consortium or determined its direction. Members, including Department of Defense officials, reached a broad consensus on technical goals and then left the consortium management to implement the program.

The industry interaction within the consortium, and between the consortium members and the suppliers, improved the dynamics between the device makers and the equipment industry, with collaboration generating new technical perspectives for the participants and encouraging the give-and-take between manufacturer and supplier necessary to expedite the technology development process.

2. Analyzing SEMATECH

Measuring the contributions of research consortia is a difficult task (Box E). As noted in this report, there have been relatively few empirical analyses of the impacts of R&D cooperation on industrial R&D.[42] For the semiconductor industry, some empirical analysis suggests that the consortium has boosted the "effec-

[41]While many believe that SEMATECH contributed to the resurgence of the U.S. semiconductor industry in the early 1990s, it was by no means the only element in this unprecedented recovery. For example, time for the industry to reposition itself was provided by the 1986 Semiconductor Trade Agreement. The U.S. industry also repositioned itself, profiting from shifts in demand, i.e., away from DRAMS (where Japanese skill in precision clean manufacturing gave significant advantage) towards microprocessor design and production (where U.S. strengths in software systems and logic design aided in their recovery.) Arguments about which of these elements were most decisive probably miss the point. The recovery of the U.S. industry is thus like a three-legged stool. It is unlikely that any one factor would have proved sufficient independently. Trade policy, no matter how innovative, could not have met the requirement to improve U.S. product quality. On the other hand, by their long-term nature, even effective industry-government partnerships can be rendered useless in a market unprotected against dumping by foreign rivals. Most important, neither trade nor technology policy can succeed in the absence of adaptable, adequately capitalized, effectively managed, technologically innovative companies. In the end, it was the American companies that restored U.S. market share.

[42]Few researchers have empirically assessed the effects of joining SEMATECH on its member firms' expenditure on private R&D. As noted by Flamm and Wang, *op. cit.*, in this report, even

Box D
Contributions of the Consortium

SEMATECH has made a variety of important contributions to the health of the semiconductor industry in the United States. For example, the consortium has:

- Played an integral role in promoting the development of effective manufacturing technology in the semiconductor industry.
- Developed industry-wide standards for manufacturing tools, notably through collaboration with the equipment industry and through industry-wide technology roadmaps.
- Fostered a shared perspective on the technological development required to maintain the industry's high growth rate through the semiconductor roadmap process.
- Aided companies in:
  -developing reliable manufacturing tools
  -creating an effective quality control process
  -understanding the needs of the industry and advancing the sophistication of the manufacturing process.

though SEMATECH is the highest-profile R&D consortium in the United States, it has been the focus of study for only three statistically rigorous papers. One study (Douglas A. Irwin and Peter J. Klenow, "High-Tech R&D Subsidies: Estimating the Effects of SEMATECH," *Journal of International Economics* 40(3-4):323-44, May 1996) found that SEMATECH firms reduced their individual expenditures on R&D by about $300 million dollars. They further concluded that the reduction in firm-level R&D of member firms does not justify public support for the consortium, since firms are essentially free-riding on federal funds and would have expended the equivalent federal funds out of their own budgets had there not been a consortium. Irwin and Klenow argue further that firms joined SEMATECH to "share" information but not to necessarily "commit" funding for high spillover R&D, which, if true, would have resulted in an increase in R&D. There are both conceptual and econometric flaws with this argument, as pointed out in Flamm and Wang, *op.cit.* Irwin and Klenow interpret the data as member firms reducing R&D expenditure that would have been conducted in the absence of the consortium, rather than reducing poorly appropriable R&D, which would not have occurred at all had it not been for the consortium's formation. Further, even a "commitment" approach to R&D among consortium members, with relatively low spillovers, would actually lower R&D, a conclusion the authors do not draw.

Box E
Organizing Successful Consortia

Because of its contributions, SEMATECH is sometimes considered a model for future public-private partnerships.[a] Some of its lessons for organizing a successful consortium are:[b]

- *Understand the Need for Cooperation*—the great range of R&D needs, from basic science to manufacturing infrastructure to whole new industries, is, arguably, best understood in terms of a process where industry works in close cooperation with universities and government research laboratories. To achieve the full benefits of cooperation, it is important to:
- *Ensure Quality Leadership*, including key leaders of the major participating industries
- *Convey your Message* publicly to leaders in the government and private sectors
- *Focus the Program* on key sectors and build on this developed strength, rather than approach the entire industry
- *Set Measurable Objectives* for advancing generic or pre-competitive knowledge
- *Set Uniform Requirements* of participation so that support is not fragmented
- *Plan first—Spend later*. Roadmaps are needed before consortia can be properly launched
- *Develop an Industry-driven process*—recent collaborative work, such as that on extreme ultra-violet lithography, shows that successful consortia are industry driven.

---

[a] National Research Council, "Government support for technology development: The SEMATECH experiment," *Conflict and Cooperation in National Competition for High-Technology Industry,"* pp. 141-51.

[b] See remarks by William Spencer in the Proceedings section of National Research Council, *Partnerships for Solid-State Lighting, Report on a Workshop,* Washington, D.C.: National Academy Press, 2002. Toshiaki Masuhara of Hitachi also suggests a number of criteria for organizing a successful consortium. See his presentation in Panel VI of the Proceedings of this volume.

tive" R&D level of its members.[43] The work of Flamm and Wang suggests that SEMATECH reduced the R&D expenditures of its membership somewhat, in part by eliminating duplication.[44] In essence, if the number of dollars spent on similar R&D projects across firms is reduced, and the yield of overall industry R&D is unchanged, then this is a better outcome from both a social standpoint, for society, and for industry since resources are freed which can be put to productive use elsewhere. This is a positive result both for the firms and for society as a whole. Moreover, this outcome lends credence to the idea that a consortium can add to the dynamic efficiency of both its member firms and the industry as a whole.

While the precise measurement of contributions is difficult, SEMATECH is widely believed, within the industry, both in the United States and abroad, to have made a positive contribution to the resurgence of the U.S. semiconductor industry. More indirectly, the consortium's activities have contributed to greater cooperation among producers, suppliers, and the government. For example, the current promising cooperation on next-generation lithography tools, (i.e. the EUV Consortium) illustrates this enhanced willingness to collaborate in innovative ways. This positive perception of SEMATECH has contributed to its emulation, notably in foreign programs to support national or regional semiconductor industries and among other U.S. industries, (e.g. in optoelectronics and nanotechnologies).[45] More broadly, SEMATECH helped sustain the rapid technological progress of the industry as projected by Moore's Law. This technical progress was facilitated by the collaborative research encouraged by the consortium, including the development of the Semiconductor Industry Roadmap.

For its part, the government partner achieved many of its objectives. The Department of Defense achieved its goal of maintaining a robust, technologically advanced manufacturing capability within the United States. SEMATECH thus helped the government achieve a key objective, namely, sustaining a U.S.-based industry able to provide cutting-edge, low-cost devices to support defense requirements[46] and thereby avoiding the risk of dependency on foreign suppliers for U.S. defense systems.[47] Throughout the decade of the 1990s, the Defense Department was able to acquire higher-performance, lower-cost components from

---

[43]See Flamm and Wang, *op. cit.*

[44]*Ibid.*

[45]See Box B in the Introduction in this volume. See also National Research Council, *Small Wonders, Endless Frontiers: A Review of the National Nanotechnology Initiative*, Washington, D.C.: National Academy Press, 2002.

[46]A healthy U.S. industry also ensures a surge capacity for the defense industrial base, should it be required.

[47]The erosion of the U.S. semiconductor industry's position was a source of growing concern to federal defense officials and was reflected in the creation of the National Advisory Commission on Semiconductors (NACS). One of the NACS' missions concerned the dependency of modern weap-

commercial suppliers than would have been available from a dedicated defense production facility. This trend contributed to dual-use defense acquisition designed to benefit from the rapid evolution of commercially available semiconductors characterized by rapidly increasing performance and falling costs.[48]

The combination of rapid gains in semiconductor capabilities and sharply falling costs has contributed to the government's capacity to carry out many other non-defense missions more efficiently. These contributions are reflected in the economy as a whole. Also, as noted above, the U.S. economy recorded substantial gains in productivity growth between 1995 and 1999, with productivity growth more than double that of the 1973-1995 period. The Council of Economic Advisers attributed "these extraordinary economic gains" to three factors, namely, technological innovation, organizational changes in businesses, and public policy.[49] Two of these factors concern information technology, in particular the simultaneous advances in information technologies—computers, hardware, software, and telecommunications—which combine these new technologies in ways that sharply increase their economic potential. Progress in semiconductor capabilities enabled advances in information technologies, driving innovation in each of these product areas. In short, the government and the economy as a whole have benefited from the contributions of a robust U.S. industry.[50]

---

ons systems on state-of-the-art semiconductor devices. Specifically, under the legislation creating the Commission, Congress notes in its findings that: "Modern weapons systems are highly dependent on leading-edge semiconductor devices, and it is counter to the national security interest to be heavily dependent upon foreign sources for this technology." The charter further states that this Committee shall "identify new or emerging semiconductor technologies that will impact the national defense or United States competitiveness or both." For the objectives set forth for NACS, see <http://www4.law.cornell.edu/uscode/15/4632.html>.

[48]See Jacques Gansler, *Defense Conversion: Transforming the Arsenal of Democracy*, Cambridge, MA: MIT Press, 1995. See also the presentation by Paul Kaminski, then Under Secretary of Defense for Technology and Acquisition, in National Research Council, *International Friction and Cooperation in High-Technology Development*, Washington, D.C.: National Academy Press, 1997, pp. 132-133. Dr. Kaminski points out that tighter linkages with commercial markets shorten cycle time for weapons-systems development and reduce the cost of inserting technological improvements into DoD weapons systems. By placing greater reliance on commercial sources, DoD can field technologically superior weapons at a more affordable cost.

[49]Council of Economic Advisers. *Economic Report of the President*, Washington, D.C.: Government Printing Office, 2001.

[50]These broader contributions may be relevant with respect to proposals for recoupment of federal contributions to government-industry partnerships. Some analysts suggest that the best means of recoupment is, in fact, the tax system. For example, many of the companies that have thrived following SEMATECH's inception have returned the government's original investment to the consortium many times over in the form of tax revenue. See testimony by Christopher T. Hill, Vice Provost for Research and Professor of Public Policy and Technology, George Mason University, before the Subcommittee on Environment, Technology and Standards, Committee on Science, U.S. House of Representatives, March 14, 2002.

As described in greater detail in the Introduction, the SEMATECH consortium's contributions to the resurgence of the U.S. industry were significant but are best understood as one element of a series of public policy initiatives that collectively provided a positive policy framework for U.S. semiconductor producers. Overall, SEMATECH's record of accomplishment was achieved in no small part through the flexibility granted its management and the sustained support provided by DARPA, the public partner, complemented by the close engagement of its members' senior management and leading researchers.

Recommendations

The Committee's recommendations outline a series of modest steps that nonetheless may prove important to the long-term welfare, economic growth, and security of the United States.

RESOURCES FOR UNIVERSITY-BASED SEMICONDUCTOR RESEARCH

To better address the technical challenges faced by the semiconductor industry and to better ensure the foundation for continued progress, more resources for university-based research are required.

The Committee believes that universities have an important role in maintaining a balance between applied science and fundamental research. This balance is key in generating ideas for future research.

The Committee suggests consideration of the development of three-way partnerships among industry, academia, and government to catalyze progress in the high-cost area of future process and design. These partnerships would:

(a) **Sponsor more initiatives that encourage collaboration between universities and industry**, especially through student training programs, in order to generate research interest in solutions to impending and current industry problems.

(b) **Increase funding for current programs.**[51] Research programs that are already operational, such as the Focus Center Research Program developed by the SRC, could usefully be augmented through substantially increased direct government funding. These centers also represent opportunities for collaborative research with other federal research programs, such as those supported by the National Science Foundation.

(c) **Create Incentives for students.** A key role for universities is to ensure the flow of technical innovation and skills that originate with students. In order to address the undersupply of talented workers and graduate students in the industry, more incentive programs should be established. Since professors typically respond to appropriate research incentives, augmented federal support for programs that encourage research in semiconductors would attract professors and graduate students.[52] In addition, specific incentive programs could be established to attract and retain talented graduate students.

[51]The president's FY 2003 budget makes important steps in this direction. It calls for a 3 percent increase, to $1.9 billion, in the *Networking and Information Technology Research and Development Program* (NITRD). This particular program could play a key role in funding the basic research that confronts the technical challenges in the semiconductor industry. The NITRD coordinates key advanced information technology research across multiple agencies to make broad advances in computing and networking. These advances manifest themselves in the development of new technologies such as computing platforms and software, which can support advances research in physics, materials science and engineering as well as biomedical and earth and space science. The 2003 budget envisions emphasizing critical areas of research such as networks security issues; high-assurance software and systems; micro- and embedded-sensor technologies; and revolutionary architectures to reduce cost, size, and power requirements of high-end computing. The budget emphasizes research on the social and economic impacts of developments in the fields of information technology. For the text of the president's proposed initiatives, see *Fiscal Year 2003*, *Analytical Perspectives*, Budget of the United States Government, Washington, D.C.: U.S. Government Printing Office, 2002, p. 164.

[52]See Paula Stephan and Grant Black, "Bioinformatics: Emerging Opportunities and Emerging Gaps," in National Research Council, *Capitalizing on New Needs and New Opportunities: Government-Industry Partnerships in Biotechnology and Information Technologies,* p. 244.

III

PROCEEDINGS

Opening Remarks

Bill Spencer
Chairman Emeritus, International SEMATECH

On behalf of the National Academies Dr. Spencer expressed appreciation for the effort that many participants had made to come "from the four corners of the world," especially those from Taiwan, Japan, and Europe. He described the assembly as "a particularly distinguished set of panels even for the National Academies of Sciences and Engineering."

Dr. Spencer said that the economic regions represented at the symposium—the United States, Europe, Japan, and Taiwan—represented over 90 percent of semiconductor manufacturing and sales worldwide. He suggested that these regions had moved into leadership positions because they had adopted the kinds of programs that the symposium would explore.

Dr. Spencer said that two weeks earlier the National Academies had held a related seminar on the New Economy and the impact of technology on it. He defined the New Economy in the United States as "the spectacular growth that we have had since 1995, which economists have not been able to explain by looking at the usual economic indicators." He said that additional growth of GDP amounted to 0.5 to 1 percent per year. Dr. Spencer further noted that a group of approximately 30 leading U.S. economists and technologists were uncertain of the existence of a new economy.[1] Nevertheless, the group unanimously agreed that the gains in productivity in computers, communications, and software were due principally to growth in productivity in the semiconductor industry.[2]

[1]For one perspective on growth and the existence of a New Economy see Paul Krugman, "How Fast Can the U.S. Economy Grow?" *Harvard Business Review*, July 1997.

[2]Researchers in the field of economics have classified semiconductors as a general-purpose technology (GPT). Historical periods of great progress and growth have been attributed to the pervasive-

Dr. Spencer said that it was widely recognized that the semiconductor industry had contributed not only to spectacular growth in that industry but also to economic growth in general. He expressed hope that the lessons shared during the symposium would help that industry maintain high growth and productivity over the decades to come, and then said that he thought the outlook was very good. He then introduced Clark McFadden as chair of the first session.

---

ness of such technologies. For an excellent discussion and treatment of this topic, which also touches on the semiconductor as a GPT, see Timothy F. Bresnahan and M. Trajtenberg, "General-Purpose Technologies: 'Engines of Growth'?" *Journal of Econometrics*, January 1995 pp. 83-108.

The U.S. Experience: SEMATECH

INTRODUCTION

Clark McFadden
Dewey Ballantine

Mr. McFadden said that the first panel would consider the experience of SEMATECH, which began in 1987 as a government-industry collaboration to promote certain technologies in the semiconductor industry. Noting that it has evolved to become International SEMATECH—a collaboration of private companies on an international scale—he observed that SEMATECH was an appropriate subject for this particular committee of the National Research Council because it has been a very visible and major technology partnership. "Extraordinary" in its ambitions, scope, and impact, it has met its technology goals and continues to attract growing investment from its industrial participants.

The Impact of SEMATECH

Assessing SEMATECH's impact on the semiconductor industry and on technological development in the United States, he said, is more difficult. A full appraisal of SEMATECH requires an understanding of its special features and the sources of its enduring and broad appeal. From the industry standpoint, he said, SEMATECH was initially stimulated by an external competitive threat—the emerging preeminence of an integrated Japanese semiconductor industry. The support coalesced around the acknowledgment that developing the technology required for semiconductor manufacturing was beyond the ability of any single company. The consortium leveraged the differing but reinforcing needs of industry and government participants to develop manufacturing tools and process technologies. It helped industry to develop more effective manufacturing technology and helped government achieve less costly and more accessible manufacturing of advanced military circuit designs.

A Catalyst for Change in the Industry

SEMATECH has also been creative, he said, in dealing with the issues that confront most technology collaborations: creating incentives for participation, establishing an appropriate technology focus, defining the bounds of shared intellectual property, and providing effective mechanisms for technology transfer. It accomplished these objectives, he said, without diminishing the intensity of the competition among the industrial participants. Remarking that SEMATECH had a transforming effect in many ways, he cited specifically the fostering of an industry perspective on technology development, leading naturally to industry-wide testing of tools and standards and to the development of industry-wide technology roadmaps.

He defined two main tasks of the symposium. The first was to evaluate the salient aspects of SEMATECH, both at its inception and in its current form. The second task was to compare SEMATECH to other consortia initiated around the world. Many of these were in turn stimulated in part by SEMATECH. He noted that launching an initiative of this scope and magnitude requires determined leadership, and that many of the leaders of this effort were present at the symposium. He introduced the first speaker, Gordon Moore, as "one of the real visionaries for and a very compelling advocate" for the consortium as it unfolded and a force behind benefits far beyond SEMATECH.

THE SEMATECH CONTRIBUTION

Gordon Moore
Intel Corporation

Dr. Moore cautioned the audience that his contribution to SEMATECH was restricted to its early years. He proposed to create a picture of the situation at the time SEMATECH was established and to present industry's view of some of its contributions. During the early years of the consortium, he said, the U.S. semiconductor industry was experiencing what Andy Grove[3] dubbed "X curves" (see Figure 1). This referred to the U.S. curves for market share going down and Japanese curves going up for a variety of manufacturing industries, including the semiconductor industry. For the U.S. semiconductor industry, he said, this was "disconcerting, to say the least."

[3]Andrew Grove and Dr. Moore are co-founders of Intel Corporation.

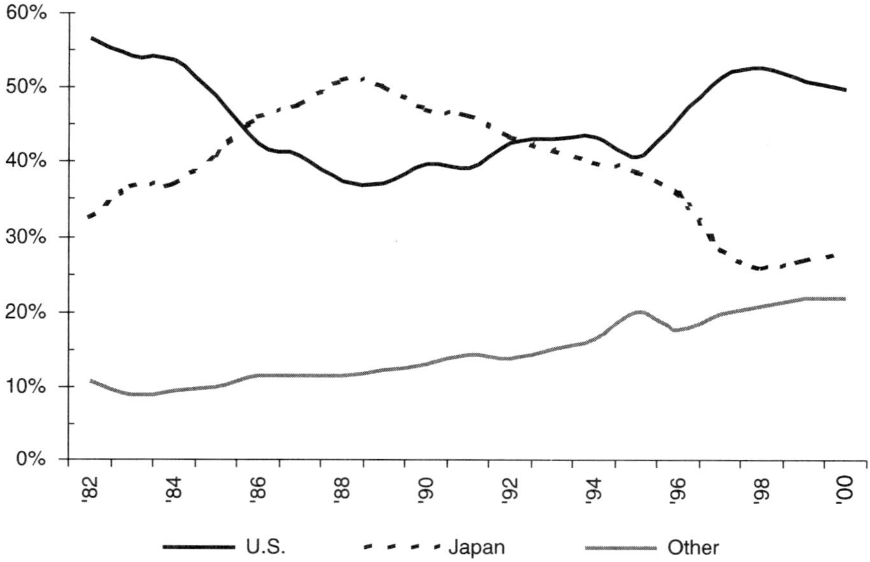

FIGURE 1 Worldwide semiconductor market share.
SOURCE: Semiconductor Industry Association

Quality Challenges and the X Curve

Dr. Moore noted that the quality levels of devices being supplied by Japanese industry and U.S. industry were dramatically different at the end of the 1980s. As one contributing element he pointed to an AQL (acceptable quality level) culture that had evolved in the United States. AQL was considered the level of quality that would pass 95 percent of the lots. He said that the industry used to argue with customers about whether 1 percent AQL was appropriate, or perhaps 0.4 percent AQL. "Nobody," he said, "ever told us they wanted a higher-quality product. I don't think the customers realized it was available until they started getting it from the Japanese industry." This discovery, he said, was "also disconcerting."

Then the U.S. industry started to compile benchmarking data that allowed comparison of yields. Of wafers that were started through the production line the industry was able to move 80-90 percent successfully to the end, whereas the available data showed that Japanese companies were succeeding with 98 percent of their wafers. Similarly, Japanese companies were achieving considerably higher overall yields—the percentage of the original silicon area that emerged as good devices. This function depends strongly on the size of the device because it is an area-dependant phenomenon. In direct labor productivity Japanese industry was roughly twice as high and in indirect productivity seven times as high, partly because U.S. plants employed far more engineers and other workers. When Japanese and U.S. plants used the same equipment, Japanese plants produced two to three times as many wafers per unit of time as the U.S. plants.

Dr. Moore said that this pattern of Japanese superiority was repeated in virtually every measure that related directly to manufacturing. Thus, another X curve could be seen developing in the semiconductor equipment industry. The U.S. market share was dropping and had nearly passed below the level of the Japanese equipment suppliers.

First Steps Toward Collective Action

That, said Dr. Moore, was the environment in which SEMATECH was established. Before that time the U.S. industry consisted of many companies that were very independent. The industry preferred that the U.S. government not be involved in industry other than as a customer. This broad set of trends, however, was a great concern to industry leaders. Charles Sporck of National Semiconductor, who has a manufacturing background, urged the industry to consider some collective activity. Eventually he succeeded and the industry looked more thoroughly into its manufacturing procedures. This examination, said Dr. Moore, showed that U.S. firms were doing an excellent job in device technology, where the industry was moving to the next generation as fast as possible; however, "we weren't doing a good job" in manufacturing technology, and the leaders resolved to create programs to improve this.

Government-Industry Cooperation

After much debate the Semiconductor Industry Association, composed of the main U.S. device manufacturers, took the unusual step of approaching the government and making the argument that collective action was necessary for the sake of long-term U.S. economic competitiveness and the national defense. Industry and government agreed on a unique arrangement in which several companies invested a combined total of $100 million per year and the federal government matched that amount. Over time the government investment in SEMATECH has totaled about $850 million.

An important feature of this agreement was that, for the first time, industry agreed to put quality people into a government-industry partnership. Participating companies included nearly all of the largest semiconductor companies in the United States and their 14 appointed representatives to work with SEMATECH.[4]

[4]The original members of SEMATECH were IBM, Intel Corporation, Motorola, Texas Instruments, National Semiconductor, Advanced Micro Devices, Lucent Technologies, Compaq Computer Corp., Hewlett-Packard Technology, Conexant Systems, NCR Microelectronics Corp., Harris Semiconductor, LSI Logic Corp., and Micron Technology.

The program was initially planned to last 5 years. At the end of that period it was extended for four years. At the end of that extension the industry decided that it was not appropriate to ask for additional government support. This decision was "much to the surprise of the people back here in Washington," said Dr. Moore, "where programs seem to have a life of their own."

Varying Motivations for Joining the Consortium

He said that the reason the program was extended to 8 years instead of the planned 5 years was that the industry had moved off to a slow start while it tried to determine the best approach. Initially the planners anticipated two manufacturing lines, using DRAM technology from IBM and SRAM (Static Random Access Memory) technology from AT&T. This did not work well partly because the various companies supporting SEMATECH had different ideas about how they would benefit individually from the consortium. Some assumed they would benefit by receiving next-generation technology, while others thought they would benefit from joint research on manufacturing. As a result, SEMATECH made little progress at the outset despite heavy initial outlays on production lines. The consortium discovered that little could be done to improve manufacturing without running a full-volume manufacturing operation. At the outset the operation also suffered from lack of full-time leadership.

Clarifying the Challenge

Eventually, SEMATECH and the industry supporting it began to clarify the problems that needed most attention. One was the developing crisis in manufacturing tools. The manufacturing tool industry in the United States was very fragmented then and still is. Many companies were one-product firms formed to build a particular kind of machine. Often the lifetime of such a company more or less matched the period during which its product met an industry need, rising and then falling on the success of a single instrument. SEMATECH worked with these companies to develop reliable tools, to teach them total quality control, and to help them understand the needs of the industry and the increasing sophistication of the manufacturing process. Leadership in tools, particularly lithography tools, was then shifting away from the United States.[5] SEMATECH began to recognize that much of the important work required to improve manufacturing equipment did not have to be done by each company individually but could be done by the

---

[5]Lithography is the process whereby a pattern is transferred to a photosensitive material by selective exposure to a radiation source such as light. As a result of the selective exposure to light, the physical properties of the photosensitive material are altered in specific areas. In the semiconductor industry, lithography is used to imprint circuits on semiconductor materials (e.g., silicon, germanium).

consortium centrally. The consortium developed a cost-of-ownership model for manufacturing tools that described the problems in detail.

A Forum for Open Communication

Another major contribution of SEMATECH was to provide a forum where companies could openly communicate. Previously antitrust concerns prevented semiconductor companies from communicating effectively about what they were doing, except at certain conferences. Now legislation allowed them to talk together on matters related to SEMATECH and, although the effect was difficult to quantify, they benefited greatly from this new avenue of communication. They discovered that they were all moving in generally the same direction and pursuing essentially the same set of problems. The internal research of companies became more efficient, as they were able to reduce duplication and the number of blind alleys.

Since then SEMATECH has succeeded in funding the development of new 300-mm tools and has taken a leadership position in pursuing the technology roadmaps designed by the Semiconductor Industry Association. It has supported initiatives on mask-making tools, lithography using very-short-wavelength (157-nm) ultra-violet light from a special laser, next-generation lithography consensus, low-dielectric-constant materials, and other innovations. It has also continued to benchmark the industry and to help improve manufacturing methods, among other contributions.

A Lack of New Members

Among SEMATECH's disappointments, said Dr. Moore, has been its lack of success in increasing the list of U.S. semiconductor companies that belong to it. Despite the rapid increase in the number of semiconductor companies during the consortium's lifetime, essentially no new U.S. companies have joined. New international companies have joined to replace the U.S. companies that have dropped out, but U.S. participation is largely confined to the founding members.[6]

New Directions

SEMATECH today has a new set of directions. One of the most important is the 300-mm wafer program, "the perfect example of the kind of place where SEMATECH can help." The industry today must move to the 300-mm standard in "lock step," said Dr. Moore, because the equipment industry cannot afford to

---

[6]International members, who joined in 1996, include Philips, Hyundai, LGSemicon, STMicroelectronics, Infineon, and TSMC. The following year U.S. government funding dropped to zero.

develop separate new generations of equipment for first 200-mm and then 300-mm wafers. SEMATECH can provide valuable help in coordinating this task.

He also raised the issue of timing, saying that some equipment suppliers had criticized the consortium for beginning to move toward the new standard too early. The equipment suppliers invested in 300-mm equipment before the semiconductor device manufacturers were willing to accept the new standard.

SEMATECH is also helping companies push the key technology of lithography to shorter and shorter wavelengths. The industry needs insulators of lower dielectric constants to improve performance, said Dr. Moore, as well as materials with higher dielectric constants for future gates. Both goals pose considerable challenges to the materials science industry.

A Positive Impact on the Industry

To illustrate the impact of SEMATECH's activities he showed some of the earlier curves extended in time past the formation of the consortium. The original X curve had become a W curve, indicating that the U.S. industry had regained significant market share in the semiconductor device industry. He said that many factors were responsible for this but that SEMATECH had certainly played a part. The same was true for the equipment industry, where the U.S. and Japanese market shares crossed and then crossed again as the United States regained share, with the effect becoming noticeable at the same time SEMATECH was beginning to "get traction." (See Figure 2) He concluded that at least from the U.S. perspective, SEMATECH had "a positive impact on the U.S. industry."

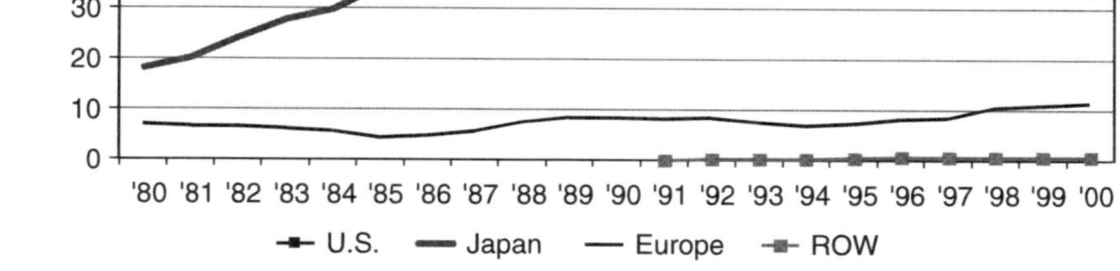

FIGURE 2 Semiconductor equipment market share.
SOURCE: VLSI Research, Inc.

To measure that positive impact he reviewed the membership over time and showed that most of the major companies had continued to participate, providing an indication of how the participants feel about the value of the organization.

"Was it a good deal from the U.S. government's point of view?" he asked. He reiterated that SEMATECH was formed during a difficult period in the semiconductor industry. Intel, for example, had lost a "significant" amount of money in 1986, as did most of the industry.[7] Today, however, Intel pays more in taxes every quarter than the entire government investment in SEMATECH, indicating that for the government and the industry a "very good investment was made."

Focusing on the Largest Challenges

In conclusion he said "a government-industry partnership can contribute." He said the challenge was to identify a problem clearly at the outset. Without a very specific charter, he said, a multi-corporation organization may not have any more impact than a single laboratory. The impact can be greatest when the problems exceed those a single company can solve. As an example he cited the challenges of the new lithography environments. As the industry moves away from optical lithography it faces billion-dollar R&D programs to reach the point where it can start operating. A challenge of this magnitude, he said, requires some form of industry or industry-government consortium.

Supporting Basic Research

Another area where government-industry cooperation is important, he said, is in the support of the basic research on which an industry depends. In effect, he said, the industry is "still mining the basic developments of the 1950s, when we developed the semiconductor industry." Since then industrial R&D has become more sharply focused on short-term, predictable results.[8] Companies have found they cannot fully capture the benefits of their high-risk, long-term research, which can produce results that are unexpected or take far longer than anticipated. Companies can no longer afford to support large central laboratories dedicated to ba-

[7]During the mid-1980s Japanese firms, because of their comparative advantage, dominated the then memory-driven semiconductor industry. As a result of their large market share in semiconductors the Japanese firms forced many of their competitors out of the memory market and into other markets. It was at this point that Intel abandoned the DRAM (dynamic random access memory) market and began focusing its production efforts on the microprocessor.

[8]According to the Semiconductor Industry Association, federal investments in pre-competitive R&D, such as programs sponsored by the National Science Foundation, have declined from 5.7 percent of the federal budget in 1965 to 1.9 percent today. During the same period many other nations have increased their investments in R&D.

sic research—even though the benefits of such work to society as a whole are significant.

Therefore, it is extremely important, said Dr. Moore, that the government continue to support long-term research across a very broad base. The whole of technology must move ahead together as a single front, without one part advancing ahead of the rest. In biological research, for example, it is poor strategy to study only what goes on in the cells themselves. Also needed are physical and chemical tools to probe and analyze these cells. He said that progress in biology owed a great deal to MRI imaging, whose uses in the health sciences no one could have predicted several decades ago. We could not have made such rapid progress in deciphering the human genome without simultaneous advances in data processing. He concluded that the one place where government has an indispensable role in working with industry is in maintaining this country's excellence in long-term basic research across the whole broad front of major disciplines.

DISCUSSION

Participation of Foreign Companies

A questioner asked, in regard to international organizations, whether foreign companies would use them as a way to draw on the U.S. base of expertise. Dr. Moore said he viewed it more as an opportunity for broader participation in what has become a global industry. "There are some general problems that we all have to solve," he said, "if we're going to continue to make progress as an industry," such as the challenge of producing 300-mm wafers.

SEMATECH Without Government?

Dan Radack of the Defense Advanced Research Projects Agency asked whether SEMATECH would have formed without the original participation of the government, provided it had not been blocked by antitrust concerns. Mr. McFadden said that in his opinion it would have been unlikely. Government participation provided a broader sense of both urgency and commitment, and made most of the companies feel as though they could not afford to be left out.

Support for a Broad Range of Basic Research

Eliot Cohen of the Palisades Institute for Research Services, referring to Dr. Moore's comment about the need for sustained government support for basic research, asked if he would advocate support for specific areas. Dr. Moore repeated that support should be "fairly broad." He suggested that the current focus of Congress on bioscience research at the National Institutes of Health should not be applied to the exclusion of information technologies and other disciplines that may also contribute to breakthroughs in health science. He said that it is very hard to pick winners and losers in research, especially in basic research, where the impact may be far removed from the locus of research.

THE IMPACT OF SEMATECH ON SEMICONDUCTOR R&D

Kenneth Flamm
University of Texas at Austin

An Economist's Perspective

Dr. Flamm said he would offer an economist's view of the impact of SEMATECH, including a review of the economic literature on R&D cooperation. He said that if one tried to evaluate the success of SEMATECH from a private perspective, a good test would be whether firms judge it to be worthwhile in the absence of government subsidies. On the other hand, to judge whether it was socially worthwhile for the government to support SEMATECH, information on the impact of SEMATECH on aggregate R&D in the industry would be needed. He said, however, that in a moment he would suggest that this is not necessarily the right question to ask.

Dr. Flamm said he would spend little time on the background of SEMATECH, which had already been described by Dr. Moore. He did note an irony in the story: The Japanese VLSI projects of the late 1970s helped shape the concepts that led to SEMATECH.[9] The Japanese VLSI projects, he said, were generally perceived as having played a significant role in bringing Japanese manufacturing technology and semiconductors up to world-class levels in the late 1970s. Subsequently that model in the semiconductor industry was basically dropped by Japan in the 1980s and, after SEMATECH, brought back again in the 1990s.

Generally Perceived as a Success

He observed that with a few exceptions SEMATECH is generally perceived as a success by the U.S. semiconductor industry, although opinions differ on the amount of credit it should be given. He said that the willingness of the industry participants to continue the program entirely on their own is the best test of whether they judge it privately to be worthwhile. He said it was also perceived as a success in Japan, influencing the formation and design of the ASET and Selete programs.

Dr. Flamm described a perception that SEMATECH played some role in the resurgence of the U.S. semiconductor industry. He said that economists generally

[9]VLSI stands for very large scale integrated circuit.

view the program as the preeminent model of a cooperative government-industry R&D joint venture. Despite its place as a model program, however, there has been only one empirical study of any significance, and the study itself has some flaws.[10]

Coordination and Cooperation

Turning to the economics literature on R&D cooperation, which he warned his audience would be "painful," Dr. Flamm said that it basically distinguished two aspects of cooperation that can logically be separated. One of them is coordination, i.e., when firms jointly commit to R&D activities, taking into account any spillover effects. The other function of R&D cooperation is to share the results of the jointly funded R&D. Much of the economic literature concerns the pure coordination model, which could be called an R&D cartel. That is, the participants agree on how much each will invest in R&D, knowing that some of the results of their individual R&D are going to spill over to the other firms. The distinction is between all companies committing jointly to individual investments in R&D and to all companies doing their R&D competitively, without commitment to others.

Internalizing Spillovers from Others' Research

The results of these two models of cooperation differ. If every firm is doing R&D on its own, the spillovers of this R&D among firms will tempt each firm to act as a free-rider. As such, the strategy of each will be to let other companies do most of the investment while the low-investment or free-rider firm attempts to garner the benefits from the others' R&D results. In the case of coordination, however, each company is internalizing some benefit from the spillovers. The industry is maximizing total profits and each participant is able to internalize some of the spillover from the efforts of other firms.

The Joint Venture Includes Information Sharing

The canonical model of the broader form of cooperation that includes information sharing is the R&D joint venture. In the cooperative mode of a joint venture, firms not only set their R&D levels together but they also agree to perform

[10]Few researchers have empirically assessed the effects of joining SEMATECH on member firms' expenditures on private R&D. As noted by Kenneth Flamm and Qifei Wang in "SEMATECH Revisited: Assessing Consortium Impacts on Semiconductor Industry R&D" in this report, even though SEMATECH is the highest-profile R&D consortium in the United States, it has been the focus of study in only three statistically rigorous papers. For the first empirical study of SEMATECH see Douglas A. Irwin and Peter J. Klenow, "High-Tech R&D Subsidies: Estimating the Effects of SEMATECH," *Journal of International Economics* 40(3-4):323-44, May 1996.

R&D together as a single entity. Thus, they completely share the benefits of their R&D. This is the opposite of pure competitive R&D by competing firms.

As a variation he suggested a competitive R&D joint venture, in which each company agrees to funnel its R&D through the single joint organization without committing to a level of spending. This case would still suffer from a free-rider problem—a situation in which each company agrees to do its R&D in semiconductors with SEMATECH without specifying what they would contribute. They would spend some amount in the interest of doing R&D, even if no other company spent money on R&D. However, this amount would be less than it would be if companies decided to set their R&D levels cooperatively.

Two Scenarios of R&D Cooperation

He then showed schematic representations of the two models. The first depicted pure coordination, where the spillover (β) between two R&D firms may vary from 0 to 1—from no spillover to complete sharing. (See Figure 3.) He also showed a schematic drawing of coordination and information sharing, picturing an R&D joint venture in which the spillover (β) takes on a value of 1, that is, complete sharing. (See Figure 4.)

He pointed out that the economics literature has examined two scenarios under different levels of cooperation and sharing. The first set-up considers the outcome of a cartel versus competitive firms, whereas the second scenario examines the consequences of a competitive joint venture versus a cartel joint venture. That is, the economics literature only considers two extreme situations; complete cooperation in R&D and no cooperation in R&D.[11]

SEMATECH as a Hybrid of the Two Scenarios

He suggested SEMATECH is a hybrid of these two pictures. A certain portion of the R&D is funneled into the organization, where it is carried out collectively and spills over to other firms. He pointed out that the spillover coefficient

---

[11]Researchers have focused on a simple model of a research cartel, in which there exist information spillovers and firms pursue R&D as a means to reduce their costs of production. Joanna Poyago-Theotoky, for example, has found that, contingent on a particular size of the spillover, the market may not provide sufficient incentives for the optimal amount of cooperation in R&D to take place. That is, not enough firms join a research joint venture, for example, to justify its existence from an efficiency perspective. This result suggests that industry-wide cooperation in R&D should be encouraged and policy should aim to foster a situation in which all firms in the industry join the research venture. For the complete analysis of this result see Poyago-Theotoky, "Equilibrium and Optimal Size of a Research Joint Venture in an Oligopoly with Spillovers," *Journal of Industrial Economics*, June 1995, pp. 209-26.

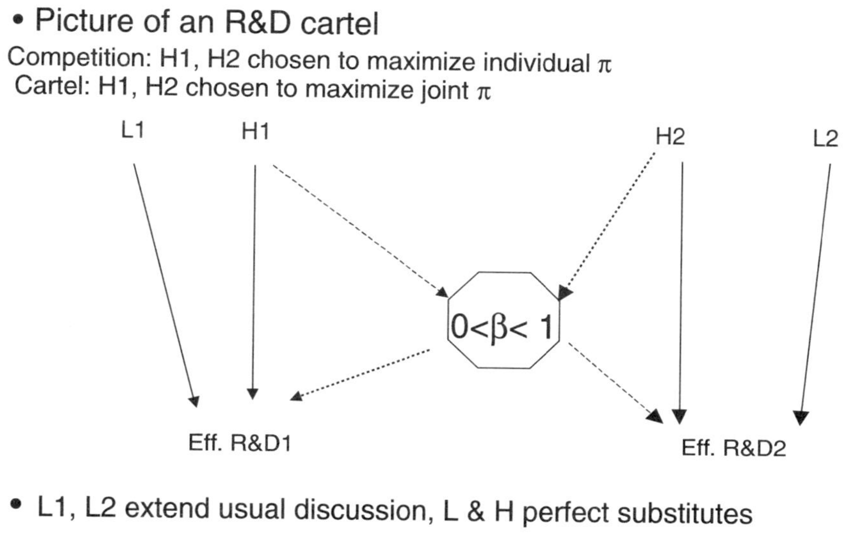

FIGURE 3 Pure coordination.

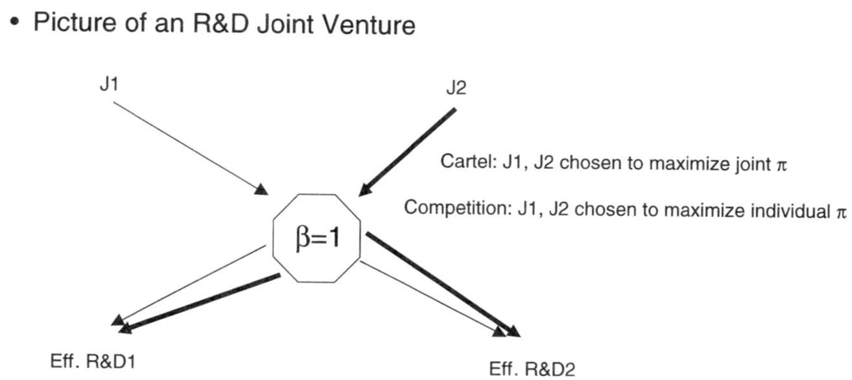

FIGURE 4 Coordination and Information Sharing.

for that is not necessarily 1. In fact, it would be somewhat less than 1 because it might be less effective than a firm's own R&D performed for private purposes. At the same time companies are also competing privately, carrying on their separate R&D programs. In addition, another factor is required to indicate the effect of the government subsidy on the consortium. Thus, there is a mixture of cooperation and competition, and the picture for SEMATECH is somewhat more complicated than the abstract, ideal cases depicted by the economics literature.

R&D Spillovers and Externalities

Differences Between High and Low Spillovers

The strongest result from these models, he said, is that with very high spillovers the aggregate amount of R&D increases with a cartel. With very low spillovers the amount of R&D decreases with a cartel. Between these two cases is a wide range of intermediate cases for which the result is ambiguous. For very high spillovers cooperating will increase the level of R&D, but for very low spillovers cooperating will reduce the amount of R&D.

He suggested an intuitive explanation for this outcome. He posed the existence of two kinds of externalities that may result from R&D. One is a competitive-advantage externality. That is, with no spillovers at all, a company continues to do R&D until the benefit of that work just meets the need of a project inside of the company. On the other hand, if there is spillover to another company, a company's R&D reduces other companies' unit costs by some amount. In the competitive model this will tend to reduce the amount of R&D the company funds relative to what would be funded without spillovers.

The Impact of Cooperation Can Be Positive or Negative

When a company cooperates, it may experience an additional externality, which might be called the combined-profits externality. In this scenario a company seeks to maximize not just individual profits but the profits of the entire industry. This effect can have a positive or negative impact on R&D. If the magnitude of the spillover is small, other companies' costs are not reduced much and their profits decrease as the first company's profits increase. Thus, the overall effect is negative and total R&D is reduced. If the magnitude of the spillover is relatively large, other firms' costs are reduced and their profits rise along with those of the first company. Subsequently, this positive effect increases the overall amount of industry R&D.

The conclusion, said Dr. Flamm, is that when firms cooperate, they can have either a positive or a negative impact on industry-wide R&D. If the magnitude of the spillover is large, the result will be an increase in the level of R&D in the industry. If the magnitude of the spillover is small, the effect will be a reduction in the overall level of R&D. If the spillovers are large enough and the two effects are combined, the effect on industry-wide R&D will be positive. That result supports the notion that cooperation actually increases R&D.

Three Motives for Consortium-Based Cooperation

Dr. Flamm then looked at R&D from a public policy view and presented three motives for cooperation. The first is to *share information*. Even without spillovers, companies that cooperate on R&D can reduce the investment required to achieve a certain level of technological progress. The second reason for cooperating is *to accomplish projects that promise spillovers so large as to inhibit an individual company from pursuing such projects independently*. The third reason is *to create an institutional structure that promises to increase spillovers*, which he suggested as a possible reason for creating for SEMATECH. For each of these motives, he said, in either a world with spillovers or a world with no spillovers the effect of cooperation can be either positive or negative. That is, the amount of R&D can rise or fall.

He gave an example to show how even without spillovers R&D can increase in a consortium because the consortium brings research opportunities: the R&D being done in SEMATECH, which is complementary to something a company wants to do for private reasons. This saves resources and might make the opportunity worthwhile.

R&D May Increase or Decrease in the Presence of Spillovers

He said that one can't infer anything about the motives or character of R&D just from the impact of the consortium on overall R&D. In particular, one cannot tell the nature of R&D being pursued, whether it aims mainly to reduce duplication or targets areas of high spillovers between companies. R&D can also increase or decrease when spillovers exist. Thus, even though spillovers are a justification for federal government support for R&D, the impact of the consortium on aggregate R&D in the industry is ambiguous, offering no information about whether the consortium is worthwhile.

Is SEMATECH Worthwhile? An Empirical Study

Empirical Evidence

Reviewing the single empirical study of SEMATECH, Dr. Flamm noted that it is characterized by a number of inherent problems in its data and analysis.[12] He suggested some improvements in this type of analysis and described a method of considering R&D as a percentage of sales for companies inside and outside the

[12]Irwin and Klenow examine the commitment hypothesis and the sharing hypothesis. The commitment hypothesis maintains that consortia such as SEMATECH obligate firms to contribute a larger amount to high-spillover R&D, while the sharing hypothesis implies that member firms will reduce the level of duplicative R&D. The researchers find support for the sharing hypothesis, which would lead one to the conclusion that government subsidization for R&D consortia was not justified. Economists, however, have criticized this study as suffering from measurement error. The results of the study are detailed in Irwin and Klenow, *op.cit.*

industry. He used a fixed effects model, which included industry conditions, age, SEMATECH membership, and possibly R&D sales as determinants of the total amount of R&D.[13] This would produce a controlled experiment for the behavior of both SEMATECH firms and non-SEMATECH firms in the same industry to compare how SEMATECH seems to affect R&D expenditure. With some modifications to the model, he said, one concludes that the increase in R&D for SEMATECH members compared to non-SEMATECH members in the U.S. semiconductor industry is approximately 0. That is, SEMATECH members do neither more nor less R&D than non-SEMATECH members.

Greater "Effective R&D"

Does that mean SEMATECH was not worthwhile? He said the answer to this question is "no". If companies in the industry cooperate and reduce the duplication of R&D, they achieve greater effective R&D, a socially and privately worthwhile result. This signifies a high rate of return even if R&D does not increase at all or even declines. In fact, a decline in R&D could be beneficial by reducing the number of duplicative projects while sharing the production of the same number of technological projects. An increase in R&D could also signify a beneficial outcome. This might indicate that the consortium was doing many high-spillover projects that would not have been done without internalizing the benefit by forming SEMATECH.

Dr. Flamm said SEMATECH had had an observable effect on suppliers and that the result may have been a slight reduction in expenditures for suppliers.

He raised an important technical question: "How were R&D contributions by the member companies to SEMATECH reported in company accounts?" He said that the answer to that question would affect his conclusion about whether R&D went up or down.

Conclusions

The long-run impact of SEMATECH on company R&D suggests little about the social value of the consortium, because it does not indicate whether R&D has increased by the duplication of company efforts. Nor does it make clear whether the consortium was doing high-spillover projects that would not have been done

[13]A fixed-effects model is a statistical technique that allows the researcher to discern the influence that the individual firm has in determining the outcome of a dependent variable, which in this case would be total industry R&D. Typically a fixed-effects model is most appropriately used, for example, when the researcher is able to observe every firm in the industry; thus he can isolate the individual role each firm plays in determining the level of industry R&D. For an in-depth treatment of fixed-effects models see C. Hsiao, *Analysis of Panel Data*. Cambridge: Cambridge University Press, 1996.

otherwise. He concluded by saying that the study does not answer whether the investment of public money in the consortium was good or bad from a social perspective. The investment seems to have been good from a private perspective, because the private companies have continued to invest in SEMATECH over a sustained period.

CURRENT CHALLENGES: A U.S. AND GLOBAL PERSPECTIVE

Michael Polcari
IBM

Dr. Polcari said he would discuss the technical challenges the computer industry is facing, the competitive landscape, and some resource issues. He called the technical challenges unprecedented. To illustrate the productivity improvement in the industry he showed the rise in the density of transistors per chip, which has approximately doubled every 18 to 24 months, a trend known as Moore's Law. He also showed the performance improvement in microprocessor speed, also trending upward and slightly ahead of its curve at about 1.2 gigahertz. This curve was plotted against lithography dimension, which had been declining by about 70 percent every 2 to 3 years. This period had been three years, but as mentioned earlier in the day, it had dropped to a 2-year cycle over the past few years.

Overcoming Technical Challenges

Improving Productivity

From now on, said Dr. Polcari, the challenge for the industry is to maintain this rate of productivity improvement. To date, these improvements have been the result primarily of scaling—the progressive shrinking of component size. The challenge of the future will be to find new solutions when scaling ends. He discussed a schematic of an MOS (Metal Oxide Semiconductor) transistor, which is the fundamental building block for the semiconductor industry. In 1974 Robert Dennard and his colleagues at IBM[14] defined how one would calculate the scaling of a transistor by means of a scaling factor alpha that could scale the relative parameters of the transistor. This scaling technique had been used for two decades, particularly for lithography, which must be scaled down to alter the tran-

[14]Dr. Dennard's invention of one-transistor dynamic RAM (DRAM) in 1966 was a core development in launching the modern computer industry. With coworkers he also developed and verified scaling theory—an orderly scientific approach to determining and dealing with the challenges posed in designing and building ever-smaller computer devices on silicon chips.

sistor dimensions. In addition, the gate must be shrunk by the alpha factor because it is one of the fundamental limitations to continued improvement.

He also described an optical lithography system, using a schematic, showing the light source, the condenser lens, and the projection lens. The other factors in the overall lithography system are the mask and the photoresist. All of these factors depend on the wavelength of the light source used to build the images.

The Limits of Scaling

He made two points to illustrate how difficult it is to continue scaling at the traditional pace of improvement suggested by Moore's Law. First, the industry has for many generations followed a path of introducing a lithography system with a certain wavelength and being able to use that system for several generations. We have now reached a point, he said, of introducing new wavelengths and changing the whole system in fewer generations, which has placed a considerable strain on the pace of engineering. In addition, these wavelengths are being used at resolutions below the wavelength of the system, which adds another tax on the system. This is done through engineering "tricks" that adjust the mask, projection optics, and other parts of the system. Both approaches place a huge tax on the industry in terms of time and resources.

Diminishing Wavelengths

He went on to discuss the progression of lithography toward smaller wavelengths. After a series of surprising improvements many people agree that this progress is likely to end at a wavelength of 157 nanometers, which would represent a gate length of 70 to 80 nanometers. Beyond that, he said, to achieve shorter wavelengths the industry would have to switch to other systems, such as electron beams or X rays. Those changes will require significant development and research. Some of this work has begun, but Dr. Polcari noted the widespread opinion in the industry is that the level and amount of research on these systems is not adequate for the expected time frames.

Where Scaling Ends

Turning to the new copper technology that has been introduced for transistors, he noted some special challenges. The performance of a chip increases with each generation of transistor, as does the performance of the overall device. However, at a certain point the performance advantage of a transistor is lost unless the material in the wiring levels is changed to yield lower resistance. Having to change materials in the system in order to maintain scaling (a process that had already begun) signified that scaling had, in effect, ended. This need adds complexity and presents additional challenges to the system.

The industry now is moving from materials such as silicon dioxide, which has been the mainstay of the semiconductor industry for several generations, into lower-dielectric-constant materials, such as organics. Eventually, to approach a dielectric constant of two, it will be necessary to introduce porosity into the system, which is a major challenge in terms of reliability. Again, such a change will require significant industry resources to continue to extend scaling.

Gate Thickness: An "Amazing Feat"

As a final technical illustration Dr. Polcari showed a micrograph of some of the atomic-level features of the semiconductor. "This shows the advances we've made in the industry," he said, "where now you have to use transmission electron microscopes to show what you're looking at." He described the gate dielectric structure, with individual grains of polysilicon above and single-crystal silicon below, separated by the gate-oxide material SiO_2. The thickness of the gate dielectric, he said, is approaching 25 angstroms.[15]

He said the tolerance of the thickness of the gate dielectric is 10 percent, which means trying to control it to within 2 angstroms, "which is quite an amazing feat of technology." In the future engineers will need to reduce this thickness to continue performance improvements and scaling. Within a few years the thickness is scheduled to drop below 20 angstroms. At around 15 angstroms, however, the gate dielectric of SiO_2 is no longer useful due to leakage and other problems, so the industry will need to switch to higher-dielectric-constant material; this is a major focus of the industry today. He emphasized how difficult it is to switch to different material. The reason for using silicon in the first place was that it is stable and has other desirable properties. The switch must be made rapidly, which presents many other challenges.

Improvements Beyond Scaling

Dr. Polcari concluded his technical discussion on a slightly more optimistic note. Because most of the improvements in transistor performance over time have been due to scaling, many people are predicting that when scaling ends, we will hit a technological brick wall in the improvement of transistors. Others, however, say that the challenge will simply shift in the direction of other kinds of improvements. Beyond materials changes, he said, will come specific improvements in Complementary Metal Oxide Semiconductor (CMOS) device performance, such as silicon-on-an-insulator, which does not require changes in the lithography dimension. Another anticipated improvement is to increase the mobility of the sili-

[15]An angstrom is one one-hundred-millionth of a centimeter, about the diameter of an atom.

con itself by enhancing the material in the gate area and developing new double-gate devices. All of these, he said, require substantial investments in research.

Too Early to Pick Winners

Beyond these changes, he said, we can expect a series of small improvements that will strengthen overall system performance. To illustrate the point he showed a list of more than a dozen such improvements, none of which depends on the familiar scaling strategy of the last 20 years.

Some of them help to significantly alter the substrate and ultimately modify the whole system structure to control the current through the channel on both sides of a double gate. Beyond this approach comes a series of new device structures and new architectures, including quantum dots and nanotubes, all of which are under investigation and will require many years of study and basic research. "It's very difficult to pick a winner from this list today," he said. "You need a broad portfolio from which to choose when you are looking at things that take 10 to 15 years to mature."

Preparing for the New Competitive Landscape

A Groundwork of Basic Science

Given the magnitude of these technical challenges, Dr. Polcari underscored the importance of building the nation's groundwork in the basic sciences, notably materials sciences and interface physics, which must be better understood at semiconductor scales. Echoing the remarks of Dr. Moore, he called on the federal government to provide more research funding for these sciences if the nation is to continue its leadership in this field.

A Sharing of Leadership

He turned his discussion from a focus on device and process technology to the competitive landscape that lies ahead. He began by discussing the number of patents being filed in the United States. The United States and Japan have dominated this activity, but he described a recent and significant increase in activity by South Korea and Taiwan. "This," he said, "is testimony to the rapid advances in the semiconductor industry in those countries." He expects the gap to continue to narrow, driven by the strength of Korea and Taiwan in memory technology and foundry-based technology. He said the trend toward a shared global leadership was illustrated by the even distribution of papers at the 2000 Symposium on VLSI Technology, one of the world's major device conferences. He also pointed out that all major semiconductor companies now belong to global alliances. These alliances handle functions from research and development through manufacturing. Considering U.S. government funding for the semiconductor industry, he said, one needs to take into account that global partnerships have become the industry norm.

To indicate further how leadership is shared among companies around the world he showed a list of the companies preparing to begin 300-mm fabrication, the most advanced manufacturing technology. The list included not only the U.S. and Japan but also Europe, Korea, and Taiwan, all advancing at about the same rate. Similarly, a chart of DRAM industry bits shipped again showed broad dispersal. Of the five major DRAM manufacturers, two are Korean, one European, one Japanese, and one American.

The Danger of Neglecting High-risk Research

He moved to the topic of funding resources, recommending especially a paper by Erich Bloch and colleagues.[16] In the 1950s about one-third of the funding for the semiconductor industry came from private sources, while two-thirds came from the federal government. Today those proportions have been reversed. He said that overall this is good news, except for one danger. Government funding goes primarily to the long-term, high-risk research programs, while industry tends to fund the more short-term, tactical activities. He cited the danger of decreasing the longer-term, more basic research. In addition, during the 1990s funding for some basic materials and physics research decreased as well. He referred to the dramatic decrease in funding by the Defense Advanced Research Projects Agency and other programs of the Department of Defense during the 1990s, mitigated somewhat by the MARCO[17] programs, as well as some potential federal dollars for nanotechnology research. Not only are the high-risk areas neglected but the graduate student population also begins to suffer as money moves away from these academic areas. Professors move to where the dollars are, and as dollars move away from the semiconductor industry, the graduate students and professors move into other fields—a recent and visible phenomenon. In this way the people needed to drive the U.S. computer industry decrease in number. Now that the industry is global, competition for the more talented workers exists in the context of a world labor market, and U.S. industry has difficulty attracting the skilled workers it needs.

---

[16]Erich Bloch, Ralph Cavin, and Kathleen Kingscott, "The Economy, Federal Research, and the Semiconductor Industry," a report prepared for the Semiconductor Industry Association, March 8, 2000, at <http://www.semichips.org>. The report calls for increased government support of university-based research and closer collaboration between government and industry. It warns that "a loss of international leadership in semiconductor technology would be economically devastating."

[17]The Microelectronics Advanced Research Corporation (MARCO) is a wholly owned but separately managed subsidiary of the Semiconductor Research Corporation. MARCO is a not-for-profit research management organization that funds and operates a number of microelectronics technology-oriented, university-based research centers as part of its Focus Center Research Program.

Widening Gaps in Research and Workforce

He pointed to a chart showing that as the cost of research has risen, industry has chosen to shift most of its research to short-term projects. SEMATECH is shown as funding more medium-term research, with the Semiconductor Research Corporation (SRC) funding much of the industry's longer-term research in universities.[18] The chart depicts a research gap: Insufficient funding at universities has the potential to erode university infrastructure and the base of graduate students.

Dr. Polcari also described a sudden, recent falloff in the engineering workforce. The SRC has surveyed the graduation rate of electrical engineering majors from bachelor's programs and found a significant decrease from 1988 to the present. It projects no recovery from the current low levels for the next several years.

He concluded by warning that the falloff in the engineering workforce comes at a time when the industry faces some of its most pressing engineering challenges. He urged the government to participate by applying funding to the area of academic research, both to stimulate research in long-term, high-risk areas and to stimulate the pool of graduate students in the field of semiconductor research. He commended the SRC for continuing its support of the basic sciences and recommended that it continue to strengthen the university system.

DISCUSSION

Jobs for Young Ph.Ds

Steve Kang of the University of California at Santa Cruz recalled that when he worked at AT&T-Bell Laboratories, he observed the steady shrinking of the chemistry and physics departments. "Where," he asked, "can the young Ph.D. go if industry no longer has R&D labs?" Dr. Polcari suggested that although central labs have been downsized, there are still opportunities at the large corporations, as well as elsewhere in the semiconductor industry. He agreed that there is less work today in pure physics and more in applied and development cycles.

Global Collaboration

Arpad Bergh of the Optoelectronics Industry Development Association asked a question about international collaboration. With SEMATECH sponsoring collaboration on pre-competitive research, would it not be equally appropriate for

[18]Over the past 15 years the Semiconductor Research Corporation has channeled some $30 million a year of industry funds into university-based research.

different countries to collaborate on basic research? Dr. Polcari answered affirmatively, pointing again to the global nature of the industry.

DISCUSSANT

David Mowery
University of California at Berkeley

Five Points About SEMATECH

Dr. Mowery said he would touch on five specific points about SEMATECH, since the overall consortium had already been described during the workshop.

The Dynamic Character of SEMATECH

First, he stressed the importance of its dynamic character. There has been considerable change in its structure since the founding of SEMATECH because some members have been replaced by others and the consortium has become internationalized. Its financing has also evolved considerably, with the termination of federal funding and the introduction of contributions of international members. The research agenda, too, has moved from a "somewhat hazy focus on horizontal collaboration" with the goal of developing a manufacturing process and diffusing the results among members to focus on a more vertical collaboration. He suggested that much of the success of SEMATECH has reflected its ability to adapt. This flexibility, he suggested, owes much to the combination of partial public funding and industry control of the agenda.

Does The Government Performance and Results Act Suppress Flexibility?

A second point he raised was the issue of how the Government Performance and Results Act (GPRA) applies to publicly funded R&D programs.[19] He reminded the workshop that SEMATECH in 1987 or 1988 created a set of goals but that those goals have changed appreciably over time. How easily could they have adapted to changing conditions, he asked, if they were forced to fit the "somewhat procrustean requirements" of GPRA? He suggested that this issue will take on added importance as the United States contemplates public funding of other collaborative government-industry ventures.

---

[19]The Government Performance and Results Act of 1993 calls all government agencies and programs to greater accountability by requiring annual performance plans and performance reports. This requirement applies to research activities as well as other government functions.

An 'Extension' Role for SEMATECH

A third issue was the importance of the breadth of SEMATECH's agenda, particularly in its emerging collaboration with the equipment industry. This is a collaborative agenda that extends well beyond conventionally defined R&D. It involves an extension role, in the agriculture-extension sense of the term, in which SEMATECH staff and member firms work with suppliers to improve quality control and address management issues. This role is not unusual in collaborative R&D, he said, but extra breadth is required in dealing with smaller, newer firms that often need assistance on a broad front. The role also extended well beyond the narrowly framed concept of R&D that economists tend to consider.

Challenges in Evaluating the Impact of SEMATECH

A fourth issue was how to evaluate SEMATECH's contributions to the revival of the U.S. semiconductor industry. He called this a difficult issue, partly because it is almost impossible to specify what would have happened to the industry in the absence of SEMATECH. Some of the most important factors, he said, are product innovation, repositioning existing firms to compete in new product areas, and exploiting new product opportunities instead of competing directly in established product lines like DRAM. The competitive revival on the manufacturing side had a great deal to do with product innovation. Manufacturing performance may have been a necessary condition, but it almost certainly was not a sufficient condition. On both the equipment side and the manufacturing side, he added, additional important factors have been the severe and lingering economic downturn in Japan and the entry of non-Japanese semiconductor manufacturers on a large scale. These events, in turn, have created substantial opportunities for equipment vendors.

An Important Catalytic Role

Finally, the SEMATECH R&D budget probably accounted for a smaller share of combined national industrial and government R&D spending in areas related to semiconductors than did the budgets of other countries' collaborative programs, such as those of Taiwan and Japan. He stressed the importance of the scale of non-SEMATECH public and private investment that contributed to the revival. He recalled Dr. Moore's suggestion that the commitment of federal financing appears to have a played an important catalytic role at the outset. It raised the expectations of prospective member firms, raised the profile of the collaboration itself, and may have encouraged the commitment of individual corporations. The question remains whether that catalytic role also required a financial contribution for eight or nine years subsequently. Would a shorter-term grant program, for example, or a longer-term loan-finance arrangement have accomplished the same purpose? The catalytic role, he concluded, was important, but no one has yet shown which were the best types of public financing mechanisms to accomplish this role.

A Plan for Comparing National R&D Support Programs

Next, Dr. Mowery proposed a possible agenda for comparing various national programs, especially collaborative R&D programs. He sees two challenges. The first is to compare the structure of these programs cross-sectionally. The second is to compare how well these programs fit with the evolving structure of the semiconductor industry. This second challenge is important, he said, because the industry structure itself is dynamic and moving rapidly toward greater vertical specialization. Current trends include the formation of new foundries and fabless firms that specialize in design, equipment producers that play a more prominent role in the development of process modules, and a great deal of learning and comparison across national boundaries. He agreed with Dr. Flamm that SEMATECH and European programs were inspired in part by their understanding of earlier successful Japanese programs. Meanwhile, the more recent Japanese programs—ASET and Selete—have been influenced by the perceived lessons of the SEMATECH experience.

He suggested three issues with respect to collaborative R&D, drawing on work that Dr. Spencer and he had begun about 18 months earlier with a doctoral student at the Haas School.

The Research Agenda of Collaborative Activities

The first issue, examining the research agenda of collaborative activities, includes the following questions:

- How is the research agenda established?
- Is it driven primarily by industry, by government, by an expert panel at arm's length from either group, or by some mix of these?
- What is the time horizon for the research agenda? The research agenda for SEMATECH has been primarily medium term; member firms and other entities have tried to complement each other's investments in longer-term, medium-term, and perhaps near-term research.
- What is the time horizon across national programs and how does the research within the collaboration complement other research activities—university research, in-house research, research in government laboratories?
- How has the agenda changed since the consortium's formation and what drives its evolution?
- Is the collaboration primarily vertical or horizontal? He said that SEMA-

TECH has become primarily a vertical collaboration, with some horizontal collaboration in planning and roadmapping.

The Structure of the Collaboration

As a second issue he mentioned the structure of the collaboration, which is characterized by the following questions:

- What is the role of suppliers within the consortium (very important within SEMATECH)?
- Who accomplishes the task of roadmapping?
- What are the various roles of non-domestic firms: as participants, non-members, or non-participants?
- Where is the collaborative R&D performed? SEMATECH originally intended to carry out most of its R&D at its central facility in Austin. Over time a more diffuse structure evolved with a substantial shift of R&D activities to supplier firms and member firms taking place.
- How is the mix of public and private funding established and how is it changing?

The Structure of Management

Finally, he considered the structure of management, which involves the following questions:

- How are member firms selected?
- How are the R&D facilities of the consortium staffed? SEMATECH was a pioneer in staffing and very innovative in getting its member firms to contribute high-quality people, who played the key role of transferring new technology back to the firm.
- What happens to member-firm assignees when they return to their parent firms? He mentioned tentative evidence that the experience of SEMATECH assignees and their parent firms may differ from the experience of some of the Japanese participant firms' assignees in earlier projects, such as the VLSI programs.
- What are the career paths in the parent firm after an assignment with the consortium? A policy of placing assignees on a fast track upon their return can have important implications for the relationship between member firms and the consortium.
- How is intellectual property managed? Customs differ among the consortia nationally and regionally. This is an issue that is likely to change over time in substance and importance, he said. In some cases consortia themselves take out patents, while in other cases consortia manage a patenting

process, so member firms within the consortium become the assignees of the patents.

In summary, Dr. Mowery characterized his talk as "a brief laundry list of issues," which could be useful as "a roadmap of sorts" to establish points of contrast and similarity. With such a roadmap, he suggested, one could think profitably about how "this unusual instrument of R&D collaboration" will likely evolve in response to ongoing changes in industry and market structures.

Current Japanese Partnerships: Selete and ASET

INTRODUCTION

Toshiaki Masuhara
Hitachi

Dr. Masuhara, who organized the session on Japanese partnerships, offered a brief introduction of the four distinguished speakers who traveled from Japan to attend this workshop. He said they would speak on the following four topics: Selete (Semiconductor Leading Edge Technologies, Inc.); ASET, a consortium between government and industry; an overview of the Japanese consortia for semiconductor R&D; and Japan's research centers for silicon technology.

He first introduced Dr. Morino, executive vice-president and chief operating officer of Selete.

THE SELETE PROGRAM

Akihiko Morino
Semiconductor Leading Edge Technologies, Inc. (Selete)

Dr. Morino began by describing Selete as a joint venture company that does R&D on behalf of the semiconductor industry. The company was established in 1996, and its shareholders include Fujitsu, Hitachi, Matsushita, Mitsubishi, NEC, Oki, Rohm, Sanyo, Sharp, Sony, and Toshiba—the major electronics companies of Japan. Its clients include all of those shareholders plus the Korean firm Samsung, and Seiko Epson. It is capitalized at roughly U.S. $42 million, has a budget of about U.S. $100 million, and is located in Totsuka, Yokohama.

The Mission of Selete

The mission of Selete is to develop process technologies and semiconductor devices that can be produced at reasonable cost. It also promotes the development of production equipment and materials for device manufacturers.

He placed its formation in the context of other industry consortia in the semiconductor industry. The first was the "very famous" Association of VLSI Technologies Development, which functioned from 1976 to 1980. During the 1980s there was virtually no collaboration among industries, but in the 1990s activity accelerated rapidly. In 1994 SIRIJ (Semiconductor Industry Research Institute of Japan), the think tank for the semiconductor industry in Japan, was formed. In 1995 came STARC, a university-industry collaboration. In 1996 came the formation of Selete, ASET, and VDEC, or Chip Implementation of Systems Designed by Universities. They were followed quickly in 1998 by VSAC, Support of System LSI Development, and STRJ, and the Semiconductor Technology Roadmap Committee.

A Program with Three Objectives

Based on the proposals of SIRIJ, Selete was established as a program with three objectives: The first was *to promote and evaluate technologies*, especially manufacturing equipment and materials for 300-mm wafers. The second was *to develop advanced technologies*, such as lithography and mask programs and technology CAD. The third was *to carry out special projects*, such as reduction of Per Fluoro Compounds (PFC) emissions. Regarding the first objective, in Phase 1, from 1996 to 2000, it worked on up to 180-nm technology and in Phase 2, from 2000 to 2001, up to 130-nm technology materials. Metrics were set up to be as specific as possible. The first year was devoted to process performance, the second year to productivity performance, and the third year to service performance. Selete also worked with suppliers to encourage development from the supplier side.

Collaboration

Collaboration with Tool Suppliers

Selete worked in collaboration with the tool suppliers. In this work Selete's first responsibility was to set up tool performance metrics, including performance metrics for process, reliability, and productivity. Its second responsibility was tool evaluation through a module process and data feedback. Its third responsibility was to supply processed wafers to tool suppliers.

Next, Dr. Morino discussed fabrication productivity improvement. Phase 1 included equipment or automatic material handling systems, communications, and control between the various production tools. In the first phase, 193-nm lithography and mask technology were developed along with electron-beam direct-writing technology. These technologies were then transferred from ASET to Selete, whose task it was to continue the first phase of development

The second phase entails Selete's focus on the execution of device process development, corresponding to the Asuka project. The four major R&D objectives associated with this phase are the development of:

1. 157-nm lithography, mask, and electron-beam projection lithography;
2. a transistor that employs a high-k dielectric as a gate insulator;
3. a multilayer interconnect that employs a low-k dielectric as an inter-layer insulator; and
4. a test fabrication line, consisting of 300-mm wafer equipment, as a platform to carry out the above three objectives.

At present, said Dr. Morino, electron-beam projection lithography (EPL) is one of the most promising techniques. Selete was focusing on 157-nm and EPL as future lithography options.

He then described Selete's role in lithography development as related to four fields: exposure tool suppliers, mask suppliers, mask defect inspection and repair tool suppliers, and resist suppliers. For Selete a critical issue is to promote R&D on the supplier side at the time required by the need for a device of a certain size; the company is now focusing on this subject.

Collaborating with Academia

As for technology CAD development, Selete has developed a framework to collaborate with academia and focus on three-dimensional process simulation and three-dimensional device simulation—a "very important field, we think." Simulation is very important in encouraging tool model development and in delivering the framework for easy implementation of each concept as a model.

International Collaboration

He turned to international collaboration between International SEMATECH and Selete. One main objective is the evaluation of equipment for 300-mm wafers. The partnership is trying to develop unified equipment performance metrics and to assist joint evaluation and data exchange. It has published a second edition of the "Unified Equipment Performance Metrics for 130-nm Technology." The partnership is also developing a unified interface for equipment or AMHS and CIM communication. The results of this work will be open to the public. The third field is to develop mask technologies, beginning with development of unified specifications. It also works on repair technology for mask defects and on mask handling for 157-nm lithography.

Following on what Dr. Morino described as "the fast pace of the restart of the industry-university collaboration," in Phase 2 Selete will focus on Project Asuka,[20] scheduled to run from 2001 through 2005 with the goal of developing system-on-a-chip technology in the 100- to 70-nm range. The original name of the project was ASCA, an abbreviation for Advanced System-on-a-Chip through Collaborative Achievement. The project's name was later changed to Asuka.

The Asuka project includes a *design technology field* and a *device-process technology field*. The device-process field will involve three technology fields on a five-year development schedule. Selete will carry out device process technology, while design technology will be carried out by the STARC. He concluded by saying that Project ASCA has been undertaken in collaboration with equipment and materials suppliers, ASET, research institutes, universities, and overseas consortia. The project is also open to other potential members.

DISCUSSION

A questioner asked how the budget of Selete compares with that of SEMATECH, which in the early years consisted of $100 million of federal money matched by $100 million of private funds. Dr. Morino said that Selete's budget is roughly U.S. $100 million, all provided by the 11 industry shareholders plus Samsung and Seiko Epson, the two client firms. It receives no government money.

THE ROLE OF ASET

Hideo Setoya
Association of Super-Advanced Electronics Technologies (ASET)

Mr. Setoya, the executive director of ASET, remarked that some people have wondered what could come after "super-advanced." He responded by saying that "super" is enough for a few decades, and then they may decide to call it "ultra- or hyper-advanced."

The Characteristics of ASET

He began by describing the characteristics of ASET, as follows:

- A research body of national R&D programs for semiconductors, hard disks, and liquid-crystal displays;
- A consortium of the electronics device industry that also includes equipment and materials suppliers (of 41 members, 6 are non-Japanese companies or subsidiaries);

[20]Asuka is the name of a historical site in Japan.

- A mission to perform research between the basic and applied levels;
- Research performed under contract to NEDO (New Energy and Industry Technology Development Organization) by the staffs of the member companies;
- Projects 100 percent financed by the national government with all results open to the public; and
- International collaboration.

Initiating New Projects

He then described the process for initiating new R&D projects, which involves trade associations, MITI (Ministry of International Trade and Industry), AIST (the Agency for Industry, Science, and Technology), and numerous research committees with members from industry, academia, and government. In the case of semiconductor research, proposals are initiated by industry through an industrial think tank organized in 1994. When a proposal is adopted by MITI, it is passed to the federal budget for funding. If the topic is purely application, it is funded by private ventures. For semiconductor research the proposal goes to Selete for the pre-commercial phase of the research. MITI requests the following year's budget through the minister of finance, who prepares the official budget draft for the cabinet. After the budget proposal is approved by the Diet, MITI assigns research funds through NEDO, which publicly announces the program for the new project.

ASET: A Research Consortium

ASET has the legal status of "research consortium," which was established under a special law. In such a consortium all members share equal partnership, and participation is open, subject to approval by two-thirds of the member companies. Research consortia are usually organized around specific research subjects, but ASET is unusual in having three major areas of research. And while most research consortia can get tax incentives for research equipment, this is not the case with ASET, because the equipment it uses is owned by NEDO. Finally, a research consortium usually terminates after the authorized research period has expired, but ASET, which was supposed to end in 2000, was extended for three more years.

The difference between ASET and Selete is that ASET is funded by the government whereas Selete is 100 percent funded by industry, although in both cases development is done by assignees from individual member companies. In the case of the universities, most funds come from the government. However, industries are encouraged to make joint agreements or fund research in universities. In the case of government-owned laboratories, there are few industry contributions, but the national laboratories are scheduled to change rather drastically, so it will have more industry funding.

Performing Pre-competitive Research

Within the spectrum of the semiconductor-related research activities sponsored by MITI, ASET's activities fall in the mid-range between basic and pre-competitive. Universities perform most of the basic research, while Selete performs pre-competitive research and companies perform competitive research. Current ASET projects include semiconductors (by far the largest), semiconductor equipment, PFC alternative technology, and electronic system integration. (See Figure 5).

The ASET research schedule for semiconductor-related projects features a variety of lithography programs, including argon fluoride lithography (ArF), proximity X-ray lithography (PXL), extreme ultra-violet lithography (EUV), and electron-beam direct writing and mask writing. All the programs started at the end of 1995 and were scheduled to finish by the end of 2000. However, several new programs were begun that have changed the schedule. These include development of semiconductor equipment, development of F2 lasers, PFC (PerFluoro compounds) alternative technology, and electronic system integration development programs.

All Research Originates from Contracts

At ASET all research is contract research with NEDO. Actual research is performed at two types of facilities. The first is the research centers, operated by ASET, where participating members assign researchers to work. The second is a system of satellite laboratories, where one member company performs the project with its own facility and staff. In the case of research centers, which are newly built or rented facilities, ASET has to provide all the equipment. In most cases 100 percent of the research funds come from NEDO, but in fact these funds do not cover the cost of the researchers. In some cases the money does not meet the operation costs of the centers. ASET then raises money from member companies. A few projects include joint research between universities and national laboratories.

ASET's head office is in Tokyo; its five research centers are distributed around the country, along with some branch offices, satellite labs, and university partners.

All ASET research results must be open to the public. Report seminars are held each year and these are open to the public. The materials are distributed in Japanese as well as in English and technical papers are submitted to the academic societies. In the course of ASET studies, more than 10 researchers from the member companies have received doctorate degrees through ASET research.

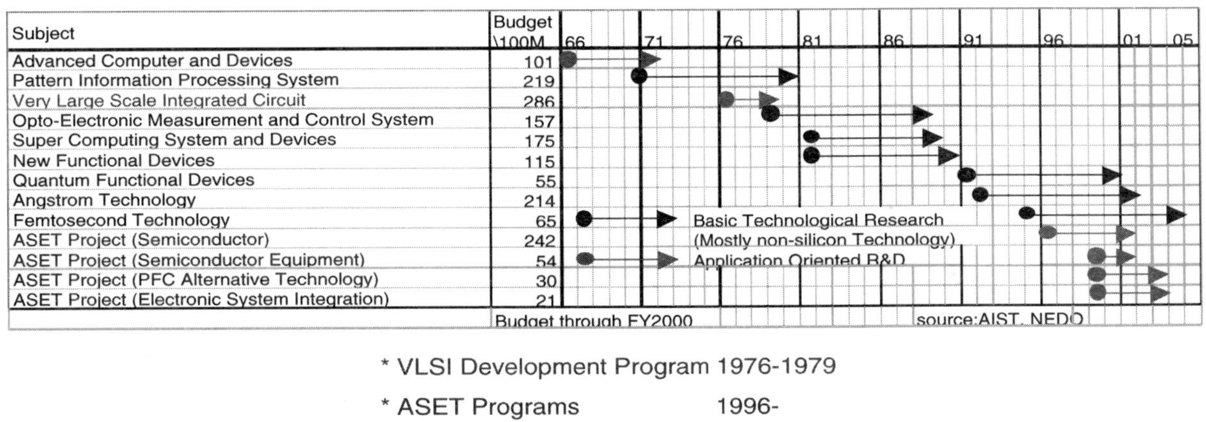

FIGURE 5 MITI-sponsored semiconductor-related R&D programs.

- ArF and EB results evaluated by Selete
- Developed technologies applied to products

Subjects	Patent Application	Articles/Reports
EB Direct Writing	52	78
PXL	20	222
ArF Lithography	9	161
EB Mask Writer	41	75
Plasma	2	171
Cleaning	11	115
EUV	6	60
EB Lithography	0	10
Sub total : Semiconductor	141	892
Equipment	3	17
PFC	0	10
Electronic SI	1	18
TOTAL	145	937

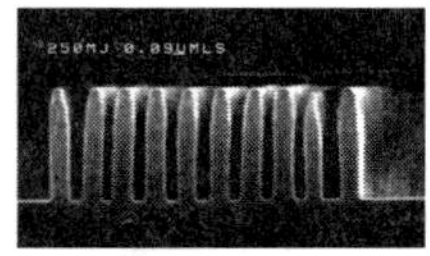

FIGURE 6 Selete achievements (as of March 2000).

The treatment of patent and other industrial property rights is in the process of change. Before September 2000 old patents and other property rights were shared by NEDO and ASET. ASET's share was transferred to the company that was the actual inventor. After five years, outside companies could use the patent with the approval of NEDO and the other owner. As of October 2000, however, the contractor could claim 100 percent ownership of the patents. This change was meant to encourage the contractor to apply for more patents through the national programs.

Mr. Setoya showed a list of the achievements of ASET, including patent applications, articles, and reports. (See Figure 6.) The research results of the ArF and electron-beam projects have been evaluated at Selete. In addition, some of the developed technologies have already been applied to commercial products. The total number of patent applications to the end of March 2000 was 145, and the, total number of articles and reports was 937.

International Cooperation

Finally, he showed the results of international cooperation in ASET. There were 41 members, including TI Japan, IBM Japan, Merck Japan and Merck KGaA, Samsung, and Intel. The first three were the original members of ASET, and Merck KGaA joined ASET soon after. Intel and Samsung joined the EUV lithography research, and Samsung also participated in the PFC alternative technology research.

ASET sponsors the International Forum on Semiconductor Technology along with ISMT, IMEC, and other research consortia. They have held three forums: two in Kyoto and one in Monterey, California. In 2001 they will hold the Fourth International Forum on Semiconductor Technology in Antwerp, Belgium. ASET has also hosted many technical meetings, including meetings on 193 nm, Hi-CP, XEL, plasma, 157 nm, and EUV. ASET has also been responsible for direct exchange of information with other companies and research consortia.

Mr. Setoya concluded by saying the future of ASET is not clear, but it plans at least to extend the EUV and, possibly, the EB (electron-beam) program for a second phase. He expressed the hope for additional international cooperation for development of EUV technology.

ASET is to start two new programs in 2001. They are the MIRAI Project to develop next-generation basic technology for semiconductor process and material, including high-k and low-k research, and the HALCA Project to develop the highly agile minifab concept. MIRAI is a joint program between the newly reorganized national laboratory, Advanced Semiconductor Research Center (ASRC) of the National Institute of Advanced Industrial Science and Technology (AIST), and ASET.

DISCUSSION

Technology Transfer

Phillip Webber of the Congressional Budget Office asked if there were technology transfer mechanisms for ASET other than international collaboration. He asked also how the technology that was developed in satellite labs would make its way into the public domain or to other members of ASET. Mr. Setoya answered that the transfer of technology was the responsibility of the patent holder if the patent was included in the technical results. The overall technological papers themselves are public, but the patent holder would decide if third parties could use the patent. In the field of semiconductors, he pointed out, most companies had cross-licensing agreements, and so patents were not a big barrier.

To a question about the budget he answered that in the case of ASET almost $370 million had been paid in by the government in the past five years. The total budget, including development of magnetic storage and LCD programs, came to $500 million.

Mingae Song, a graduate student in the economics program at Harvard, asked about the different market structures of DRAMs and microprocessor chips. For DRAMs the market is globally very competitive but for microprocessor chips Intel has about 80 percent of the whole market share. The question was how much weight ASET puts on microprocessor chips in R&D. Dr. Setoya said that ASET does not do research on specific products. It supports fundamental technology for semiconductors overall, such as lithography and basic materials.

JAPANESE CONSORTIA FOR SEMICONDUCTOR R&D

Yoichi Unno
Semiconductor Industry Research Institute of Japan (SIRIJ)

Dr. Unno said that much had already been explained about Japanese consortia for semiconductor R&D but that he would add his own perspective. SIRIJ was founded in 1994 as a think tank by 10 Japanese semiconductor companies to promote joint R&D of silicon technology. The objectives of SIRIJ were to plan and promote the development of next-generation semiconductor technologies, to study the future of the semiconductor industry, and to implement projects for international cooperation within the semiconductor industry. Corporate members include the leading members of the semiconductor industry from Fujitsu to Toshiba.

A Spirit of Cooperation

Activities of SIRIJ were helped by a spirit of cooperation between academia and industry in 1995, when the Japanese were lagging behind in the semiconductor industry. (See Figure 7.) SIRIJ planned and proposed joint research systems with Japanese universities and industry, and STARC (Semiconductor Technology Academic Research Center) was established in 1995. In 1996, he said, the Japanese semiconductor industry was still in decline. Many semiconductor in-

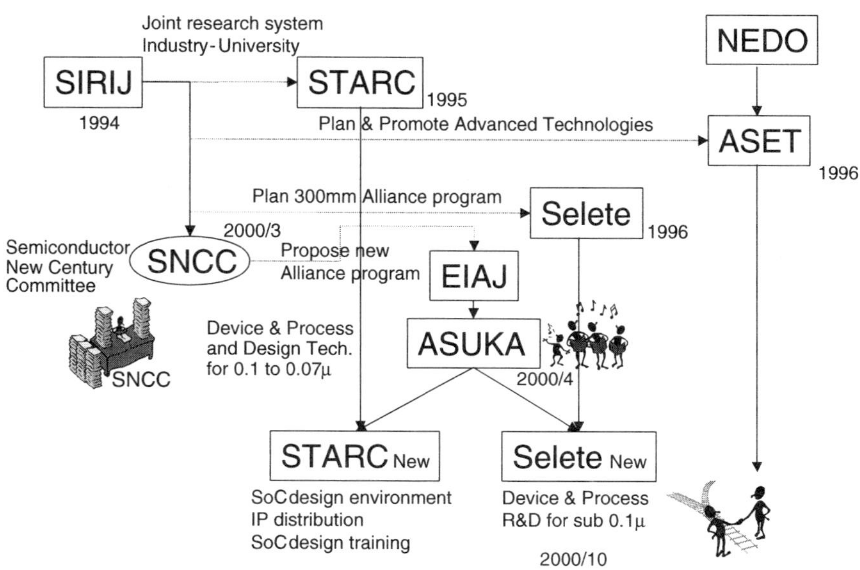

FIGURE 7 Historical flow of Japanese consortia.

dustries could not afford the development expense of 300-mm equipment, so SIRIJ planned a new alliance to develop leading-edge semiconductor technologies. The result was Selete, founded in 1996 by 10 companies. Also in 1996 SIRIJ planned and promoted ASET, the Association for Super-Advanced Electronic Technologies.

Other activities were educational. The universities were already educating students, but there was a need to educate more and to provide continuing education for people in small design companies and venture businesses. Therefore, in 1998 SIRIJ started an organization called VSAC (Venture System LSI Assist Center) to promote LSI design by small design companies and to provide support.

In addition, in 1998 SEMATECH and the SIA asked all major semiconductor firms to help produce an international roadmap for semiconductors. SIRIJ planned and organized a new committee called STRJ, the Semiconductor Technology Roadmap Committee of Japan, in 1999.

Revitalizing the Industry

The final step for SIRIJ was to organize a team to study the needs of the industry for the new century with the objective of revitalizing the Japanese semiconductor industry. The corporation proposed a new strategic plan for global competitiveness called "The Revitalization of the Japanese Semiconductor Industry," published in April 2000. He summarized the recent history of Japanese consortia with a flow chart from 1994 to 2000.

Evaluations

Evaluations by Managers

After reviewing the Selete program (described earlier), he discussed a SIRIJ program designed to evaluate the production of 300-mm equipment. The evaluation was done by managers of the 10 client companies, who assigned grades of blue (good), green (acceptable), or white (poor). For LP/AP-CVD (low-pressure/atmospheric pressure-chemical vapor deposition), 90 percent of the managers rated the performance acceptable. For ox-diffusion-RTP (rapid thermal processing), 80 percent judged it acceptable. However, for lithography and etching, only 60 percent of managers judged it acceptable. AMHS (automated material handling systems) was given a grade of acceptable by 60 percent of the managers, and for cleaning—CMP (chemical mechanical polishing or planarization)—only 50 percent assigned a grade of acceptable. He said that in his personal opinion, much more money is required to develop advanced etching process technologies. For cleaning, he said, more progress is needed. As for AMHS, he noted that he did not altogether understand the problem but thought that many companies had different policies in this regard.

Evaluation by Engineers

Next, he reviewed SIRIJ's advanced technology program evaluation, carried out by engineers of the 10 client companies. Research and development activities had the highest scores, with almost 70 percent rated acceptable. Argon fluoride resist needed more support, he said, and optical mask was a very challenging technology, but he said that the industry seemed to be finding its way. EB (electron-beam) direct writing disappointed many people, with only 8 percent acceptable, and TCAD (technology Computer-Aided Design) also "left our head down," with only 17 percent acceptable. Finally, he mentioned a comprehensive evaluation by the managers in 1999 of timing, investment efficiency, and development results. Almost 80 percent of the managers gave a grade of acceptable, which Dr. Unno said was "not bad, but maybe an easy path."

Turning to ASET, he mentioned the interim evaluation of laboratories performed in 2000 by both technical committee members and ASET department managers. Illustrating the results with humorous cartoon figures, he summarized that EUVL and environmental technologies did well, PXL did fairly well, plasma needed help, and cleaning needed "a lot of help."

UNIVERSITY RESEARCH CENTERS FOR SILICON TECHNOLOGY

Masataka Hirose
Hiroshima University

Government-sponsored Research

Dr. Hirose, who was director of the Research Center for Nanodevices and Systems at Hiroshima University, said that he would offer an overview of Japanese university research centers for silicon technology. He noted that he would focus on three centers and mention some recent research activities. The three research centers are all sponsored by Monbusho: Ministry of Education, Science, and Culture.

The first center is VDEC, the VLSI Design and Education Center, established in May 1996 at the University of Tokyo. VDEC services include:

- Distributing the latest technology information on VLSI design and education;
- Providing media, licenses, and training courses for CAD (Computer-Aided Design) tools; and
- Supporting VLSI chip fabrications and measurement for academic use.

By 1999 VDEC had 120 institutional users, including national and private universities, and a total of about 350 professors making use of VDEC activity. By

1999 nearly 300 chips were fabricated by client companies NEC, Motorola, Rohm, and Hitachi.

Techniques for Mobile Networks

The second center is the New Industry Creation Hatchery ("incubation") Center located at the Department of Electrical Engineering of Tohoku University. A major program, called Monbusho Scientific Research on Priority Areas, aims to produce new image-processing techniques for mobile network applications, including applications for the motion picture industry. From 1996 to 1999 the center's program featured "ultimate integration of intelligence on silicon electric systems." This was a joint research project involving about 20 university groups focusing on applications-oriented research. From 2000 to 2003 the center will focus on "mixed integrated systems for real-time intelligence processing," again involving about 20 university research groups.

Another part of the work at Tohoku University consists of the MITI R&D Program, supported by NEDO. The major purpose of this program is to develop strategic manufacturing tools. The major projects include the following:

- A microwave-excited plasma tool using a radial-line slot antenna (this tool features a high-density, uniform, and low-electron-temperature plasma and also a high-accuracy plasma process without metal contamination or surface damage);
- A balanced electron drift magnetron plasma-etching tool;
- A multi-target, long-throw balanced electron drift sputter tool; and
- A vertical integrated cluster tool.

The third center is the Research Center for Nanodevices and Systems at Hiroshima University. Research projects include advanced metrology for high-k gate dielectrics, modeling of gate tunnel leakage current, 30-nm gate-length MOSFETs,[21] and Cu drift in low-k dielectrics.

To illustrate one focus of the work at this center he showed a graph of a direct tunnel regime and a second graph of a 30-nm gate length MOSFET, with drain voltage plotted against drain current. Finally, he showed a series of energy band profiles of MIS structures with high-k gate dielectrics. A second emphasis at Hiroshima University is CREST, the Core Research Program on Science and Technology, supported by the Agency of Science and Technology (AST). CREST sponsors an effort to understand the self-assembling of silicon quantum dot and its application to floating gate MOSFETs.

[21] A MOSFET (metal oxide silicon field effect transistor) is a device for electronic systems.

DISCUSSION

Greg Linden of the University of California at Berkeley asked whether the research programs described were part of STARC. Dr. Hirose said they were completely separate, supported basically by the Ministry of Education and Sciences.

Concerns Over Research Support

Dr. Wessner recalled the concern of several American speakers about the level of basic research supported by the U.S. government and asked whether this was also a concern in Japan. Dr. Hirose said that the universities should indeed focus on fundamental research, but that they should also understand the practical problems of industry, so that near-term work also requires university involvement. He said that good balance between applied science and fundamental research work will be one of the important missions of the university research involvement, which is a basis for startups, new ideas for the future, device technology, and material technology.

Funding Levels at Universities

Genda Hu of Taiwan Semiconductor Manufacturing Company asked Professor Hirose about the general funding level of the three university programs. Professor Hirose said it varies year by year, but the base level is at least a few million dollars annually, provided by the government for learning centers. Additionally, depending upon the application, programs would receive more than $5 million per year to support new equipment or meet replacement needs.

Jeffrey Gren of the U.S. Department of Commerce asked if the current Japanese focus on DRAMs might change in the future as a result of the activities outlined today. Professor Hirose answered "maybe," and noted that the question is very delicate. He said that in Japan many companies cooperate in developing next-generation technologies, such as Hitachi and Toshiba, because they want to reduce development costs. That, he said, is a new fashion in Japan.

Urging Consortia to Collaborate

Mr. Gren then asked why there were multiple R&D consortia in Japan, in addition to International SEMATECH, and whether they might be combined in a single international framework. Professor Hirose agreed that this is desirable and said that there have been several examples of international collaboration between consortia. He said that a higher level of international collaboration is needed to both conquer technical barriers and reduce costs. "The issue," he said, "is how to collaborate effectively on specific items." He called for more experiments in collaboration, some trial-and-error, to learn the best ways of doing it.

William Joyner of the Semiconductor Research Corporation said that he was under the impression that STARC in its previous incarnation would support university research at several Japanese universities. Now it has moved primarily toward training designers for system-on-a-chip design. He asked whether another organization had stepped in to continue funding the projects that used to be funded by STARC. Toyoki Takemoto of STARC said that the organization continues to be funded by industry to do collaborative research with universities but that more assistance was needed for education, especially curriculum reform.

An Expertise Gap

Dr. Moore recalled the need for more technical graduates in the United States and asked whether such a huge supply-and-demand gap existed in Japan as well. Dr. Masuhara said that the total number of students graduating from electrical engineering programs had not fallen in Japan, but he said that several years ago a study of expertise needed by industry revealed a large gap. In the field of designing VLSI chips company demand was four times larger than the number of students, a ratio that has not improved and is probably increasing. He said that was the reason Dr. Takemoto had mentioned the need for improved university education, particularly for the design of the silicon chip system.

European Partnerships

WELCOME

Dr. Wessner introduced the third panel to "what I think has so far been a remarkably rich discussion of some of the challenges facing partnerships around the world." He welcomed the European colleagues who like those from Asia had made substantial efforts to attend and introduced moderator Michael Borrus, who was a member of the steering committee under which this study was being carried out.

INTRODUCTION

Michael Borrus
The Petkevich Group, LLC

Mr. Borrus said that the panel might prove to be "a bit of a surprise to some of you" in light of the rapid strengthening of the European presence in semiconductors in recent years. A decade ago, he said, many people were focused on the United States and Japan in assessing the struggle for leadership in semiconductors and did not anticipate a significant European presence. Recent developments, he said, had proven this belief to be misguided. Indeed, he said, Europe has reemerged as a very significant player in information technology and a dominant one in some sectors of communications services and equipment, notably wireless, embedded industrial electronics, and certain parts of the semiconductor industry.

Europe's Resurgence

He described a "complex constellation of reasons behind Europe's resurgence in information technologies," including the slow but persistent progress to a truly common market, the development of the euro, and an increasingly more integrated European financial services marketplace. The latter has meant greater liquidity and a wider range of financial institutions, including venture capital and growth-issue stock exchanges like the Neue Markt in Germany. Europe's financial marketplace can now fund innovation in technology industries, he said, in ways it could not a decade ago. Another reason for Europe's emergence is the choice of common standards in wireless communications, which has helped to boost European manufacturers to the leading edge. And finally, the European cooperative programs, such as ESPRIT and RACE, which were formed more than a decade previously, had been important in fostering collaboration among European companies and between Europe and the rest of the world. To discuss the first of these programs he introduced Dr. Jürgen Knorr, chairman of MEDEA, who had spent many years at Siemens in a variety of senior management roles.

THE MEDEA PROGRAM

Jürgen Knorr
Micro-Electronics Development for European Applications (MEDEA)

Dr. Knorr, who began by characterizing himself as "one of those European dwarfs who would like to play a small role in semiconductors," confessed that it is "very strange" to be leading a program to support the multinational semiconductor industry. The tradition had always been to support one's own industry and nation, but now his organization was saying "no." The semiconductor industry has become global and MEDEA's objectives have to be shaped accordingly.

Some Special Difficulties for Europe

He then said that the national features within Europe do cause some special difficulties. Europe, he noted, is neither a nation nor a republic. He observed the political economy of this federative system to be the "analog of a semi-custom integrated circuit (IC), a non-optimized, coordinated conglomerate of different functions which try to shoot for one target—a target, though, that is interpreted in different ways." French English is different from German English, he said, and both differ from what the Italians may understand when using the same words. As a result this reality caused the Europeans to think together about the best way to close the gap between Japan and Europe in manufacturing, about intellectual property creation in the United States, and about what is possible politically and financially.

He also noted that Europe's background is different from that of the United States or Japan, Korea or Taiwan. The differences show up at the World Semiconductor Council, where long-term perspectives vary and are worth reviewing for a moment.

The Context of Competition

Entering a New Cycle

The development of industries forming our society, he said, is relatively well described by the so-called Kondratiev cycle.[22] Earlier in this century, resources such as steel, oil, and electricity defined engineering technologies and their application in railways, bridges, skyscrapers, and automobiles. Companies focused mainly on the competition in the context of specific regions and of companies in those regions. These different regions used different standards and protected their companies and their employment.

After the 1970s, he said, the so-called fifth Kondratiev cycle began. The resources being used were now silicon technology, semiconductors, and design software. The applications were to the military, consumer electronics, data processing, data communication, and especially mobile communication. The competition was increasingly determined by intellectual property, not by location or hardware, as it was in the past. IP (intellectual property) includes the experience needed to make something using the knowledge of engineering, physics, and chemistry. This is transferred through brains, through human resources. Human brains in today's economy are the most flexible resource and they can be applied globally. The forces of globalization, along with World Trade Organization rules, have pushed away, first, the political barriers and, second, the economic barriers.

In this way, said Dr. Knorr, policy entered the electronics industry, especially the microelectronics industry. Global competition developed between nations, such as the United States and Japan, and led to the famous U.S.-Japan trade agreement.[23] The Europeans entered global competition not as a single European nation but as a patchwork of nations.

---

[22]The Kondratiev cycle, articulated by Nicolai Kondratiev in 1935, argues for the existence of an economic cycle of boom and bust that lasts approximately 60 years. See Nicolai D. Kondratiev, "The Long Waves in Economic Life," *Review of Economic Statistics*, 1935, 17(6).

[23]In 1986 the United States and Japan agreed on a five-year accord known as the Semiconductor Trade Agreement (STA), which aimed to open Japanese markets to foreign producers and end the dumping of Japanese semiconductor products in U.S. markets. By 1990 the foreign share of products in the domestic Japanese market had increased significantly. For a further analysis of the impact of the STA see National Research Council, *Conflict and Cooperation in National Competition for High-Technology Industry*, Washington, D.C.: National Academy Press, 1996, pp. 133-41.

Competition Is Now Between Companies

That, he said, is the background to why we are here today. It is important to understand that the real competition is no longer among nations but among multinational companies, including Siemens, Philips, Motorola, and others. Globally competitive companies need to form transborder joint ventures, corporations, and mergers. These activities, he said, will determine the future structure of the industry more than programs such as MEDEA.

However, the necessary infrastructure in education, human resources, research, and technology can and should be driven by global competition, which in turn is stimulated and supported by national programs. So competitive structures are an amalgamation of both, and sometimes contradictory because governments try to spend taxpayer money only on activities that benefit their nation.

Responding to the Value of Information Technology

An important force behind MEDEA is that in all the countries of Europe the politicians and parliamentarians who write budgets have learned that the driving force of industry in the next 15 years will be electrical engineering and electronics. He noted data from the U.S. Semiconductor Industry Association illustrating that semiconductors add more value to the U.S. economy than any other manufacturing industry. He also showed data from Germany showing that information technology creates more jobs than any other employment sector. Europe has a special need to cooperate, he said, because most of its countries are too small to compete alone against regions such as the United States and the Asian Pacific. Moreover, he noted, they must cooperate specifically in IT, which will offer the most jobs and industrial strength—as long as the countries remain globally competitive.

This shows up in the regional trade balances. During 1992-1997 Europe and the United States consistently had a negative balance of trade with Japan, while Japan and other Southeast Asian nations individually experienced positive trade balances with Europe and the United States. One of Europe's disadvantages, he said, is that it is a patchwork of nations with different cultures, different languages, and different national policies. To develop synergies through cooperation is more difficult than it is in the United States. Even different national behaviors add problems that should be solved before claiming that Europe is a unified market.

The Benefits of Cooperation

The development of the common wireless standard GSM is a success, he said, not because it is the most advanced technical system but because at least a few big countries have agreed upon it. Cooperation can help create a critical mass, share risks, reduce the costs per partner, share know-how, and shorten lead times. If countries work together, even if they have spillovers, sharing is the most effective way for those who, for whatever reasons, cannot do it alone . For these reasons and to stimulate employment the European Commission and the national governments stimulate transborder, cooperative R&D.

He reviewed the microelectronics support programs in all the major manufacturing regions, including the United States, Western Europe, Japan, Taiwan, and Korea. He then turned to MEDEA, an industry-initiated and industry-driven program supported by the national governments of 12 participating countries.[24] Its objectives are to stimulate transborder R&D cooperation and to strengthen the global competitiveness of the European microelectronics industry in technology and in future applications.

Focusing on Core Competences

In semiconductors, he said, there are three core competences for technology: design techniques and libraries, CMOS-based technology platforms, and manufacturing technologies. There are also three core competences for applications: multimedia technologies, communications technologies, and automobiles and traffic applications, or automotive electronics. The outcomes of these six core competences are system-on-a-chip domain and IC technologies domain capabilities. These are intended for the globally competitive industry segments: ICT companies, semiconductor companies, and equipment and materials companies.

MEDEA was a 4-year program that started in 1997 and terminated at the end of December 2000. The total cost was about 2 billion euros for 45 projects. These projects were distributed among 25 major partners and a total of 105 smaller partners, including SMEs, universities, and institutes. The 12 participating companies were Austria, Belgium, Switzerland, Germany, Ireland, France, Finland, Greece, Italy, the Netherlands, the United Kingdom, and Sweden. The United Kingdom made the smallest contribution, having had its own policy.

Criteria for Inclusion in MEDEA

The main criteria for project selection, as defined by industry, include the following:

- Innovation in fundamental research, industrial research, and pre-competitive development;
- Existing European R&D capacity—a condition of the national governments for using taxpayers' money;

[24]In MEDEA 46 percent of the partners are small companies, 36 percent are large companies, and 18 percent are research institutes.

- Competences of the partners;
- Complementary strengths of the consortia;
- World market potential; and
- A leveraging effect on employment.

A board manages MEDEA, and a support group oversees steering groups for both technology and applications. The technologies include CMOS technology and manufacturing technology. The applications include multimedia technologies and communications, both of which share a design function, and automotive applications.

Four Principal Results

After its 4 years of supporting cooperative R&D between European microelectronics companies, summarized Dr. Knorr, MEDEA has achieved four principal results.

1. Above-critical-mass collaborative projects. Each project involved a minimum of two countries and two companies. He also cited collaboration with national programs in France, Germany, Italy, Belgium, and the Netherlands. He noted that in the future there may be a 300-mm program in France with Philips and STMicroelectronics. There was also a German program with Infineon and Motorola that was first in manufacturing 300 mm. He called this effort a parallel, not overlapping, program.
2. Better understanding between semiconductor suppliers and system houses. This was expedited by choosing English as the common language.
3. Better understanding about where to focus resources.
4. Closer cooperation along the whole "food chain." Dr. Knorr remarked that this result was acknowledged by both industry and governments, even though the consortium had not created quantitative criteria to evaluate outcomes. The main objective, especially between different countries in a particular market, was to learn to cooperate insofar as competitors can do so. The second important result was that the whole society gained awareness that collaboration is desirable and a productive use of taxpayer money, and that it also tends to increase employment.

Moving to MEDEA Plus

As a consequence of these results the participants decided to initiate a MEDEA Plus program in 2001 under the guideline, "system innovation on silicon." The objective of this sequel program goes beyond system-on-a-chip to include system-in-a-package. The overall mission of MEDEA Plus includes the following goals:

- Address the main microelectronics challenges in silicon application platforms and enabling technologies following the international technology roadmap. He expressed hope that the consortium would be effective in drawing together the large amount of decentralized, sub-critical expertise needed in many institutes.
- Focus on key priorities and build on European industrial strengths to improve Europe's competitiveness in system innovation.
- Adapt to the fast-moving technical and marketing environment and expedite global cooperation. This objective is not normally in line with national government support, he said, but he expressed satisfaction at being able to forge international cooperation with IBM and Motorola.
- Offer opportunities for participation by SMEs, institutes, universities, and large groups.

He closed by noting that the CEOs of international companies might have to withstand pressure emanating from their own national governments for joining MEDEA Plus. "Otherwise," he said, "life would be so easy."

DISCUSSION

Dr. Wessner asked about the level of funding from the MEDEA program for actual work and about contributions from the larger member nations. Dr. Knorr answered that cost of the four-year program for companies and institutions participating in projects was roughly 2 billion euros, or 500 million euros per year. National governments supported the parts of the project that were done in their countries. The amount of funding for near-to-market research was less than 30 percent; funding for more basic research was about 50 percent. In the national programs, he said, it is sometimes harder to differentiate semiconductor research within ICT. He estimated that in France, Germany, Italy, and the Netherlands the public support ranges between 20 and 50 million euros per year. In other countries the amounts would be smaller and harder to estimate.

GOVERNMENT-INDUSTRY PARTNERSHIPS IN EUROPE I

Peter Draheim
Submicron Semiconductor Technologies

European Strategies in the Global Semiconductor Industry

Dr. Draheim, CEO of Philips Semiconductor, said he would explain the reason his company chose to join the submicron consortium, "even though cooperation in Europe is not easy." He began by illustrating some of the differences between the world's regions with respect to technology.

- In Asia memory drives microelectronic technology.
- In the United States processors and memory drive microelectronic technology.
- In Europe application-oriented circuits drive semiconductor technology.

Europe Has Diversified Its Semiconductor Business

Dr. Draheim said Europe cannot focus on one particular area. Instead, Europe in recent years has diversified and been able to take a leading position in several areas: communications, automotive electronics, smart cards, and multimedia. For applications in these areas, system innovations involving silicon are the driving force, requiring embedded technologies. As examples he said that communications and cellular radio need RF (radio frequency) integration, portable systems need flash integration, smart cards need non-volatile memory, multimedia need DRAM integration, and automobiles need high voltage. Specifically, automotive applications need more than 60 volts and high temperatures of about 200°C—a difficult challenge but not essential just yet. For telecommunications, RF integration is needed at high gigahertz and low power so that a device can run for weeks instead of just hours. For smart cards one needs non-volatile embedded memory, while multimedia demands chips for real-time processing.

Drawing on Particular Strengths

Therefore, he said, even though Europe and the United States are following the same roadmap, Europe has some particular strengths in the applications-oriented fields. In particular, Philips expects major breakthroughs in four fields: portable infotainment, third-generation mobile communication, home networks, and enhanced digital TV. He elaborated on one example, the home network. To develop such systems requires many cooperating partners: research institutes, equipment suppliers, semiconductor manufacturers, system houses, SMEs, business groups, and technical centers. For home networks one needs microelectronic components and software modules. The particular components include high-bandwidth internal networks, a high-bandwidth access network to the Internet, storage systems, and enhanced image sensors. For software the needs include personalized services, multimedia Internet services, and easy information management. All of these platforms have to be concurrently developed. He pointed out how rapidly the placement of microcontrollers is growing, especially in home applications, including coffee makers, telephones, garage-door openers, microwaves, sewing machines, and cameras. The total number of ICs in the home is expected to grow from about 60 in 1990 to 300 in 2005.[25] In the office the total number of ICs for 2005 is expected to reach 110 and, in the automobile, about 95.

---

[25]Source: Motorola Electronics.

Why in the home? He made the point that all appliances are not only going digital, but they are also being connected. "The home will become so intelligent," he said, "that I fear I will not be intelligent enough to live in my own home."

The Necessity of Cooperation

As systems become more complex, we need more cooperation. He said that Philips participates very actively in MEDEA, which is oriented toward hardware development, and with ITEA (Information Technology for European Applications). These are complementary programs that are "absolutely important for our future development." He listed two primary reasons why cooperation is necessary.

1. For complex systems, the increasing cost of technology development calls for horizontal cooperation to share costs, reduce risks, and shorten time to market.
2. The increasing costs of complex-systems engineering calls for re-use of intellectual property and for more vertical and horizontal integration with the customer.

A company cannot, he concluded, do everything on its own. As an example of the cooperative development of a technology, he cited SiGe-BiCMOS (silicon germanium Bi-polar Complementary Metal Oxide Semiconductor) for RF applications. Cooperation has begun with nine partners, including telecom companies, semiconductor companies, institutes, and universities. He said that the next big cooperative challenge will be to develop systems-on-a-chip with the goal of achieving the same functionality in one-fiftieth the space. To meet this challenge, he said, Philips plans to cooperate in both MEDEA Plus and ITEA.

DISCUSSION

Funding for ITEA

A questioner asked if ITEA is financed by the European Union or by the private sector. Dr. Draheim answered that ITEA has the same financing structure as MEDEA, which is supported by national governments. The difference is that MEDEA is oriented toward hardware and ITEA is oriented toward software. The budget is not based on specific requirements from each company. Money is raised on a per-project basis from the different countries. He did not know ITEA's budget for its first, startup year, but said it will grow to more or less the same size as MEDEA for its 4-year planned lifetime.

Dr. Wessner asked if there is any linkage with the European Union framework program. Dr. Draheim said there is no official linkage but rather close cooperation and information exchange with the IST program of the European Union.

GOVERNMENT-INDUSTRY PARTNERSHIPS IN EUROPE II

Wilhelm Beinvogl
Infineon

Dr. Beinvogl began by noting that with the internationalization of SEMATECH the three major information technology players in Europe are all now members of SEMATECH. He said the three are not only financial contributors to International SEMATECH but also significant technical contributors. He gave two examples. The first was 300-mm technology, which is a key agenda point in International SEMATECH. He suggested that Europe has achieved a world leadership position in 300 mm that in turn benefits International SEMATECH and its members. Another example, he said, was that the IMEC institute in Belgium, a world leader in cooperative research, has already achieved very close cooperation with International SEMATECH on one major project.

Strategic Challenges in the Semiconductor Industry

He then turned to some aspects of government-industry R&D partnerships in Europe. He showed a chart illustrating the time required for various technologies to reach one million users, ranging from 20 years for black-and-white TV to fewer than four years for the analog cellular phone. "The message is," he said, "that everything in this world seems to be accelerating." He said that the ITRS (International Technology Roadmap for Semiconductors) has been moved ahead by nearly a year over the last couple of years.[26] In discussing government-industry partnerships, he said, we have to take this speed into account and be prepared to accelerate.

He emphasized the outstanding importance of semiconductors, which have "huge leverage" on all kinds of electronic equipment, creating more value added and contributing to faster growth and many new jobs—especially highly skilled jobs. The downside is the scarcity of highly skilled specialists. Based on these facts, he said, microelectronics is considered to be a major strategic industry by national governments worldwide.

Partnerships as a Response to Challenges

Despite continuing competition between multinational companies, regions, and even countries, there is an increasing tendency to collaborate and to form partnerships. These partnerships, especially in Europe and particularly in Germany, can be divided into three major domains.

---

[26]For a description of the ITRS and its goals see its Web site at <http://public.itrs.net/>.

1. Industrial cooperation with publicly funded R&D institutions, such as a university (e.g., IMEC and the Fraunhofer).
2. Public funding of industrial research, which can go up to 50 percent in Europe generally. A complex set of rules defines the actual percentage, which can be considerably less than 50.
3. In most cases public funding is combined with funding from other sources, such as institutes or companies from different countries.

In Europe there is no agency comparable to the Defense Advanced Research Projects Agency for microelectronics—meaning no non-civilian support from governments. In addition, there is no indirect public funding for R&D, only fiscal incentives. He said that the network of contracts of the corporate research group is complex, that research conflicts are common, and that it is hard to maintain an overview of so many activities and relationships.

A "Mental Gap" in Technology Transfer

He showed a depiction of the various technology transfer mechanisms in Europe, the United States, and Japan. In Europe he noted a gap between non-industrial research and industrial research and labeled this a "mental gap" in order to indicate that it was more imagined than real. He said that IMEC, GRESSI (Grenoble Sub-micron Silicon Initiative) and FhGmbh are good examples of how the mental gap can hamper cooperation with industry.

Dr. Beinvogl said that the overall funding structure in Europe is basically top-down, from the commission level through framework programs, which defines the contents of funding for individual projects and funding. By contrast, the national authorities, which provide most of the actual funding, cooperate at the European level through a formal certification process and by labeling national, bi-national, or multinational products with the so-called EUREKA label (European Network for Industrial R&D). This label helps to obtain funding.

Successes and Lessons

A Major Success Story

He emphasized as one "full-blown success story" the joint venture between Infineon and Motorola, called the SC300, for semiconductor 300, which had ended about three weeks before the workshop. This joint venture was formed by two companies whose ambition was to be poised at the leading edge in the transition to the next wafer size. The company, set up in Dresden, had met its goals without encountering any major obstacles. The first fab was already fully constructed and was due to start operation in the second quarter of 2001, with a capacity of 5,000 wafer stocks per meter at 300 mm, starting with DRAM manufacturing.

He summarized another major collaboration between Infineon and Motorola, which had cumulatively spent $300 million in the previous 2½ years. Both the German federal government and the state government of Saxony agreed that the project had excellent potential to create jobs and they provided substantial funding.

For comparison he mentioned a decade-long collaboration with IBM, and to a large extent also with Toshiba, to develop DRAM technology in New York state. They have expanded this partnership to add a logic program. The cumulative cost to all partners has been $1.1 billion.

Partnerships Have to Be Global

Finally, he said that these kinds of partnerships now have to be global. He mentioned Infineon partnerships with Intel, which was a strategic investor in Infineon at the initial public offering. There is also a three-way partnership now among Infineon, IBM, and UMC (United Microelectronics Co. of Taiwan) to develop and manufacture advanced logic processes. The company also participates in a newly founded advanced DRAM consortium which brings together the major companies in the field, along with Intel.

Dr. Beinvogl closed by saying that "we are very deeply convinced that there is no way around globalization in our business today." Everyone has to find and fund their own best partnerships—those that are consistent with the globalization of their development efforts and of their business.

DISCUSSION

A Decrease in the Number of Engineers

In response to a question about the sufficiency of human resources in Germany, Dr. Beinvogl described a major decrease in the supply of engineering graduates since the early 1990s. "This is absolutely dramatic," he said, "it's not just a little effect." He said Infineon is now forced to look all over the world for skilled people. He said that about eight years ago young people were discouraged by predictions that electrical engineers would not be able to find jobs. He conjectured further that career "fashions," which may dictate that it is more desirable to be a lawyer than an engineer, have had an inhibitive effect.

Bill Long of Business Performance Research asked whether the relatively long history of multinational cooperation in Europe gives it an advantage over other countries in this symposium. Dr. Beinvogl said that they had indeed learned to cooperate across borders, which is an advantage. He said the cooperation was necessary, moreover, to achieve a critical mass to compete with a country as large as the United States.

The Taiwanese Approach

INTRODUCTION

Patrick Windham
Windham Consulting

Mr. Windham called the short history of the Taiwan semiconductor industry "a rare success story." The principal Taiwanese companies in semiconductor production are already world leaders in their specialties: Taiwan Semiconductor Manufacturing Co. (TSMC),[27] established in 1987, and UMC, originally founded in 1979.[28] He called Taiwan's journey to becoming the fourth-largest producer in the world a "remarkable" one and voiced the hope that today's discussion would reveal some of the reasons behind the country's success.

Clearly, he said, Taiwanese entrepreneurs deserve and receive credit for the lion's share of that progress. The government as well deserves credit for pursuing a unique set of policies at the right time. Policies pursued by the government not only helped to reform the capital markets in the direction of equities but also contributed substantial R&D support through ITRI (the Industrial Technology Research Institute) and its chief R&D facility, ERSO (the Electronic Research

[27]TSMC changed the semiconductor market by specializing in the manufacture of custom wafers under contract to chip designers. This frees the designers to concentrate on making and marketing the integrated circuits formed on the wafers to form microchips; it helped spark an explosion of fabless microchip companies, such as those that populate Silicon Valley.

[28] United Microelectronics Corp. began as a designer and producer of integrated circuits. In the mid-1990s founder Robert Tsao changed it into a contract manufacturer like TSMC. Recently it has expanded into high-end chips, a segment in which TSMC is not dominant.

and Service Organization). The government also helped to set up industrial parks in an effort to build industrial clusters where they did not previously exist.

He then introduced the first speaker, Genda Hu, vice-president for advanced technology development at the Taiwan Semiconductor Manufacturing Co.

GOVERNMENT-INDUSTRY PARTNERSHIPS IN TAIWAN

Genda J. Hu
Taiwan Semiconductor Manufacturing Company (TSMC)

Dr. Hu, who worked at IBM, Xerox PARC, Cypress Semiconductor, and other U.S. companies before moving back to Taiwan, played a major role in setting up Taiwan's partnership programs. He said that many people have wondered what the secret was behind Taiwan's rapid progress in the semiconductor industry. He dismissed the idea of a secret, saying that it was certainly not a miracle and that it had taken Taiwan 25 years of hard work to reach its current position. Nevertheless, he did consider a number of factors to be critical to the successful story in Taiwan. He said he could not cover the financial aspects, which were complex, but would confine his remarks to the story of the technology R&D.

An Overview of the Industry

He began with an overview for those not familiar with Taiwan's industry. (See Figure 8.) The total revenue generated by the industry in Taiwan in 1999 reached about U.S. $14.3 billion; for the year 2000, revenues were expected to reach about U.S. $22 billion, a 57 percent increase, compared with a world growth rate projected at about 37 percent.

He said that Taiwan divides its industry into four sectors: design, which consists primarily of fabless semiconductor or design houses; fabrication, which includes foundries and also IDMs (integrated device manufacturers), such as Winbond and some DRAM companies with fabs and other products; packaging; and testing. He said the Taiwanese industry is a "vertically dis-integrated" infrastructure where many companies focus on a particular expertise rather than trying to do everything. The definition of revenue is different from the revenue quoted by the WSTS (World Semiconductor Trade Statistics). The WSTS figure for revenue is based on the final integrated circuit as a product. In Taiwan revenue may relate only to a service value, such as packaging, which is considered a value added. Likewise, a foundry produces wafers but not integrated circuits per se. For example, Taiwan has 127 design houses without fabs, 21 companies with fab facilities, 42 packaging companies, and 33 dedicated testing houses. To many people, he said, this is a "really amazing" number of companies. In addition, there are supporting companies, such as those that provide starting substrates, chemicals, leadframe, and substrates for the packaging industry. Most companies are of small to medium size. Only a few are very large, including TSMC, a foundry with 1999 revenues of about U.S. $2.2 billion, forecast to be in excess of U.S. $5 billion in 2000. UMC is the second-largest company, with revenues of U.S. $1.6 billion, and is also growing rapidly. The rest, by international standards, are small.

R&D Activities

Funding and Execution

He divided R&D activities into two categories: funding and execution. Basic funding comes from the National Science Council (NSC) and the Ministry of Education, which sponsors primarily basic and applied research. The institutions or organizations that execute the activities are universities, research institutes, and industries. Among the more prominent labs are the Nano Device Laboratory under the NSC and the Chip Implementation Center, which is supported by NSC but managed by ITRI. Most of the R&D activity in Taiwan in the past has been concentrated around ITRI and most of the funding has come from the Ministry of Economic Affairs.

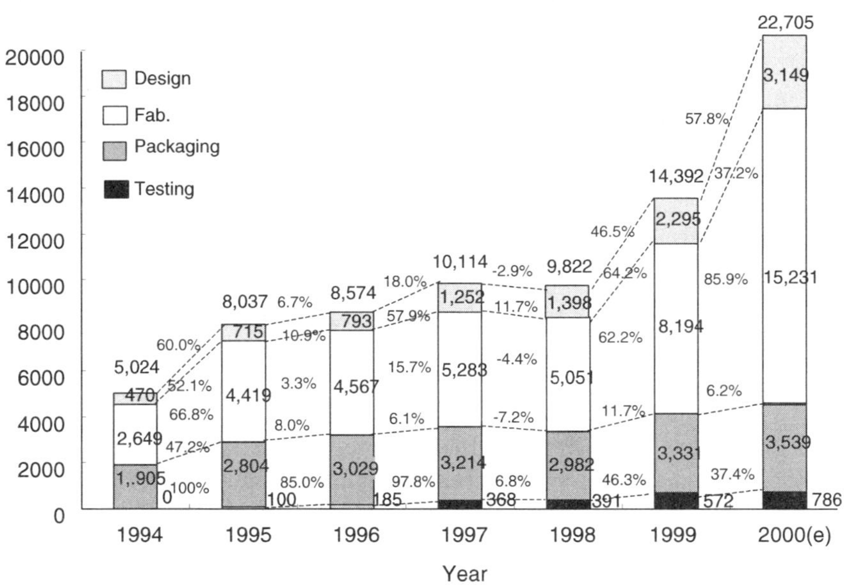

FIGURE 8 Revenue of Taiwan integrated circuit industry.
SOURCE: IEK ITIS project/ITRI, October 2000.

A small part of the activity takes place at the Hsinchu Science Park. Its activities cover a wide range but the funding is quite low. On the industry side most R&D work is in technology development and commercialization of the technology.

Low R&D Funding for Universities

He then focused on the university program, in which the number of projects, the budgets, and the workforce have increased steadily since 1996. Most of the researchers are students and professors. The National Science Council and the Ministry of Education fund most of the university research, at quite low levels, despite the large number of projects (236 in 2000). In the year 2000 the total budget is only about U.S. $5 million, which he called a "minuscule amount of research money by any standard," but it was very important because of the number of people involved and because of the students trained in the universities who eventually support the industry.

Funding levels have increased toward the applied research and development end of the spectrum. The NSC supports two main laboratories. The first is the Nano Device Lab, located in Chiao Tung University but totally financed by NSC. This lab's R&D is mainly associated with silicon-based semiconductor devices and material, with a special focus on deep sub-micron MOS devices. This lab concentrates on process-technology development. An important mission is to support silicon-related research in universities, providing much of the equipment used by the professors and students. The Nano Device Lab's annual budget runs roughly NT $300 million, with extra amounts sometimes needed for equipment. This amount has compensated, in some sense, for the low direct funding for universities.

Help for Students and Professors

The second lab, the Chip Implementation Center (CIC), concentrates on design. It functions much like the MOSES program in the United States in that students and professors execute design projects and then take them to the CIC to put them on a "multiproject wafer." This process, pioneered at Xerox PARC many years ago, makes it possible to assemble 20 to 30 different designs on a single mask. For 0.13- and 0.15-micron technology, it costs close to half a million dollars for tape-out, which professors cannot afford. Instead, they use resources at the CIC, where the average budget is about NT $140 million. That program, managed by ITRI, has been very successful.

The majority of the development work sponsored by the government has been financed through MOEA, the Ministry of Economic Affairs. The level of funding was very high before 1994. Since then it has been trending down, primarily because industry has shown sufficient vitality to take over more of the funding and perform the R&D itself. He noted that the amount of MOEA spending has been far larger than spending by the Ministry of Education, National Science Council, or other agencies. The MOEA has allocated between hundreds of millions to over a billion new Taiwan dollars each year for R&D. The MOEA is the funding agency and the work is done through ITRI by ERSO, which is the primary laboratory doing semiconductor research under ITRI.

A History of Taiwan's Rapid Rise

Dr. Hu illustrated how this funding has been used to help the industry during its brief history. He said it has not been a matter of taking the government's money to do the R&D and then hoping for technology transfer to industry. It began in 1974, when the Taiwanese government decided to focus on the semiconductor industry as a key industry. At that point Taiwan's economy was primarily based on agriculture—"nothing but paddy fields, sugar cane, and pineapples." The Ministry of Economic Affairs chose semiconductors as the industry to work on. In the early 1970s the government established ITRI and, under it, about 10 different laboratories, including ERSO, which focused on semiconductors. ITRI was not part of the government and its employees were not government employees. The organization was more like ASET (described earlier in the proceedings) in that it was an independent entity doing contract work for the government. Initially there was some grant money from the government, but once ITRI was established the government money stopped, and ITRI survived by obtaining contracts. The strategy was to make it more efficient.

Creating a Company

With ITRI as the interface, ERSO contacted RCA, and the government paid RCA several million dollars for its 7-micron metal-gate CMOS process. RCA transferred that technology to Taiwan and helped ERSO build Taiwan's first 3-inch-wafer fab, in 1975, which started semiconductor activity in Taiwan. After ERSO had worked for six years on developing its own technology, the government decided to create an industry to use the technology. It needed a commercial company, so it created UMC (United Microelectronics Corporation). At that point, no one in the private sector wanted to invest. Therefore, the government supplied the initial funds and did some arm-twisting to get the banks to put up some money as well. This was before there was any venture capital in Taiwan. The government did not try to run the company and later sold all its shares, so the company became purely private. Once the company was formed, ERSO transferred technology to UMC, along with skilled people, and helped it build its 4-inch-wafer fab. UMC, established in 1980, was Taiwan's first commercial semiconductor company.

Creating an Industry

The government came to realize that it did not yet have an industry—only one company. The private sector was still fearful of risk and could not raise enough money to start new firms. Once again the government put up money, this time to start TSMC. And this time it applied a condition: The government's share would be less than 50 percent. Lacking enough funds from the private sector, the government made an offer to Philips, which put in almost 35 percent of the initial investment; this pushed the private sector's share above 50 percent, and TSMC was born. Again, the technology and all the people came from ERSO, including about 130 engineers, the 2-micron CMOS developed at ERSO, and its latest 6-inch-wafer fab. TSMC had no product, only the fab technology, which matched well with the idea of a pure-play foundry.

That same year another group felt that it could do the same thing without government money, so nine months later a third company, Winbond, was formed. Again, most of the people came from ERSO, along with licensed technology and some old integrated-circuit (IC) products, such as wristwatch chips and other low-end consumer items. At first no one wanted to put money in, but once TSMC was formed some companies thought it probably was doable and put enough money in to create the third company. After that a new company started almost every year, and many were spinoffs from ERSO. ERSO continued to be the source of human resources for the industry.

The Power of Spinoffs

He skipped ahead to 1994, when a similar company formation gave rise to Taiwan's DRAM industry. The story began in 1990, when the government awarded a major project contract to ERSO to make the half-micron CMOS, including the 8-inch wafer and two kinds of products. One was a 4M SRAM and the other a 16M DRAM. With that technology and product a new company called Vanguard was formed. This time the government did not have to invest additional money; rather, it owned part of Vanguard when Vanguard was spun off. The government had contributed about $100 million to the DRAM project, which was converted to stock in Vanguard, and the stock was worth about $400 million two years later. "No one had ever heard of a government making money on R&D," said Dr. Hu, "but this one did." Because of that success the private sector was able to create four or five companies to make DRAMs. He said the important message here was how much could be done by government to benefit industry from this kind of R&D work, primarily through spinoffs.

Government and Industry Funding

The Strength of Small Grants

Dr. Hu next introduced the other, smaller R&D projects sponsored by the government through the Industrial Development Board (IDB), an agency under MOEA. The IDB made R&D grants to companies with very specific projects that were close to commercialization. Companies proposed projects, but the funding amounts were not large. The Hsinchu Science-based Industrial Park (HSIP) administration also allocated R&D funding each year to support the dozen or so companies inside the park. The amounts of funding were small but they helped to create many new products.

The industry itself has been growing since its inception, starting with UMC in 1980. In the past 20 years the industry has grown quickly, and in 2000 was estimated to reach about $22 billion in total revenues. The industry has been spending about 6 percent of revenues on R&D.

Industry Assumes the Primary Funding Role

Beginning in about 1990 the government was underwriting over 44 percent of total spending on R&D to the benefit of the private sector. Since then the government contribution has remained relatively constant, but its percentage has dropped rapidly. By 1999 the government's share had fallen to only 6.5 percent. This was a clear indication that industry had moved in to take over the primary role in funding semiconductor R&D in Taiwan's industry.

Dr. Hu summarized by saying that the government has played a proactive and pivotal role in establishing the IC industry over the past 20 years. The turning point came around 1994, when Vanguard was spun off and industry started to play a leading role in R&D. The government received criticism for making so

TABLE 1 Government Versus Industry R&D Investment

Year	$ from Gov't (MNTD)	$ from Industry (MNTD)	Gov't/Ind
1994	2,160.0	4,832.0	44.7%
1995	1,413.0	8,936.9	15.8%
1996	1,296.0	9,669.4	13.4%
1997	1,439.0	16,158.8	8.9%
1998	2,079.0	23,537.6	8.8%
1999	1,750.0	26,834.8	6.5%

SOURCES: NCS, HSIP, IEK/ITRI.

much money on its investment, and there have been no more spinoffs from ITRI since 1994. For its part industry does not want more spinoffs, which now would create more competition in the market place. As a result the government's role in R&D has gradually evolved into a more traditional activity and is expected to continue to decrease in the future.

DISCUSSION

A Consortium in Taiwan?

Glen Fong of Thunderbird, the American Graduate School of Management, asked whether Taiwan's spectacular startup story would evolve to include more collaborative R&D, like in the United States, Europe, and Japan. He asked about the "so-called Taiwan SEMATECH" called ASTRO. Dr. Hu said there has indeed been an attempt to establish collaborative research in Taiwan, because the benefits of collaborative activity are well understood. As in other countries government funding levels are inadequate to support an effective R&D effort, simply because the cost of manufacturing equipment has increased so fast. Taiwan has studied the models of SEMATECH, Selete, ASET, IMEC, and other consortia, and it plans to combine industry money with government money to do R&D for the common good. For 2 years ERSO has been chartered to create that kind of consortium, which was moderately successful until the beginning of 2000. With a new political party in power, industry policy has become less clear, and all further action is on hold until plans are firm and new funds are committed for the consortium.

Another questioner asked whether the Taiwan semiconductor companies are doing their own research. Dr. Hu said they are, and more noticeably at the international level. TSMC is a member of International SEMATECH, and UMC has formed an alliance with IBM and Infineon. The plans of smaller companies are not as well known.

THE SCIENCE PARK APPROACH IN TAIWAN

Chien-Yuan Lin
National Taiwan University

Providing Good Soil

Dr. Lin of the Institute of Building and Planning at National Taiwan University said that he would introduce another approach to R&D, the science park. He pointed out that even though Taiwan is small, the government has promoted economic development very heartily, allowing the country to rise from "almost nothing after the Second World War" to a point of considerable accomplishment. He said that one could think about promoting R&D or technology in the same way we might try to design a kind of tree that could produce good fruit. The science-park approach is designed to provide good soil that will grow healthy trees so that we can enjoy the fruit.

He traced the beginning of the IC industry in Taiwan to the packaging industry that started in 1976. By the end of 1999 Taiwan had 237 IC firms, including IC design and manufacturing. The industry is still growing at a rapid annual rate of about 50 percent and has moved into third or fourth position globally in market share, depending on the segment. Most of the jobs created have been in the fabrication and packaging segments. Both the IC industry as a whole and the science parks have created many job opportunities as well.

The Government's Active Role

He compared the government's role in Taiwan and the United States, saying that the U.S. government maintained a primarily free market in which the industry had to grow and compete and survive. He contrasted this with Asian countries, notably Japan and Taiwan, where the government has been very active in promoting economic development. One of the ways the Taiwanese government has done that, beginning in the late 1970s, has been to identify the most promising industries and then develop an attractive environment in which high-tech companies in those industries could become established and grow. This succeeded even though the country had virtually no technology to build on—only a labor force.

The Concept of the Science Park

In 1980 this government policy was augmented by the concept of the science park. The government provided major venture capital as well as some tax deductions or exemptions for companies that moved to the park. In addition, it provided the infrastructure, one-stop business service as well as other services such as R&D and education. This complete package was considered to be a partnership between the government and the semiconductor industry.

Incubating Factories

Dr. Lin showed a map of Taiwan and the location of the Hsinchu Science Park plays. (See Figure 9.) He pointed out that most of the country's IC factories are distributed in the northern part of Taiwan—either within the science park itself or nearby. He said that the Hsinchu Science Park played a key role in incubating these factories.

The park was planned in the 1970s and began operation in 1980 under the administration of the National Science Council. That the park is managed by the central government directly rather than by local government is a measure of its

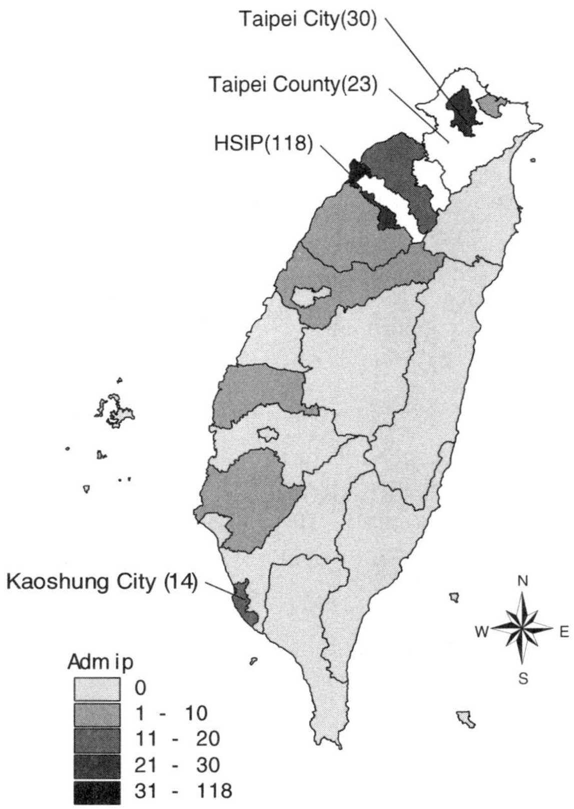

FIGURE 9 Spatial distribution of integrated-circuit factories.

importance. Park land is leased to tenants—not sold—thereby allowing the government to maintain control over the use of the park.

The park is located close to the Chang Kai Shek (CKS) International Airport and near Taipei City, so it enjoys locational benefits of transportation and human resources. Another important feature is that it is located close to two universities, the Tsing-Hua University and the Chia-Tung University. These are the leading universities in IC industries. It is also near ITRI, the Industrial Technology Research Institute, which has the important function of helping to incubate the industries in Hsinchu Science Park.

Like a New Town

In a sense, he said, the Hsinchu Science Park is like a new town. More than just a place for manufacturing, it also includes services such as restaurants, clinics, banks, housing, and bilingual schools. He said that Taiwan does not have sufficient human resources, especially high-tech people, who are the key to a successful R&D operation. The park has had to recruit talented people from the United States. When it does manage to bring "those high-tech families" to Hsinchu, it has to provide the education programs for their children.

In addition to those services the park also has support businesses, like the one-stop services, on-the-job training, and a range of domestic and international information services. Because this park is manufacturing-oriented it requires automated customs services for shipping.

Rising Employment and Capital

Employment growth at the park has been strong, rising from 19,000 workers in 1989 to 33,000 in 1994 to 83,000 in 1999. In the beginning most of the companies came from the United States, but by the end of 1999 more than 80 percent of them were local or other domestic companies.

Total paid-in capital in the park has surged from U.S. $3.5 billion in 1994 to U.S. $20.4 billion in 1999. In the beginning most of this capital came from the government, the United States, and other countries; now 92 percent of it is domestic money, and only 4 percent is from the government. The park has a total of 291 units, including PC manufacturing, IC design, and other IC-related industries. The most important trading partners are the United States, Japan, and Europe.

Dr. Lin noted that many in the labor force do not have a baccalaureate degree, so there are many opportunities for high-school graduates. This is because the park is very manufacturing-oriented. The production growth rate has slowed somewhat from the early years but was still 43.1 percent in 1999.

Because of the success of Hsinchu Science Park, both TSMC and UMC have requested additional space for expansion. The government has planned new sites in Chunan, 30 km south, to focus on biotechnology, telecommunications, and opto-electronics, and other sites in Tonglou, a short distance farther south, which would focus on telecommunications and opto-electronics. In the southern part of Taiwan the new Tainan Science Park is developing 638 hectares for four new TSMC factories and other companies. Tainan Science Park had 15 IC factories committed to operations in 2000.

The Strains of Rapid Growth

On the downside Dr. Lin acknowledged worsening environmental problems caused by the sudden growth of the science park. Hsinchu Science Park has a "terrible" traffic situation, he said, and wastewater has become a critical and irritating issue for the local government. This is a politically difficult situation because the local government does not experience benefits from the park, which is managed by the central government. These issues will become more acute as other areas, seeing the success of HSIP, ask for science parks of their own. In addition, competition is becoming a problem as other regions, including Mainland China, build their own parks. Dr. Lin concluded by saying that competition in general will have to be considered more closely in the future, now that the idea of science parks has become popular.

DISCUSSION

Dr. Flamm asked if Dr. Lin could expand on the nature of the tax deductions and exemptions for semiconductor producers. Dr. Lin said there are two kinds of exemptions. First, for the first five years of operation IC companies can deduct the cost of all equipment or investment, which is "a nice deal for them." Second, an investment incentive allows the creation of a tax shelter for any money invested in the IC industry. Dr. Hu added that many Southeast Asian countries, unlike the United States, tend to have these tax incentives.

Discussant

Michael Luger
University of North Carolina at Chapel Hill

Dr. Luger began with a comment about tax incentives. The U.S. federal government, he said, provides general tax incentives, such as R&D tax credits, while state governments and some local governments have location-specific credits.

He then congratulated the STEP Board for putting together such a rich program and, like Dr. Mowery, noted that he would limit his comments to what others had talked about during the program. He noted that most of the presentations had focused on the importance of national semiconductor consortia to national and global competitiveness.

Three Conditions Favoring Consortia

He noted three conditions favoring consortia: First, new science is increasingly expensive to develop and requires multiple partners to pay the bills; second, R&D has spillovers that invite free-ridership that consortia can internalize; and third, the absence of cooperation and coordination can lead to duplication of effort and what Schumpeter called "destructive competition." He said he would expand on two points made in many of the presentations. One, national consortia like SEMATECH and SIRIJ are one of several models along a continuum that are intended to enhance competition. And second, all of these models have important geographic dimensions that make them variably important as economic development strategies, not just as science and technology strategies. Here, he noted that as a regional economist, he was especially interested in this aspect of the topic.

A Taxonomy of Collaborative Vehicles

First, there are the national industrial consortia in advanced economies—SEMATECH, SIRIJ in Japan, perhaps MEDEA in Europe—whose goals are advances in fundamental science that can be used by industry.

Second, there are national-government-funded labs and demonstration centers, such as the U.S. national laboratories. In transitional economies, such as Thailand, these would be science and technology centers that are used to develop new industries with strategic national importance. The Nano Devices Institute would fall into this category as a focused, leading technology center. An example at the state level in the United States would be the North Carolina Biotechnology Center.

Third, there are laboratories with direct government support for industrial research. In the United States these are funded by the Department of Defense, National Science Foundation, and other agencies, including through the SBIR program. These are supported in the belief that public funds, if strategically spent, can unleash the inventiveness of the private sector.

Fourth, there are what the National Science Foundation calls "virtual co-laboratories," vehicles to foster collaboration among experts around the world who share research interests. These vehicles may use mail, teleconferences, and other information-sharing mechanisms like membership and trade organizations.

Fifth, there are university-based R&D centers such as those in Japan discussed by Dr. Morino. Many of the engineering schools in the United States have targeted centers for research in semiconductors, materials science, or information technology hardware. In general these centers have three common characteristics.

- They serve as recipients of federal, state, and foundation funding.
- They reach out to industry for active and passive partnerships, joint ventures, and funding.
- They create a focus within a university to attract faculty, students, and internal resources.

With dozens or hundreds of universities doing the same thing, they tend to compete rather than cooperate for resources, for personnel, and for industrial interest. Recognizing this, states have responded by developing multi-university or regional centers as a way to reduce competition and to create a critical mass. That is an activity of the EPSCoR[29] programs in some states. Local industries are important focal points for all of these centers.

[29]EPSCoR, the Experimental Program to Stimulate Competitive Research, is a joint program of the National Science Foundation and several U.S. states and territories. The program promotes the development of the states' science and technology resources through partnerships involving academia, industry, and state and federal government.

A Continuum of Consortia

Dr. Luger placed these programs on a continuum. On one end are the large, national consortia that emphasize basic research; they also have a high degree of spillover and consequently fewer direct local applications. Next along the continuum are programs supported directly by federal funds, which also emphasize basic research, including university R&D centers. Beyond them are state-funded R&D centers, which feature more applied research, fewer spillovers, and more concentrated spatial effects.

The Importance of Clusters

Dr. Luger then elaborated on the importance of spillovers in the context of government-industry partnerships. The kind of research that economists call appropriable science has fewer spillovers and leads to more localized geographic effects. Businesses want to locate near this research or to do it themselves so that they gain the rights of first refusal of intellectual property. The more appropriable the science, the more important physical proximity becomes. As a new technology develops commercially in a particular place, spatial agglomeration occurs, bringing what are called localization economies. As in the case of Taiwan, these economies lead to clusters of firms—not only from single industries but also from industries that are related through input-output linkages and other relationships. These clusters stimulate growth in the local economy. This has been the logic behind the development of the Hsinchu Science Park and other science parks around the world.

The Case of Research Triangle Park

Dr. Luger mentioned Research Triangle Park in his own state of North Carolina, which he has studied extensively. In addition, he discussed his trip to Taiwan several years ago to visit Dr. Lin and HSIP. In North Carolina in the 1970s and early 1980s there was virtually no semiconductor research or industrial presence outside the engineering school of North Carolina State University. Research Triangle Park was built in 1959, with IBM the first corporate client, and in the 1980s the state invested $6 million dollars—a good deal of money at the time—in the Microelectronics Center (MCNC), expecting to attract semiconductor firms.

Like in the case of Vanguard in Taiwan, the investment paid off in several ways. Recently a company called Kronos spun out of MCNC. Though the deal was private, the investors gave the state of North Carolina $30 million of the proceeds to build infrastructure in rural areas. MCNC has also served as an anchor for the recruitment of major IT companies to the park and region, including Nortel, Ericsson, and Cisco Systems.

The Importance of Being Adaptable

The focus on information technology was not initially envisioned in Research Triangle Park (RTP). After IBM located there and the Microelectronics Center was built, the state expected to develop a large presence of semiconductor companies. During the 1980s RTP attracted Harris Semiconductors, Mitsubishi, and a few smaller microelectronics companies. The industry then began to restructure and RTP failed to attract additional investment in the semiconductor area. It changed its recruitment strategy. It renamed the Microelectronics Center simply MCNC and added programming for information technology firms. As noted above, the state was successful in attracting major information technology companies. "Adaptability," said Dr. Luger, "was important."

By most standards RTP is judged to be highly successful. He showed a cluster diagram that indicated a 20 percent greater presence of information technology, communications, and software in the region than one would expect from the national averages. The location quotient has been growing at the same time these sectors have been growing nationally. On the applied side, semiconductor research is becoming the foundation of other industries, a point Dr. Knorr made earlier. So the earlier development of infrastructure to attract the semiconductor industry became useful later for attracting the two sectors that have come. He concluded with the speculation that the work of SEMATECH in helping to enhance the semiconductor industry also had impacts in related industries for many areas of the country.

DISCUSSION

Questions of Human Resources

A questioner returned to Dr. Hu to ask about the supply of human resources in S&E, and he replied that there is a shortage of engineering talent. From TSMC's point of view, he said, it will be very difficult to sustain the company's growth rate as projected by the government unless the supply increases. The company is supporting university programs and encouraging professors to generate more graduates, but he affirmed that, "on a large scale, there is a problem."

Dr. Hu was also asked about the H-1B visa: Could young Chinese engineers still come to the United States, go to graduate school, gain work experience, and then return to Taiwan? Dr. Hu did not address this directly but said that most of the people TSMC attracts *back* from the United States are very seasoned people, many of whom are U.S. citizens. He said that the people who had visa problems working in Taiwan were the Mainland Chinese. Dr. Luger added that some parts of the United States are trying to duplicate the repatriation efforts of Taiwan. He said that Pennsylvania and Georgia both offer college loan payment exemptions to young, college-educated people to return.

Panel V

Challenges Facing the Equipment Industry

INTRODUCTION

Erik Kamerbeek
Semiconductor Equipment and Materials European Association

Dr. Kamerbeek introduced the members of the panel. He noted that the technical challenges to the equipment industry constitute a challenge to the global information technology industry as a whole, and he praised the panelists as a distinguished group uniquely qualified to discuss these challenges.

CHALLENGES I

Kalman Kaufman
Applied Materials

Sustaining Technological and Business Viability

Dr. Kaufman, corporate vice president for strategic planning and new business development, began by summarizing ideas. The first one, he said, was obvious: the semiconductor and electronics industry represents an increasingly significant force in the economy—both because of the sheer size of the industry and because of its influence on the information age. Second, he said, equipment suppliers play an increasingly crucial role in enabling the technological and business viability of the industry. He presented the following imperatives for ensuring sustained success:

- Equipment suppliers need to continue to invest heavily in technology development and commercialization.
- The government must promote the supply of highly skilled people, particularly researchers. He called education "an absolute imperative."
- The government must also ensure fair access in every country to markets and technologies.
- The semiconductor producers must create a win-win environment to reduce risk and improve the overall efficiency of the industry. Even though the industry is incredibly successful, he said, it is far from using its resources at maximum efficiency.

A Need for Capital Investment

He reviewed the "food chain" of the semiconductor industry. In 1999 total revenues of the electronics industry were close to $1 trillion, or 3.1 percent of worldwide GDP. The revenues for the semiconductor components of the electronics industry were close to 16 percent of that, and capital spending for wafer-fabrication equipment about 21 percent. He stressed the magnitude of capital investment in this industry in relation to other industries. A substantial portion of this capital investment is for technology, not just for capacity. Five years from 1999 electronics revenues were estimated to grow 1.4 times, to $1.3 trillion. Semiconductor revenue is estimated to more than double, to more than $300 billion dollars. Capital spending and spending on wafer-fabrication equipment are estimated to more than double.

A Need for Alternative Technologies

In lithography, pushing the limits of technology brings a steep price in the increased cost of masks and additional requirements. The ability to extend this technology down to 0.13 microns from 248 nanometers and, subsequently, further may give the illusion that this can happen repeatedly. For example, 193 nanometers has not been attained, even though it is viable, and 157 is still in the future, so 70 nanometers will certainly present a serious dilemma. So far there is still no viable alternative to scaling. He said that the world should be concerned about the future of this technology because of its global importance. He suggested that EUV (extreme-ultraviolet) and electro-projection technologies are viable opportunities, and urged support for research on such alternative technologies.

A Comparative View of the Semiconductor Industry

He compared the semiconductor industry with other major industries. (See Figure 10.) The steel industry has grown 6 percent a year over the past 30 years, the plastics industry 10 percent, and the semiconductor industry at the "extremely

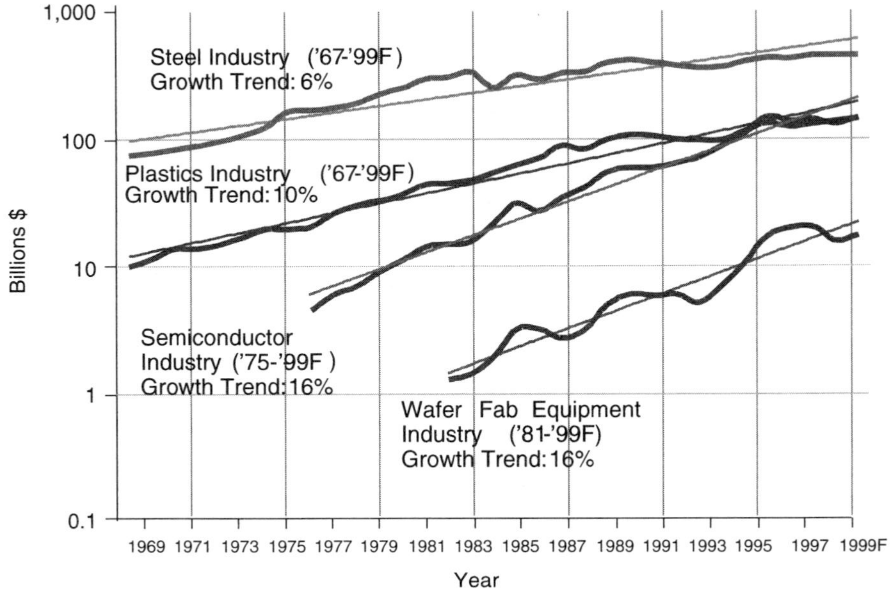

FIGURE 10 Electronics and semiconductors: Key economic forces.

fast" rate of 16 percent. Semiconductors will pass the steel industry by the year 2004. The wafer-fabrication equipment industry also grows at about 16 percent, but the cyclicality of the industry creates tremendous inefficiencies.

The manufacturers of semiconductors are outsourcing more of the process in order to create new products. In the 1970s the equipment suppliers shipped only poorly automated machines and the customers were left to learn how to use them. In the 1980s the process specifications were being defined and sent to the customers. In the 1990s the process specifications were guaranteed and the process became much more efficient. In 2000 companies began to deliver process module integration, which Applied Materials and other suppliers are now seriously promoting, as well as new solutions that improve efficiencies.

Challenges Facing the Equipment Industry

He used the transition to 300-mm equipment to illustrate the expanding responsibility of the equipment industry. In previous transformations, to 150-mm and 200-mm wafers, key customers such as Intel and IBM led the transitions, taking the responsibility and assuming the risks. This was not the case for the

300-mm transition, when the equipment industry for the first time assumed all financial and technical responsibilities. This demonstrated a much higher maturity level.

Along with its responsibility, Applied Materials itself has grown, from just over 10,000 employees in 1995 to over 20,000 in September 2000. Mr. Kaufman said this growth was typical for good companies in the sector. He estimated the size of the semiconductor equipment industry, including suppliers, at six or seven times the size of Applied Materials, or about 130,000 workers. At an average salary of $65,000-$75,000, the combined workforce represents a large economic sector.

A Need for Continuous Investment in Technology

He said one of AMAT's primary responsibilities is continuous investment in new technology. From 1981 to 2000 the company introduced at least one new product every year. In this field, he said, new products were *not* like new models of cars, for which changes can be cosmetic. New equipment models represent entirely new technologies. In addition, the number of products has soared recently, to 30 new tools in the year 2000, which he called representative of the innovation rate in the industry. He said that innovation goes beyond the tool itself to include the mode of operation and the interaction between the supplier's equipment and the industry itself. Capabilities such as integration of software and tools and productivity have to be an inherent part of tool design. This allows the two industries to converge and puts even more pressure on the semiconductor industry to invest continuously. He said that in one year at the beginning of the 1980s AMAT spent three-and-a-half times its cash flow on R&D for new products.

A Need for Both Research and People

To continue its growth, he said, the industry needs a continuous flow of technical innovation and continuous training of people. He expanded on the topic of investment in R&D, which increases every quarter in spite of downturns and fluctuations in the market. He said that companies with small market share cannot maintain this pace of investment. Applied Materials believes that a company needs at least a 15 percent market share to sustain significant R&D. Below that share a company can function and introduce valuable technologies, but it is financially impossible for it to offer research support to its customers on a long-term basis.

Mr. Kaufman said that the necessary role of the universities is to ensure the flow of technical innovation and skills. This is a dual responsibility. The first responsibility is to provide long-term research, while the second and more urgent obligation is to produce motivated and skilled researchers for the worldwide industry.

Meeting New Challenges

The Roles of Universities and National Labs

He looked at the research itself from two perspectives. The first is that universities have a role in teaching and motivating the next generation of researchers and engineers. The second is that the national labs should bridge the widening gap between academic research and the next-generation industry requirements in generic, pre-competitive research. Such research should be funded and defined mostly by industry, but the government has a role to play in inspiring and catalyzing the research.

To ensure the flow of technical innovation, he again stressed the need to prevent national protectionism. Any barriers from any country, he said, would have a negative impact on the whole industry. He praised the potential of International SEMATECH and called for great care not to create national boundaries that would impede the ability to grow.

The Role of a Consortium

He turned to supplier-customer cooperation and discussed the need to reevaluate the role of SEMATECH based on lessons learned during the 300-mm transition. The best role for a consortium, he said, is to bring people together in order that they may cooperate and provide the semiconductor industry with valuable information so "we can change our roadmaps and learn how to serve customers." Problems arise when the group attempts to pressure the semiconductor equipment manufacturers toward certain guidelines or attempts to dictate pricing. Those problems did arise in the case of SEMATECH's planning for the 300-mm transition, he said, when Applied Materials invested some $300 million in unnecessary equipment because they received the wrong targets when they introduced the tools.

He concluded with the message that such decisions should be made together in order to best promote the industry, and he called for more cooperation and better communication among all elements of the industry.

DISCUSSION

Questions About SEMATECH

Dr. Kaufman was asked whether, despite his criticism of SEMATECH, the consortium has contributed to the growth of AMAT. He responded that SEMATECH has benefited the industry by promoting better cooperation and opened doors to understanding. He said he thinks AMAT's own position has not changed much because of SEMATECH, but he said his company strongly promotes working with SEMATECH and that International SEMATECH is creating excellent tools for cooperation and communication.

Another participant asked Dr. Kaufman to clarify the role that SEMATECH played in pricing the 300-mm wafers. Dr. Kaufman said that SEMATECH did not set prices but tried to use a formula in regard to new tools for 300 mm. The formula was based on large-scale economic estimates of the affordable costs of the transition from 200-mm to 300-mm equipment.

A New National Lab?

Dr. Wessner recalled Dr. Kaufman's reference to the role of the national labs and asked if he was suggesting an expansion of a laboratory such as the National Institute of Standards and Technology or a modified role for a weapons lab like Sandia. Dr. Kaufman said that it should be a very focused lab, perhaps modeled on IMEC in Europe. He said that its relationship with the universities should be a close one but that the people in the lab should be dedicated to solving precompetitive, generic research problems much like those the semiconductor equipment companies work on.

CHALLENGES II

John Kelly
Novellus Systems

Three Critical Needs

A Need for More Talented People

Dr. Kelly has had the unusual advantage of having worked both on the IC side of the line, at Hewlett-Packard, and for the last five-and-a-half years as a supplier. He said that problems facing the supplier industry are "fairly simple and straightforward." The first is the undersupply of talented graduate students. He said that the good news is that many of the students they have hired trained through the Semiconductor Research Corporation (SRC) and were prepared to "hit the ground running." The bad news is that there are not enough of them, and the situation seems to be worsening. Many graduates have moved away from the semiconductor industry into other areas, such as nanotechnology, he said, and the professors have been going "where the money is." More students might be drawn to semiconductor-related programs, he said, through various incentives. The SRC has only limited funds to spend, however. He said that Novellus is very involved with the SRC and uses it as a primary source of new talent.

A Need for Resources for Long-term Research

Another, more complex issue, he said, is the problem referred to by Drs. Moore and Polcari as shrinking resources for long-term research. The technological "brick wall," he said, could be very real "if we don't work on the right problems fast enough." He said that the industry has good momentum and good people, and that "with some help" it could either get over the wall or around it.

A Need for More Fundamental Research

He said that the largest change in the equipment industry that poses serious challenges is that it is no longer acceptable simply to deliver a tool to the customer. It has to be delivered with a process or as part of a process, and the process has to be perfect. This makes the supplier's life more complex, because more knowledge is required and no one can tell in advance what problems may arise. There is less knowledge production being done in the absence of the longer-range industrial labs. With an already large research load the company has limited ability to maintain its own basic research. This means a limited ability to answer fundamental questions quickly.

"If a company such as Novellus wants to move quickly," said Dr. Kelly, it needs more work on "fundamentals, materials—the real basics." The most significant gap is in the area of two-year to five-year research—the long-term work that brings understanding of materials and interactions and allows the integration of the processes previously described by Dr. Kaufman. If a company starts a development program on a new process or piece of equipment and does not understand the science, then developing the process usually takes twice as long. If it does know all the basics, however, the company can move fast and develop a good tool the first time.

Growth of the Research Burden

Even without doing this fundamental research, the suppliers' research burden has grown enormously over the last 5 to 10 years. During a downturn companies have to spend large amounts of money to maintain their research level. In growth years they are pressed to find enough good people to expand the company's activities.

Dr. Kelly said there are several sources of help. One is MARCO, the Microelectronics Advanced Research Corporation run by SRC, which provides a modest long-term knowledge resource for companies. At the other extreme, SEMATECH helps with testing and evaluating tools that are in the development stage. Between these two extremes, however, lies a chasm with little ongoing research.

CHALLENGES III

Papken Der Torossian
Silicon Valley Group

A Brief History of Silicon Valley Group

Dr. Torossian said that enough people had asked him about his company, which produces high-end lithography systems, that he decided to give a brief review of its history. In the late 1980s, 90 percent of the lithography business belonged to Nikon and Canon, with ASML in Holland supplying 9 percent, primarily to TSMC. The company that was making scanners, Perkin-Elmer, left that business, which was losing money, and Silicon Valley Group (SVG) bought it, much to the dismay of Wall Street, which saw nothing but risk. SVG raised a small amount of federal funding and sold stock to raise most of its financial capital. Over the last 10 years, he estimated, SVG has spent over $1 billion in R&D and capital. At present the company has about 8 to 10 percent of the global lithography market, and probably 40 to 50 percent of the advanced technology market. "We make [the equivalent of] Ferraris and Porsches," he said.

The Need to Cooperate

The Role of SEMATECH

He acknowledged that SEMATECH, IBM, and Intel helped SVG become established in the early 1990s. He said that an essential fact about SEMATECH is that it has created an environment in which a buyer and seller can work together. In all its successes, whether in the etching business or measurement or lithography, SEMATECH helped by being the midwife in cases when a customer really wanted to buy and a supplier really wanted to manufacture. He said that the manufacturer alone cannot perfect the complex equipment of this industry. Someone has to buy it, work with it, and improve it. So there is mutual benefit when buyer and seller work well together. In such cases, said Dr. Torossian, "government money doesn't help; you need to have a customer who buys, who beats you up and makes the machine better."

Why Cooperation Is Required

He said that some companies joined SEMATECH on the assumption that they could harvest other companies' pre-competitive technologies. Some of those companies were sorely disappointed, he said, because that was not the purpose of SEMATECH. In the equipment business, he said, pre-competitive work is very difficult. The only area now that is pre-competitive is 300-mm research, because no single company can sell 300-mm machines. Unless all the parts (lithography, etchers, positioning) work together, companies will not buy a 300-mm factory. No one can invest in this specific equipment. Customers want suppliers to share technologies because they want two strong suppliers to compete over price. But suppliers do not share technologies. Instead they have to figure out how to work together pre-competitively.

Mounting Challenges

A Huge Investment in Research

He then complemented earlier comments about the technical challenges of lithography, copper, and low-k. He reinforced the point that moving from a three-year cycle to a 2-year cycle requires a huge investment by the industry. The lithography industry already invests 12 to 15 percent of its sales in R&D, he said, or about $5 billion to $8 billion a year; SVG has invested $612 million in R&D over the last 10 years and $350 million in capital expenditures. Research spending would have to increase by almost 30 percent to accelerate the equipment cycle.

The second challenge is that some of these investments are very long term. For example, he said that it has taken 10 years to go from 248 in 1989 to 193 (130 nm), which is just being introduced now. This required an investment of about $500 million to $800 million. The next generation, 157 (100 nm) and EUV, will take another $500 million. These huge investments cannot be borne by a single company, no matter how large. So, he said, we have to find a way to work together. He said that the EUV consortium is one simple way of working with competitors, which is "the only way we're going to advance the science in the next few years."

A Need to Strengthen Supply Lines

He added that the semiconductor equipment industry today includes about 3,000 suppliers globally. Only about 50 of them are integrated suppliers, like Applied Materials, Novellus, and SVG. Most of them make all the numerous components required for systems. A tremendous challenge for the industry is to work better with customers to strengthen the supply lines, especially in regard to a "tremendous problem" with optics.

Changes in Market Share

He showed a slide of 1999 semiconductor capital spending depicting Intel in the lead and said that TSMC, then ranked fourth, would almost equal Intel within the following year. (See Figure 11.) He also showed a chart of the worldwide

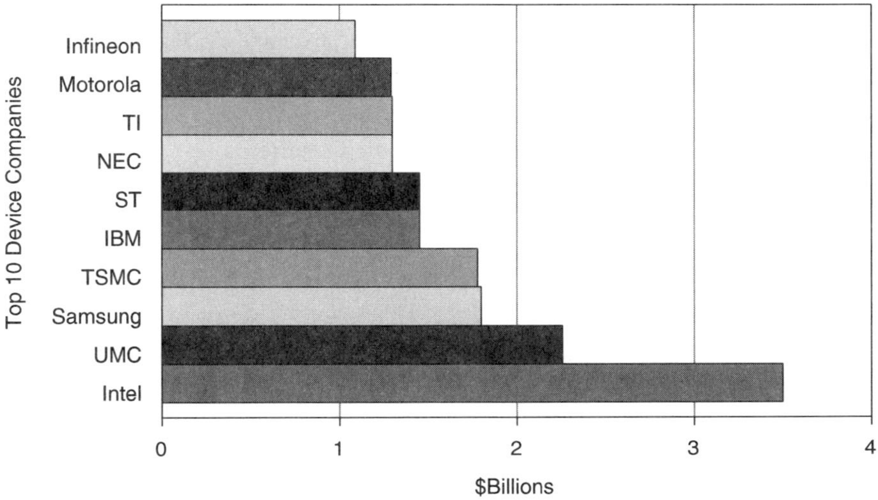

FIGURE 11 1999 semiconductor capital spending.

lithography market, which was a $5 billion business in 2000 that is projected to grow to $8 billion by 2004. (See Figure 12.) "A significant point," he said, is that Taiwan, Korea, and Singapore have sales almost double those of Japan. He then predicted that Japan was going to come "roaring back" very strongly. Finally, he showed a chart of worldwide market share for lithography depicting both AML

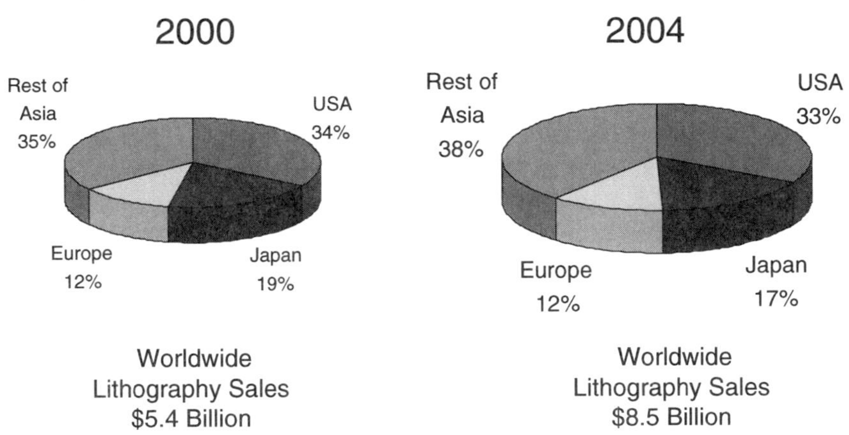

FIGURE 12 Worldwide lithography market.
SOURCE: Dataquest, July 2000.

and SVG to be growing at the expense of Nikon and Canon. He pointed out that neither ASML nor SVG had been able to sell any machines in Japan.

Finding Ways to Work Together

He said that SVG had spent almost $1 billion dollars of its own money in the last 10 years. It has received about $50 million from the government and $25 million from SEMATECH for R&D. Most of its investment capital came from stockholders.

To conclude, he reemphasized that research is expensive. He showed a SEMATECH chart on lithography funding that showed over $900 million in annual R&D expenditures by the industry to advance lithography. "I think it's important to realize," he advised, "that no one, including some of our customers, has that kind of money, so we have to find a way to work together on it."

The Internationalization of Cooperation— New Challenges

INTRODUCTION

Edward Graham
Semiconductor Industry Suppliers Association (SISA)

Dr. Graham pointed out that all the members of the previous panel were from the Semiconductor Industry Suppliers Association (SISA), which is headquartered in Austin, Texas. He praised the organizers of the workshop for the quality of the presenters and the program. He echoed the comment by Ken Flamm that success has many fathers and applied it to the semiconductor industry, underlining just how successful it has been. He then introduced the first speaker, George Scalise, president of the Semiconductor Industry Association, past chair of the Semiconductor Research Corporation, and a former member of the board of SEMATECH.

A U.S. PERSPECTIVE

George Scalise
Semiconductor Industry Association (SIA)

The Unusual Value of Information Technology

Mr. Scalise said he would try to build on what had been said so far, beginning with a comment by Dr. Knorr that the semiconductor industry is the greatest creator of wealth in the United States. He said that roughly 70 percent of semiconductor product goes to work in information technology, a segment of unusual value. IT, he said, represents about 8 percent of the U.S. economy but nearly 40 percent of the growth in the economy. Information technology is also a deflator, reducing inflation by 0.5 to 1 percent per year and increasing productivity by a point to a point-and-a-half per year.[30] Because these are major contributions to the economy, he said, we have the responsibility to "keep this thing going" in order to maintain the growth of the global economy.

A fact of great importance, he said, is that the number of transistors produced each year increases by about 55 percent. This year [2000] the industry will make roughly 40 million transistors for every man, woman, and child on Earth; by 2008 the per-person figure will rise to about 1 billion transistors. This, he said, implies a "major, major transition" that will require significant preparatory steps. One step is to decide whether current industry structures work well enough.

Accomplishments of the Semiconductor Research Corporation

He offered the example of the SRC, the Semiconductor Research Corporation, as a structure that does work well. It was founded in 1982 for two major reasons. The first reason was concern about the insufficient number of engineers coming out of colleges. The second reason was that not enough engineers were being trained in the new solid-state technology. Since then SRC has helped to increase the number of engineering graduates and to enhance technology training.

It has also created an "integrated, virtual semiconductor research laboratory" that funds projects at about 65 universities across the country. One goal has been to find the best principal investigators from any part of society, not just at the major universities. SRC now supports about 275 faculty members and over 800 graduate students, creating a pattern of broad participation that could be imitated in other countries. He advocated extending the SRC model to other associations around the world as a means of both increasing funding and encouraging even broader participation. The current funding level is about $35 million per year without restrictions on the kind of work or the nationality of the students funded. The outcomes of the research are available to all users without restriction.

Another important accomplishment in which SRC has participated, along with the SIA and SEMATECH, is production of the semiconductor roadmap,

---

[30] Dale Jorgenson has asserted that the decline in prices of IT equipment, and thus the subsequent large-scale investment in IT, has played a major role in the economic growth of the 1990s. Jorgenson states that "a consensus is building that the remarkable behavior of IT prices provides the key to the surge in economic growth." In particular Jorgenson attributes the "development and deployment" of the semiconductor in different industries as the engine for this surge in productivity. For in-depth analysis of his research on semiconductors and economic growth see Dale W. Jorgenson. "Information Technology and the U.S. Economy." *The American Economic Review*, March 2001.

which guides the industry today. The domestic roadmap has now been expanded to an international version involving information technology leaders from around the world.

Addressing the Research Gap

He turned to the research gap that had been discussed earlier and acknowledged that the MARCO program designed to address this gap is limited in resources. Two projects were under way to extend this: one for semiconductor design and testing, another for layout. The next steps were to enhance MARCO and add more programs, two of which were then in the planning stage. He conceded that those plans would probably not be enough, considering the proximity of the "brick wall." He agreed that the effort to get over or around the wall must be international in scope, and he said that SEMATECH could be a key mechanism to promote international research programs on materials structures and devices, circuits systems, and software that would begin to fill a part of the research gap. He recalled that the decision to end government funding to the consortium has helped to move SEMATECH in an international direction. For the SRC as well, he suggested, internationalization is the next logical step.

The Need for a Super-consortium

To pull together a "consortium of consortia," said Mr. Scalise, would be a more complex exercise than to ask SEMATECH, Selete, MEDEA, the SRC, and others simply to collaborate. One meeting of these groups had already been held, in April 2000, under a structure started by Japan called the International Forum for Semiconductor Technology. Two cooperative first steps were identified: one in environmental safety and health, the other in lithography. At a further meeting in midsummer, however, the group realized it was not sure how to implement the idea. They did agree that a consortium of consortia was the right approach to pursue, because collaborative action is needed to make more efficient use of the R&D dollar. In the United States, for example, public funding over the last decade has declined by 20 percent each for mathematics and physics, while for engineering public funding has declined 30 to 40 percent. The number of electrical engineering graduates has declined by 40 percent. "We're going to have to create a sea change," he said, "that will allow us to be far more efficient, far more effective."

He said that cooperative research was needed to bring about a linked series of necessary steps—more research can drive market competition; more competition can allow the industry to stay on Moore's Law; the continued improvements implied by Moore's Law can bring greater functionality and lower cost. "With the issues that are ahead of us in the next five- to seven-year period," he said, "we don't have a choice. This has to happen."

DISCUSSION

Enrollment and Funding Levels

Dr. Wessner asked two related questions. First, he asked for Mr. Scalise's impression of how much funding levels for engineering and related disciplines were dropping and to what extent this was responsible for the fall-off in enrollments. Second, he asked if it was feasible to double the size of the SRC's budget.

Mr. Scalise answered the second question with "absolutely yes." For engineering and the hard sciences, he said, the SIA is trying hard through a number of programs to get students more involved, including an effort to develop 50,000 K-12 teachers across the country as mentors for young people in math and science. The goal is to foster and reinforce the interest of talented students in those subjects.

Dr. K. C. Das, director of Office of Science and Technology, Department of Technology, Commonwealth of Virginia, observed that the statistics on scientists and engineers are "very frightening," with little increase for basic research in the universities. He asked at what level of government this issue might most appropriately be addressed.

Funding Engineering and the Hard Sciences

Mr. Scalise said that basic research in the hard sciences and engineering is primarily the responsibility of the federal government and a responsibility that it has backed away from. Funding for the National Institutes of Health has risen by about 47 percent while the funding level for engineering and the hard sciences has declined by about 17 percent. Overall the United States is spending about 3 percent of GDP on basic R&D. In the short term, industry has filled the gap; but this creates an unstable environment, because when economic conditions decline, industry reduces its spending. There has to be a more stable environment, which only the federal government can provide. Encouraging long-term increases in federal funding for basic research is an ongoing major objective of the SIA.

A JAPANESE PERSPECTIVE

Toshiaki Masuhara
Hitachi

Dr. Masuhara said he would try to summarize the Japanese view and then focus mostly on international collaboration. He began with an illustration of the 1999 International Roadmap for Semiconductors, showing "quite a few so-called brick walls which begin around 2005 and 2008." (See Figure 13.) In the process area the gate CD (critical dimension) control is the major issue in lithography, as

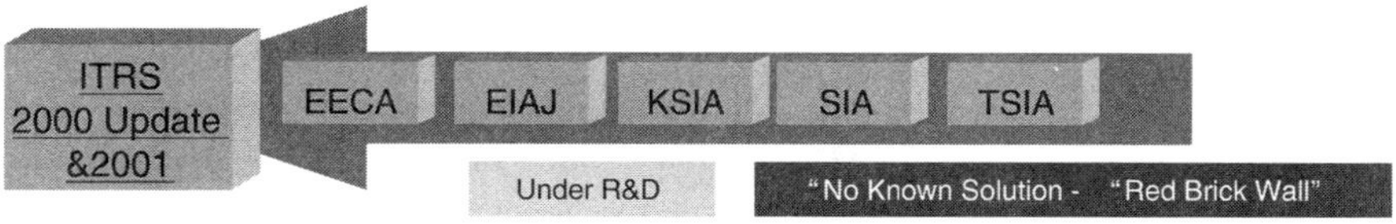

Items	Unit	1999	2002	2005	2008	2011	2014
Technology Node	nm	180	130	100	70	50	35
MPU Isolated Line	nm	140	85-50	65	45	30-32	20-22
Gate CD (3 Sigma)	nm	14	9	6	4	3	2
Wafer Diameter	mm	200	300	300	300	300	450
Gate Insulator	nm	1.9 - 2.5	1.5-1.9	1.0-1.5	0.8-1.2	0.6-0.8	0.5-0.6
Gate Leak (100C)	pA/μm	5	10	13	40	80	160
Clock Freq.(CP-HP/MPU)	GHz	0.6-1.2	0.8-1.6	1.1-2.0	1.4-2.5	1.8-3.0	2.2-3.6
Voltage (LP-HP)	V	1.5-1.8	1.2-1.5	0.9-1.2	0.6-0.9	0.5-0.6	0.3-0.6
Power (LP-HP)	W	1.4-90	2-130	2.4-160	2-170	2.2-174	2.4-183

FIGURE 13 1999 international roadmap for semiconductors.

well as gate insulator and gate leak. In the design area a major challenge is to achieve a low-power operation at a very low supply voltage. He said that the international roadmap committee, in discussing the 2000 and 2001 updates, had concluded that the node 130 nm might be accelerated to 2001. He then showed a chart that summarized the major issues discussed by the roadmap technical working group, beginning with the challenges in lithography—not only the tools but also the types of resist, CD control, and gate stack material, equipment, and database. He also summarized the design challenges for developing systems-on-a-chip, including design reuse, system level design, physical synthesis, and testing. Here designers expect the brick wall in 2008.

A Japanese View of U.S. Collaborations

He then gave the Japanese view of government-university-industry collaboration in the United States. In the Japanese view there has been a good balance of support for research by government and industry through the universities. Industry support goes to SEMATECH and the SRC, and university support comes from industry partly through MARCO and the Focus Center Research Project, and partly through SEMATECH, with "very good balance between design and processing." He said that the overall success of U.S. industry appears to have come from the contributions of five overlapping efforts.

1. The use of the SIA roadmap to determine the direction of research;
2. The planning of resource allocation by SIA and SRC;
3. The allocation of federal funding through Department of Defense, National Science Foundation, and the Defense Advanced Research Projects Agency;
4. The success of SEMATECH and International SEMATECH in supporting research on process, technology, design, and testing; and
5. The Focus Center Research Project.

He said that one indicator of the success of U.S. research is the number of papers delivered at the ISSCC and IEEE Field Award events: Nearly twice as many papers originated in the United States as in Japan. As discussed, he said, Japanese industry generated a relatively small number of papers.

The Research Gap

Dr. Masuhara reviewed the research gap discussed by several speakers and the Japanese view of government projects in Europe, Japan, and the rest of Asia. In regard to Europe he emphasized the success of the applications-related technology research center IMEC, created and supported by both national governments and the European Union. This, he said, had led to success in LSI in application areas such as GSM, cell phones, DSP, A/D mixed signal, and SoC. He also mentioned a new regional model created in Scotland called Alba.

The Picture in Japan

Summarizing the picture in Japan, he said it was significant that the semiconductor industry has been generously supported by both STARC and Selete, as well as VDEC. He said that forming partnerships between industry and academia in both process and design would be more important in the future and would require a great deal of funding for development because of changes in technology. A national plan for a government-academia advanced semiconductor research center to study systems, SoC design, and advanced process is needed. STARC has supported industry-academia collaborations in design and testing and Selete has supported advanced process and device R&D.

For international collaboration the Japanese view began with Japan's participation in the international roadmap. Inter-consortium collaboration is also needed in Japan and on a global scale. Finally, future technology standardization is an effort to achieve global, joint guidance for 300-mm R&D.

Consortia Tradeoffs

He balanced this list by suggesting that participating in consortia does bring tradeoffs that can have negative effects on industry.

First, R&D by individual companies can be more efficient and faster—as long as it commands the best R&D people, good management, and sufficient funds. Recently, he said, these three conditions have become increasingly difficult to meet. In addition, the openness of R&D can be limited by patent policy.

Second, government-funded consortia can be bureaucratic. Once a target is determined, it is difficult to change direction and achieve progress quickly.

Third, consortia of private companies are suitable in commonly shared, basic technologies. It is difficult for industries in different sectors, such as devices and equipment, to collaborate on more specific applications. Too often industry managers are reluctant to send their best R&D people to work in a non-competitive area.

Academia-industry consortia often seem to be the best solution because of openness, availability of top R&D people, and clear benefit to industry; but the value of academic accomplishments is traditionally measured by numbers of papers, degrees, and awards, whether or not the work is relevant to industry's needs. Often, he said, the R&D tends to be too academic. He suggested the example of Dr. Kilby's Nobel Prize in physics as excellent academic work with practical results.[31] Dr. Masuhara further recommended a prize be named after Dr. Kilby for work in the technology area in order to inspire academic people to work in technology.

Five Criteria for Organizing a Successful R&D Consortium

He offered five criteria for organizing a successful R&D consortium.

1. *Business merit*: Is the technology applicable to industry? Can the market accept the new technology? Can the concept lead to new business?
2. *Technical merit*: What is the technical merit of each company? Can new patents or technologies be generated? Are there pitfalls in application, technology matching, suitability, or reliability?
3. *Participant merit*: Does the consortium provide good opportunities for participants and a good career path?
4. *Academic merit*: Can the consortium lead to research papers, master's and doctoral degrees, and faculty success?

[31]Jack St. Clair Kilby shares with Robert Noyce the credit for inventing the integrated circuit, or microchip, which made possible the development of the modern computer. Kilby's "monolithic idea" stated that reliability and miniaturization could be improved if all circuit elements—resistors, capacitors, distributed capacitors, and transistors—were made of the same material and placed together in a single chip.

5. *Industry manager merit*: Are managers willing to send the best R&D people from industry? Dr. Masuhara said that this is a "negative value" in many cases.

He suggested the use of "Masuhara's success equation" to measure whether a consortium would be successful, giving a weighting to each function, with the highest weighting for "A" and the lowest for "E." (Dr. Masuhara suggested that the value of "E" is "maybe zero" in most cases.)

Dr. Masuhara concluded by suggesting three points about successful government-industry cooperation. First, semiconductors will continue to be the key technology in the twenty-first century for information, communications, and consumer technologies. Second, the industry faces major challenges in the years immediately ahead if it is to avoid hitting the brick wall. Third, reaching solutions will require the industry's best efforts in both competitive R&D and inter-consortium collaboration.

A TAIWANESE PERSPECTIVE

Genda J. Hu
Taiwan Semiconductor Manufacturing Company

Dr. Hu began with the message that Taiwan has always been a player that believes in international cooperation. Even as a late-comer to the industry, in whose R&D activities it has not played a significant role, Taiwan has tried not only to improve its internal R&D capabilities but also to participate in international activities, especially in the International Technology Roadmap for Semiconductors. Taiwan has been an active participant in ITRS since its inception in 1998, and last December Taiwan hosted the annual conference for updating the year 2000 roadmap.

The Industry Consortium Called ASTRO

He also discussed in more detail the planned Taiwanese consortium called ASTRO, which has been placed on hold due to issues beyond the control of the industry. The attempt to form that organization, he said, is a clear demonstration that Taiwan intends to participate in R&D consortia. Part of the objective of ASTRO, he said, is to facilitate participation in international R&D activities when there is an opportunity. To him the important message is that virtually all the companies in Taiwan are willing to join international-level R&D activities.

Without ASTRO, the most feasible strategy for the time being is for individual companies to join consortia activities. He said that trying to participate internationally through the various Japanese research institutes would be difficult because they are funded primarily by the government. He suggested that international R&D activity is best handled through either individual companies or consortia of private industries.

Industry Must Play a More Active Role

The plan for ASTRO was to start out with some government funding, but less than 50 percent, so that decision making would be primarily in the hands of the industry. Individual companies would have to play a more active role to make the consortium a success.

In the semiconductor industry in Taiwan almost every company is engaged in some type of international collaboration. TSMC has a major working relationship with Philips and is an active member of International SEMATECH. TSMC participates in ITRS activities and has numerous programs with U.S. as well as Japanese equipment manufacturers. UMC is involved with SRC and IBM-Infineon, and with Hitachi in Japan. Winbond is working with Toshiba.

At the present time, Dr. Hu concluded, even though Taiwan does not yet have a true consortium similar to SEMATECH in the United States or Selete in Japan, there are still "very high levels of intention within our industry" to collaborate internationally. He said that the industry will watch for any opportunity to do so in the near future.

A EUROPEAN PERSPECTIVE

Erik Kamerbeek
Semiconductor Equipment and Materials European Association (SEMEA)

Dr. Kamerbeek began by explaining that the name of his association had recently changed. The name was once Semiconductor Materials Association, but in 1997 it was decided that a combined activity was needed for both the materials and equipment industries; the two were combined under a single title. It used existing organizations in Germany, France, and the United Kingdom. The Dutch companies joined to create an umbrella European association.

Cooperation as a Way of Life in Europe

In discussing support measures for semiconductor technology, he said that internationalization is a common concept for any European. He said that if he drives for more than one hour from his home, he is in Germany, Belgium, or the sea. In other words, anything a company does is likely to involve a form of international cooperation. European countries are relatively small, so that if a company wants to develop new equipment, new materials, or new processes in this industry, it will probably find its partners in other countries.

He said that European cooperative programs can be divided into two categories: national projects, which are very country-specific and too numerous to address, and international projects.

International Projects

He mentioned first the IST (Information Society Technologies) Programme of the European Union, which has the longest history.[32] The European Union has developed the IST into a single integrated research program that builds on the convergence of information processing, communications, and media technologies. The other major international effort is EUREKA—a decentralized effort that is divided into a variety of sub-categories or subjects.[33] The largest EUREKA project is MEDEA (Micro-Electronics Development for European Applications).

In all European collaborative projects the rules call for international cooperation, which is defined as a minimum of two partners coming from at least two countries. Usually the areas where this work will take place are described and the nature of the research must be advanced or pre-competitive R&D. The work is either centralized, as it is for IST and other projects under the European Framework 5 Program, or partly decentralized, which is the case for EUREKA projects.

The IST, a Centralized Effort

The European Union IST program is planned and organized by the commission with the support of industry. The representatives of the 15 countries participating in the European Union approve the programs. The commission issues regular calls for projects which are evaluated by commission experts who are employees of the European Union in Brussels, independent consultants, or industry representatives. The program is centralized, and the commission issues contracts from its own funds. Commission employees or hired independent experts and consultants review R&D progress. The typical funding level for each European Union IST program and project is 50 percent of eligible costs.

The interests of the equipment and materials industry in this program are

---

[32]The objective of the the Information Society Technologies (IST) Programme is, according to its Web site, "to realise the benefits of the information society for Europe both by accelerating its emergence and by ensuring that the needs of individuals and enterprises are met." It is managed by the European Commission, has expanded steadily for several frameworks, and now has a budget of 3.6 billion euros.

[33]EUREKA, founded in 1985, is designed to develop and exploit technologies for industry. Now representing 29 countries, it helps industry and research institutes find partners and funding. Research partners receive financing from their national governments, which support 100 to 200 new EUREKA projects per year. Projects range widely in size from the Joint European Submicron Silicon Initiative project (JESSI), which has 100+ partners and a budget of 3.8 billion Euro, to the two-partner feasibility projects involving less than 1 million Euro.

spelled out in subchapters of the IST program. These include microelectronics R&D, microelectronics semiconductor equipment assessment, and microelectronics SEA300, which is the 300-mm support program. This subchapters program was created because it was clear at the outset of 200-mm development that the costs related to equipment testing, wafer supplies, and other components would be extremely high for the next technology and very difficult for the smaller companies to support.

EUREKA, a Less Centralized Effort

The less centralized EUREKA supports numerous projects on very different topics, from transportation to agriculture; MEDEA is devoted to information technology. Each EUREKA project must be accepted by the ministers' conference of the EUREKA countries, and each country has its own EUREKA secretariat. An important point is that all projects are initiated and guided by industry.

EUREKA funding differs from European Union funding in that each country decides to participate in a project. Industrial partners in different countries participating in a particular project will enter into contracts with their own national government. This means that partners from different countries must each have a contract with their national government. National governments have different rules for issuing contracts, so the countries are still learning how to accommodate their rules to the various EUREKA projects. The learning process has helped the authorities learn to cooperate and coordinate contracts in order to start projects smoothly. One difficulty, and a difference from the European Union initiatives, is that funding for EUREKA cannot cross borders. This is because national governments want their funding to be used within the country of origin, whether by institutes or companies.

He reviewed the MEDEA program, which ran from 1996 through 2000 with a budget of nearly 500 million euros. It worked as an umbrella program for projects in semiconductor R&D and was run by a MEDEA board consisting of top industry executives. The overall program was controlled by industry.

Other EUREKA programs in information technology are ITEA and PIDEA. ITEA, or Information Technology for European Advancement, supports the development of software technology for European industry and has a planned budget of some 3.2 billion euros for 1999-2006. PIDEA, or Packaging and Interconnection Development for European Applications, supports R&D on high-density interconnection and packaging technologies. PIDEA is a EUREKA cluster program with a total budget of 400 million euros for the period 1998-2003.

A Need to Enhance Core Competences

Dr. Kamerbeek said that the European equipment and materials industry has been interested in MEDEA because it needs help in seven core competences.

1. The 300-mm development platform;
2. Lithography;
3. Innovative manufacturing processes;
4. Next-generation gas and chemicals;
5. Metrology development;
6. Product testing; and
7. Environmental issues.

The Motivation to Create Partnerships

Europe has a greater need for this kind of support than does a single large country like the United States. Europe is a patchwork of small- and medium-size economies, and each of them is more or less sub-critical in IC activities. However, Europe is not an easy place to form partnerships because of different languages, currencies, and cultures. Countries therefore need incentives to create partnerships, and MEDEA, serving as an information clearinghouse, provides those incentives by helping to arrange funding and partners. Most of the equipment and materials companies participating in MEDEA are small- and medium-size enterprises that can benefit from this kind of support.

As the follow-up program, MEDEA Plus begins in 2001, and the industry is once again supportive. The important research feature for MEDEA Plus, concluded Dr. Kamerbeek, is the enabling technologies charter. Most work in MEDEA Plus for the industry will focus on sub-100-nm research and development to help its participating companies toward the next generation of information technology.

Closing Remarks

Dr. Graham concluded the panel session and the symposium by thanking all participants, especially those who had traveled long distances to attend. He thanked the organizers again for a rich and educational program featuring excellent material and the participation of the leaders of the international information technology community.

IV

RESEARCH PAPERS

Competing Programs: Government Support for Microelectronics

*Thomas R. Howell**
Partner, Dewey Ballantine LLP, Washington, D.C.

EXECUTIVE SUMMARY

Government promotional policies have played an important role in the development of the semiconductor industry in every country in which such an industry has emerged.† In the United States, the proper relationship between government and industry has long been controversial, and the measures taken by the U.S. government in microelectronics have proven no exception. However, the collaborative government-industry-university programs that have been implemented in the U.S. semiconductor sector since the 1980s—of which the SEMATECH consortium is emblematic—are widely admired abroad and are being emulated on a large scale in Japan and the European Union.

---

*Thomas R. Howell is a partner with Dewey Ballatine LLP in Washington, D.C. During the course of his law practice in the field of international trade, Mr. Howell has devoted 20 years to the study of foreign industrial trade policies and is the author of numerous studies, sectoral analyses, and books examining policy measures and business practices in countries such as Japan, Korea, Taiwan, and Russia, as well as the European Union and Eastern Europe. He has served as a counsel to U.S. semiconductor producers at intervals since 1982. Mr. Howell is a graduate of Harvard College and Boston University Law School. The views expressed in this paper are his own and should not be construed as representing the views or positions of any semiconductor producer or group of producers. The author acknowledges the contributions made to the research and field interviews underlying this article by his colleagues Maki Hishikawa, Brent L. Bartlett, Hui Yu, Melissa E. Murphy, Gregory I. Hume, and Lawrence E. Rothenberg.

†For an update of the program developments in this industry, particularly the emergence of China as a semiconductor manufacturing site, see the Postscript.

At a moment when U.S. government support for the semiconductor industry is significantly reduced, the level and scope of government involvement in the industry outside the United States is increasing substantially, with various elements of the U.S. economic system and industry-government relationship frequently cited as models for foreign promotional programs. The divergence between U.S. and foreign government policies toward the semiconductor industry occurs at a time when the U.S. industry enjoys a position of undisputed world leadership but also as it confronts unprecedented technological challenges with "no known solutions" and no clear plan for mustering the resources needed to surmount those challenges.

Foreign government measures to support the semiconductor industry are larger in scale and broader in scope than anything currently under way in the United States. In Japan and the European Union, as in the United States, the principal form of government support for microelectronics is the provision of funding and infrastructure for industry-government research and development projects. However, the Japanese and European programs are funded at a much higher level and place a greater emphasis on R&D with immediate commercial applications, in contrast to U.S. programs, which are normally limited to precompetitive R&D. In addition, in the European Union substantial funding is being provided by national and regional governments to individual companies for investment in semiconductor manufacturing facilities, and in both the European Union and Japan, direct government funding is being used to stimulate new "venture" businesses in the microelectronics field.

In the growing number of newly industrializing countries promoting an indigenous capability in microelectronics—Taiwan, Korea, Singapore, China, and Malaysia—government policies emphasize the acquisition and diffusion of advanced semiconductor technology from the industrialized countries rather than pursuit of leading-edge R&D. The principal forms of government support in these countries are direct provision of capital to domestic firms (including funding of small- and medium-size venture companies); funding of research institutes that assist in technology diffusion; technology acquisition and transfer to industry; tax holidays; programs to train personnel and attract scientists and engineers from other countries; establishment of industrial zones with incentives for firms locating in these zones; and outright creation of new semiconductor enterprises.

While it is difficult to predict the precise effects that the various government programs in this sector will have over the long run, government measures are clearly contributing to several of the most significant observable trends in this industry.

- Japan is pursuing a national revival of its competitive position in microelectronics, driven by an array of large new government-sponsored R&D projects.
- The European Union has reversed its declining position in the semiconductor sector and improved its relative competitive standing, a development sub-

stantially attributed to European Union- and national government-supported R&D projects, most notably JESSI and MEDEA.

- Both Japan and the European Union are now pursuing comprehensive strategies designed to challenge U.S. leadership in microelectronics by leveraging their present and anticipated advantages in mobile communications and digital home appliances.
- Taiwan has emerged as a major production base for semiconductors and—reflecting a sustained government promotional effort and with initial capitalization from the government—has pioneered a business model, the dedicated foundry, that many believe will revolutionize the industry.
- China is emerging as a potentially significant competitor, reflecting its government's efforts to attract inward foreign investment and foreign technology and measures to promote indigenous producers. The dramatic ongoing movement of Taiwanese information technology manufacturing functions to Mainland China—encouraged for differing reasons by governments on both sides of the Fujian Straits—will almost certainly accelerate the maturation of the Chinese semiconductor industry.

U.S. economic thinking places a high value on independent entrepreneurship and emphasizes the need to circumscribe carefully the government's role in the market. While U.S. federal, state, and local governments have frequently worked in close collaboration with the private sector in a diverse array of industries—agriculture, aerospace, biotechnology, and others—government intervention in specific sectors or on behalf of individual enterprises is controversial and therefore usually limited.

The United States, however, has generally not proven successful in dissuading other governments from intervening heavily on behalf of strategic sectors like semiconductors, where governments often have the explicit objective of promoting the commercial success of individual indigenous firms. Reflecting that fact, since the early 1980s the U.S. government has been drawn into a series of limited market interventions to counter the adverse effects of foreign government measures on the U.S. semiconductor industry, most notably the Semiconductor Trade Agreement and federal funding of the SEMATECH consortium. These and other, similar measures were improvisations devised by the U.S. government and industry working together in response to challenges that arose out of foreign industrial policies. These measures, while sometimes the subject of criticism, did not represent a fundamental departure from U.S. economic values and succeeded in addressing the problems that made the measures necessary in the first place.

Given the challenges confronting the U.S. industry in the coming decade, which reflect both difficult technological obstacles and dramatic increases in foreign government support measures, it is likely that additional industry-government improvisations—and continued cooperation—will be required if the U.S. industry is to sustain its current position of leadership.

INTRODUCTION

This paper surveys government policy measures that are influencing the international competitive environment in the semiconductor industry. The role that governments should play in this industry has always been controversial, and the impact that current programs will ultimately have is difficult to predict with much precision. Looking back it is clear that government policies have played a major role in the evolution of the global semiconductor industry. Indeed, to date no country in the world, including the United States, has established and sustained a world-class semiconductor industry in the absence of a very substantial government promotional effort.[1]

Many of the most significant developments of the past 15 years in this industry are substantially attributable to government policy actions; these include the rebound of the U.S. semiconductor industry from the crisis of the mid-1980s; the erosion of Japanese leadership by new competitors arising in Taiwan and Korea; the resurrection of the European position in microelectronics; and the advent of China as a potentially significant competitor. Government programs to support the semiconductor industry have shaped the competitive environment and will continue to do so for the foreseeable future.

The most striking aspect of the current pattern of government support for microelectronics worldwide is the level at which such support is declining in the United States while increasing substantially in the key semiconductor-producing regions outside the United States. Ironically, at the moment the United States appears to be curtailing direct government support for this industry, promotional programs abroad are attempting to replicate what is seen as a highly successful example of government-industry collaboration in the United States. Some argue that U.S. industry leads the world today precisely because market-based competition has shaped its evolution to a far greater degree than has been the case anywhere else, and that the steep decline in U.S. government financial support and other forms of involvement is therefore a positive trend.

But, U.S. leadership in microelectronics reflects the success of a system of innovation based not only on the activities of private firms but on government institutions, universities, and research associations and consortia in which the government plays a role. Significantly, during the next 10 years this system will confront fundamental technological challenges that simply have no precedent and

---

[1] "The semiconductor industry has never been free of the visible hand of government intervention . . . [T]he semiconductor industry, *wherever it has developed*, has been an explicit target of industrial policy—whether in the guise of military policy in the United States or in the guise of commercial policy elsewhere in the world." (Original emphasis) Laura D'Andrea Tyson, *Who's Bashing Whom? Trade Conflict on High Technology Industries*, Washington, D.C.: Institute for International Economics, 1992, p. 85.

will require a level of human, financial, and infrastructural resources that the industry, standing by itself, will be hard pressed to achieve.

At present the U.S. semiconductor industry stands at the pinnacle of success. It holds by far the largest share of the world market and is the undisputed world leader in many key areas of semiconductor technology.[2] It has fought off a major competitive challenge from Japan and has seen its Japanese rivals' market share progressively decline for a decade. Its productivity has consistently grown at a rate that far outstrips most other sectors of the U.S. economy.[3] Its methods and business culture are widely studied in academia and increasingly emulated abroad. It continues to attract extraordinarily talented people from every part of the world.[4] Its revolutionary contributions to the economy and to the society as a whole are universally acknowledged and acclaimed.

Challenges Facing the U.S. Semiconductor Industry

The present sanguine state of the U.S. semiconductor industry masks trends that could jeopardize the U.S. position in the coming decade. These challenges are both structural and technological in nature.

Structural and Technical Challenges

Despite continued high rates of growth in long-term demand the industry remains sharply cyclical—a dynamic that results in erratic levels of funding for R&D. The industry faces a growing shortage of trained scientists and engineers, reflecting such trends as a decline in U.S. graduate electronic engineering degrees and an increase in the number of foreign students who return home after graduat-

---

[2]In 1998 producers based in North America accounted for 53 percent of all semiconductor revenues earned worldwide. Japan-based producers held a 26 percent share and all other producers held 21 percent. Technecon Analytic Research, Inc., *1999 Annual Databook: Review of Global and U.S. Semiconductor Competitive Trends 1978-1998.* (Semiconductor Industry Association, 1999).

[3]The average productivity of the U.S. merchant semiconductor industry, measured in terms of sales revenues per employee, more than doubled between 1990 and 1998, from $93,000/employee to $245,000. Semiconductor Industry Association.

[4]A 1997 Israeli perspective on Silicon Valley made the following observations: "Silicon Valley swarms not only with Israeli engineers, but also with no few Indian and Oriental (mainly Chinese) engineers and designers: Altogether, the Valley seems to attract émigrés, with engineers and programmers and entrepreneurs from all parts of the world populating its every nook and cranny. Migrants have played an important role in the Valley's florescence. Intel President Andrew Grove, for example, emigrated from Hungary, and his contribution to the sector is inscribed in the Hall of Fame and on every computer. Other less well-known emigrants took part in the set up of high-tech corporations such as Sun Microsystems, Cirrus Logic, and many others." U.S. Foreign Broadcast Information Service ("FBIS"), May 9, 1997, reprint of Nir Nathan, "We've Taken Silicon Valley," *Tel Aviv Globes*, May 6, 1997 (FTS19970509000829).

ing from U.S. universities. The capital investments required for semiconductor manufacturing have become so large that a significant and growing proportion of U.S. production is outsourced to "foundries," by far the most advanced of which are operated by foreign companies based in East Asia.

The U.S. industry also confronts a formidable array of technological hurdles as it pushes miniaturization to the molecular and atomic level—in the next 5 years it will encounter problems "for which there are no known solutions."[5] These physical limits may herald the end of further miniaturization using known silicon-based technology and require a radical leap to some other form of technology to achieve further advances; but research and development on the scale necessary to develop such a replacement technology is not taking place.

Competition for New Markets: PCs to Wireless

Finally, the U.S. industry faces competition from abroad, a factor that has receded as a perceived challenge but which now should be receiving heightened attention. The U.S. industry's present dominant position is based in significant part on its success in developing products for PCs and PC equipment. European and Japanese industry leaders believe that they can dominate what they see as the main semiconductor growth markets of twenty-first century—wireless and wired telecommunications and digital home appliances.[6] The Japanese semiconductor industry, with an unprecedented level of government backing, is embarking on an intensive series of leading-edge R&D projects with the objective of recapturing world leadership in microelectronics from the United States, based on improved capability with respect to systems-on-a-chip. A Japanese semiconductor analyst recently commented on this effort as follows:

> The 21st century will be the end of the PC era, and the arena of competition will change. Instead of the PC, cell phones and digital consumer equipment, two areas in which Japan is dominant, will pull semiconductor technology forward. A favorable wind is blowing for Japan, which was defeated in the PC era.[7]

Technological Parity?

Similarly, the European industry is building on European strengths in key, rapidly growing end-use markets—most notably telecommunications—to carve

---

[5]Semiconductor Industry Association, *International Technology Roadmap for Semiconductors*, 1999, p. 1.

[6]FBIS, January 2, 2001, translation of "From Stagnation to Growth, the Push to Strengthen Design," *Nikkei Microdevices*, pp. 106-124 (JPP20010125000012) and FBIS, January 2, 2001, translation of "Continuing with Large Investment Similar to 2000," *Nikkei Microdevices*, pp. 88-93 (JPP20010131000003).

[7]*Ibid.*

out a significant market segment based on the design and production of specialized devices for these markets, which are growing rapidly as a proportion of the total end-market for semiconductors. The principal European and Japanese R&D programs are major, sustained, long-term efforts of five or more years' duration that are often succeeded by follow-on efforts of comparable or greater length.[8]

Industries in Taiwan, Korea, and Singapore are seeking to achieve and sustain technological near parity with U.S. producers by acquiring technology through foundry relationships and other collaborative arrangements. These arrangements involve the cession by U.S. firms of key segments of the manufacturing process that they once performed entirely by themselves. China, while not a major factor today, is aggressively pursuing acquisition of foreign technology and know-how, and in the view of many will emerge as a major U.S. competitor within 10 years. Increasingly, all of these countries are competing with the United States for the same limited pool of human resources and, to the extent that this rivalry remains a zero-sum exercise, it will inevitably intensify.

These developments abroad are not necessarily negatives for the U.S. industry, and in some cases they will actually enhance competitive opportunities for U.S. firms. The emergence of government-supported semiconductor foundries in Taiwan and Singapore, for example, may well work to the competitive advantage of U.S. producers, and U.S. firms have been beneficiaries of foreign government subsidy programs.[9] But, in assessing the prospect that the U.S. semiconductor industry will surmount the challenges it will confront in the new century, the government-driven competitive strategies that are being implemented abroad and their potential effects on the U.S. industry must be taken into account. The intensity and scale of these national efforts are changing rapidly, generally in the direction of a larger government commitment.

The first segment of this report surveys U.S. government programs in microelectronics. The remainder examines programs outside the United States, reviewing national strategies in Asia and Europe.

---

[8]The European Union's flagship R&D project, JESSI, began in 1988, ran through 1996, and was succeeded by MEDEA, a four-year project ending in 2000; the latter was succeeded by MEDEAPlus, which will run through 2009. Japan's new Asuka sub-0.10-micron project will run for 7 years beginning in 2001.

[9]In the mid-1990s the U.S. firm Advanced Micro Devices established a manufacturing and development center for microprocessors in Saxony, a state in the former East Germany. The German federal government and the state government of Saxony reportedly provided this project with DM 500 million in investment subsidies ("standard assistance for the East"), DM 200 million in tax breaks, and DM 100 million to set up a research facility. In addition, a loan guarantee of DM 1 billion was provided, 40 percent borne by the state government and 60 percent by the federal government. See FBIS, February 26, 1996, translation of Casper Busse, "Billions in Support Smooth Tough Road," *Handelsblatt*, January 2, 1996 (FTS19960226000662) and FBIS, January 12, 1996, translation of "Dresden Becomes Capital of Microelectronics – Biederkopf's Dreams," *Markt und Technik*, January 12, 1996, pp. 1-3 (FTS19960112000528).

GOVERNMENT AND INDUSTRY IN THE UNITED STATES

The success of the U.S. semiconductor industry—with its now legendary tradition of dynamic and colorful entrepreneurial initiative, ferocious competition, and dramatic technological breakthroughs—is sometimes held up as the very embodiment of the virtues of the market-driven U.S. economic system. The reality, however, is more nuanced. U.S. government policies have played an important role in the evolution and survival of the semiconductor industry, and the industry's current success is at least partially attributable to a close cooperative working relationship that has grown up between the industry, universities, and the U.S. government.

As a result foreign countries seeking to create their own indigenous versions of the U.S. semiconductor industry are trying to replicate not only the best features of Silicon Valley and the U.S. venture capital system but also the U.S. industry-university-government research triad. The chairman of the Nippon Electric Company, one of Japan's largest semiconductor makers, has commented on how the U.S. industry recaptured world leadership from Japan after the mid-1980s—in a manner that he now urges Japan to emulate:

> In the U.S. there have been various [government] measures that recognize the importance of the semiconductor industry, which is the basis of defense and all industries; a semiconductor revival resulting from those measures; the activation of private-sector investments, along with that; and ideal cooperation between industry and universities that supports the semiconductor industry . . . [T]he U.S. activated cooperative semiconductor-related efforts such as SEMATECH that involved industry, government and universities from the viewpoint of the importance of semiconductors as a key technology for all industries. In order to win in the 21st century, we must, at all cost, rework the strategy we have lost.[10]

The Legacy of Government Support

The U.S. semiconductor industry was given its initial impetus from U.S. government funding of research and development and procurement for military and space exploration programs.[11] While the creation of a thriving commercial semiconductor industry was a byproduct rather than an objective of these early government programs, the U.S. defense community grew to recognize the in-

---

[10]See FBIS, October 31, 1995, translation of Emi Yokata, "Japanese Manufacturers to Launch Joint Research on New Chip Technology to Counterattack U.S. Manufacturers," *Ekonomisuto*, October 31, 1995, p. 42 (FTS19951031000245).

[11]The initial impetus for the growth of the semiconductor industry in the United States came from the Apollo and Minuteman programs, which featured government procurement of integrated circuits in large volume at high prices. Government procurement enabled U.S. firms to improve yield and efficiency through volume production and encouraged wider application of integrated circuit technology, first in military and then in commercial technologies. National Bureau of Standards, *The Influence of Defense Procurement and Sponsorship of Research and Development on the Development of the Civilian Electronics Industry*, June 30, 1977.

creasingly central importance of the commercial industry to national defense, and in subsequent decades took major steps to preserve and enhance the competitiveness of that industry.

The most significant initiative was the Defense Department's *sponsorship of SEMATECH*, a research and development consortium established to ensure U.S. leadership in semiconductor-manufacturing technology. This initiative and many others in the microelectronics field benefited from support from the Defense Advanced Research Projects Agency, a small Defense Department agency that supports long-term R&D with commercial as well as military value.[12]

During the 1980s the U.S. Congress enacted legislation *extending copyright protection* to semiconductor designs, relaxing antitrust rules for joint research, and providing tax credits for research and development, all of which substantially benefited the U.S. industry. In 1986 the U.S. government negotiated a bilateral agreement with Japan, the U.S.-Japan Semiconductor Arrangement, with the objective of ending Japanese dumping and improving market access for U.S. semiconductor producers in Japan.[13]

In addition to these individual government actions a system of ongoing *collaboration between the industry, the government, and U.S. universities* has evolved with respect to long-term research and development. In 1982 the Semiconductor Research Corporation was formed by U.S. semiconductor device firms to undertake silicon-based R&D in U.S. universities.[14] While U.S. companies contributed most of the Semiconductor Research Corporation's funding, four U.S. government organizations participate in and contribute to its funding, and a number of its key leaders have backgrounds in Department of Defense microelectronics R&D programs.[15]

---

[12]Created in 1958, DARPA has an impressive record of technological achievement. Its budget in FY 2000 was about $2 billion.

[13]For a review of the problem of dumping in high-technology markets and specifically the Semiconductor Trade Agreement, see National Research Council, *Conflict and Cooperation in National Competition for High-Technology Industry*. Washington, D.C.: National Academy Press, 1996. The trade agreement is summarized (p. 111) and a discussion of Japanese-American competition in semiconductors and the agreement is included in Supplement A, pp. 131-141. For an historical view of the problem of dumping see "Dumping: Still a Problem in International Trade" in National Research Council, *International Friction and Cooperation in High-Technology Development and Trade*. Washington, D.C.: National Academy Press, 1997, pp. 325-377.

[14]As of early 2000 the Semiconductor Research Corporation (SRC) had channeled approximately $520 million into semiconductor research at over 60 universities. SRC's annual revenues in 1999 were $36.7 million from all sources. Custom research funded by individual customers accounted for 10 to 14 percent of SRC's R&D. The remainder is core R&D funding from the general pool of member funds.

[15]The four government participants are the Defense Advanced Research Projects Agency, the National Science Foundation, the National Institute of Standards and Technology, and the U.S. Army Research Office. Larry Summey, who has headed the Semiconductor Research Corporation since 1982 was previously the head of the Department of Defense's Very High Speed Integrated Circuit Program, the largest U.S. defense-related microelectronics program of the 1970s.

Beginning in 1992 the Semiconductor Industry Association, representing U.S. device manufacturers, brought together representatives of the private sector, the government, and U.S. universities to develop a National Technology Roadmap for Semiconductors, a description of and timetable for achieving technological targets necessary to ensure continued advances in the performance of integrated circuits.[16] SEMATECH, while no longer federally funded, continues as an active participant in joint industry-government activities such as the ongoing development of the roadmap and R&D partnership with the National Laboratories.

Support: Past, Present, and Future

This multifaceted collaborative industry-government-university R&D effort is widely credited abroad with playing a major role in reversing the relative competitive decline of the U.S. industry in the late 1980s and its return to world leadership in the 1990s and into the present.[17]

Yet the U.S. government's involvement in microelectronics has remained a subject of controversy. The end of the Soviet military threat raised questions about the continuing need for defense spending in this field, at least on the same scale as during the Cold War.[18] The Defense Advanced Research Projects Agency's annual funding of microelectronics R&D—the principal channel of direct federal financial support—is expected to decline.[19] In addition, critics, including some executives in the U.S. semiconductor industry, have blasted federal

[16]See Semiconductor Industry Association Web site at <http://www.semichips.org>.

[17]"A major factor contributing to the U.S. semiconductor industry's recovery from this perilous situation [in the 1980s] was a U.S. national policy based around cooperation between industry, government, and academia." See Hajime Susaki, chairman of NEC Corporation, in FBIS, January 12, 2001, translation of "Japanese Semiconductor Industry's Competitiveness: LSI Industry in Jeopardy," *Nikkei Microdevices*, December 2000, pp. 245-48 (JPP20010112000011). The U.S. industry undertook a comprehensive effort to "incorporate scientific methods into semiconductor production management technology, on-site maintenance, and the like . . . . I would like to point out that the earnest national effort made by the U.S., in particular, was a very important motive force behind the American revitalization." See FBIS, August 1, 2001, translation of Michio Mizogami, "Prescription for Japan's Revival," *Nikkei Microdevices*, August 2000, pp. 207-09 (JPP20000817000061). National Research Council, *Conflict and Cooperation* reviews the SEMATECH program (p. 141) and provides an early review of the then emerging Japanese and European programs (pp. 19-24). For a review from the perspective of the U.S. industry, see Andrew A. Procassini, *Competitors in Alliance: Industrial Associations, Global Rivalries, and Business-Government Relations*. Westport, Conn.: Quorum Books, 1995.

[18]Conrad W. Holton, "Federal Funds for Chip Research Dwindle Under Congressional Pressure," *Solid State Technology* (July 1996), p. 86.

[19]Scott Nance, "Broad Federal Research Required to Keep Semiconductors on Track," *New Technology Week*, October 30, 2000. Sonny Maynard, Semiconductor Research Corporation, cited in presentation by Dr. Michael Polcari, "Current Challenges; A U.S. and Global Perspective," National Research Council, *Symposium on National Programs to Support the Semiconductor Industry*, October 2000.

programs in microelectronics as corporate welfare.[20] Thus, even though federal funding for SEMATECH ended in May 1997, debate has continued within the Congress and the Executive branch as to whether and to what extent the U.S. government should continue to invest federal funds in supporting R&D in microelectronics.[21] Many observers argue that if anything, the role of the government should be further curtailed. This trend, if it continues, will run directly counter to those in Europe and East Asia, where governments are dramatically escalating their levels of direct and indirect funding in this sector.

Challenges Facing the U.S. Semiconductor Industry

Confronting the "Brick Wall"

For three decades advances in semiconductor technology have followed Moore's Law, which predicts that transistor performance and density will double at a predictable and relatively constant rate (Moore originally postulated a doubling every three years, but the rate of doubling has been adjusted over time). This dynamic, which has resulted in a relentless driving down of the cost of electronic functions and concurrently an exponential increase in performance, underlies the information revolution that is credited by many analysts with the U.S. economic resurgence of the 1990s. As the economist Robert Gordon has noted,

> [The] Clinton economic boom is largely a reflection of Moore's Law. [T]he recent acceleration in productivity is at least half due to the improvements in computer technology.[22]

For many years it has been recognized that there are limits to the miniaturization of semiconductor components; at some point optical lithography—the predominant mode of semiconductor manufacturing—will no longer be workable. Although scientists have repeatedly succeeded in pushing optical lithography beyond what were once seen as its absolute limits, a more fundamental array of technological barriers may portend the end of Moore's Law. In September 1999

---

[20]T.J. Rodgers, "A CEO Against Corporate Welfare," Testimony before the House Subcommittee on Government Management, Information, and Technology of the Government Reform and Oversight Committee, September 6, 1995.

[21]In 1990 DARPA Director Craig Fields was dismissed by the Bush Administration over initiatives seen as aimed more at enhancing U.S. commercial competitiveness than at improving U.S. military capability. One of these programs was SEMATECH. "DARPA Looks Ahead," *Computerworld* July 28, 1997, p. 76.

[22]Moore's Law—"more like an engineer's rule of thumb" than a basic law of nature—was articulated in 1965 by Gordon Moore, then the research director of Fairchild Semiconductor, who was asked by *Electronics* magazine to predict the future of the semiconductor industry. His estimate that the doubling of performance would occur at a predictable rate has proven remarkably accurate. Charles C. Mann, "The End of Moore's Law? Technology Information," *Technology Review*, May 2000, p. 42.

Paul A. Packan, a highly regarded Intel researcher, warned that miniaturization of electronic components had been taken to such extremes that arcane physical effects at the molecular and atomic level now loom as a major obstacle to further miniaturization.[23] Moore's Law, he said,

> seems to be in serious danger. Fundamental thermodynamic limits are being reached in critical areas, and unless new, innovative solutions are found, the current rate of improvement cannot be maintained.[24]

"No Known Solutions"

Packan pointed out that "solutions for these problems have not been found," and the 1999 International Technology Roadmap for Semiconductors concurred, indicating that industry will face technical challenges in the next five years "for which there are no known solutions."[25] These multiple technological obstacles have come to be known collectively as "the brick wall." U.S. companies have developed techniques that may enable them to squeeze one or two more generations of miniaturization from refinements using conventional methods, but this will merely defer the confrontation with the brick wall.[26]

---

[23]Packan identified some of the challenging technological problems facing the industry. The gates that regulate the flow of electrons within semiconductor devices have become so small—now less than 2 nanometers in thickness—that electrons can tunnel through them even when they are shut. Dopants, which are impurities mixed with silicon to increase its ability to hold localized charges, must be present in progressively higher concentrations as device size shrinks in order to enable them to hold the same charges; but at a certain level of concentration the dopant atoms begin to interact with each other, forming clusters of dopant atoms that do not serve the function of holding a charge. In addition, transistor dimensions have become so small that small changes in the exact number and precise distribution of individual dopant atoms "can cause appreciable changes in the device behavior." Paul A. Packan, "Pushing the Limits: Integrated Circuits Run into Limits Due to Transistors," *Science*, September 24, 1999, p. 2079. See also "1999 Roadmap: Solutions and Caveats," *Solid State Technology* May 2000, p. 18; Ralph K. Cavin III, Daniel J.C. Herr, and Victor Zhirnov, *Semiconductor Research Needs in the Physical Sciences*. Semiconductor Research Corporation working paper, undated; Pieter Burggraaf, "A Closer Look at Some of the Most Difficult Processing Challenges," *Solid State Technology* September 2000, p. 76.

[24]Packan, *op. cit.*

[25]Semiconductor Industry Association, *International Technology Roadmap*, p. 1. Paolo Gargihi, Intel's director of technology strategy and the chairman of the committee that prepares the roadmap, said at the beginning of 2000, "For the next 5 years, we see no limitations, but beyond that there are formidable obstacles to the industry's classically silicon, silicon dioxide, polysilicon scaling approach to continuously improve performance." "2000 Begins with a Revised Industry Roadmap," *Solid State Technology*, January 2000, p. 31.

[26]Intel scientists recently announced a series of technological successes that suggest that no major departure from today's CMOS process techniques and materials will be needed to create transistors with 70-nm design rules. "Lab Looks to Wring Two More Generations from Today's Techniques, Materials—Intel, CMOS up to 70-nm Task," *Electronic Engineering Times*, December 11, 2000 (Nexis Reprint).

Some U.S. scientists are experimenting with non-conventional solutions to some of the problems identified by Packan and the roadmap, such as the use of non-silicon and biological materials, but the scale of these efforts is grossly inadequate for the task.[27] As Gordon Moore has expressed it, the U.S. industry is currently living off the benefits of investments in basic microelectronics R&D that were made in prior decades.[28]

Shortfalls in Research

Because the nature of the technological solutions to the brick wall is not known, the financial, human and infrastructural resources needed to surmount it cannot be quantified with precision, although it is generally agreed they are substantial. A 1995 working group, established under Semiconductor Research Corporation auspices, examined this issue based on circumstances prevailing at the time. It identified an annual shortfall of $492 million in basic microelectronics R&D. It concluded that while $153 million of this shortfall consisted of topics suitable for university R&D, U.S. universities had the capacity to perform only $48 million of this total.[29]

In a 1999 "back of the envelope" analysis the Defense Advanced Research Projects Agency concluded that the basic research funding gap had widened to an estimated annual shortfall of $1.2 billion. The growth in the research gap between 1994 and 1999 reflected not only an increase in research funding needs from the $750 million range to about $1.4 billion but also a drop in the level of funded basic research from $270 million in 1994 to $155 million in 1999.[30]

Shortfalls in Human Resources

The looming shortfall in human resources is probably even more serious than the research-spending gap. U.S. universities are graduating progressively fewer students with degrees in electrical and chemical engineering and in the physical sciences, notwithstanding the immediate demands of the private sector and the need to find solutions to the technological brick wall that will require a substantial increase in the number of trained people committed to R&D. To compound

---

[27]Scientists at Bell Labs have developed strands of synthetic DNA that will self-assemble into a motor analogous to a CMOS transistor and may provide the basis for creating electronic circuits integrating billions of devices. "DNA Motor Drives Nanotechnology Toward the Post-CMOS Era," *Electronic Design,* September, 18, 2000, p. 40.

[28]Nance, *op. cit.*

[29]George Bodway, Sunlin Chou, Mark Melliar-Smith, and Peter Verhofstadt, *SRC White Paper on Research Investment Gap Analysis, mimeo*, 1995.

[30]Defense Advanced Research Projects Agency, *Focus Center Research Program*. Undated view graph presentation.

the problem the proportion of foreign students in these fields at U.S. universities has increased dramatically, and a growing percentage of these graduates are returning home rather than remaining in the United States. Japan now produces 75 percent more engineers than the United States, and China produces over twice as many.[31]

Existing U.S. Industry and Federal Efforts

The magnitude of the technological challenge and the looming resource shortfalls confronting the United States in microelectronics should not obscure the fact that the private sector and the U.S. government continue to devote very substantial resources to microelectronics R&D, and that a number of successful or promising industry-government R&D efforts are under way. These programs and perhaps others like them could provide a foundation for a national effort to surmount the brick wall.

Industry R&D

The U.S. semiconductor industry devotes a comparatively large portion of its revenues to research, spending about $9 billion annually on research and development. The industry's annual R&D expenditures account for 10 to 15 percent of sales revenues—a rate of spending that is substantially higher than that of such technology-intensive sectors as aerospace, computers, telecommunications, and precision instruments.[32]

The vast preponderance of total private-sector R&D expenditure is directed toward the development of commercial products or specific objectives of immediate concern to the companies involved.[33] A few of the biggest companies (IBM,

---

[31]Mary L. Good, president of the American Association for the Advancement of Science, cited in "S.F. Conference Opens With Plea for Cabinet Position," *San Francisco Chronicle*, February 16, 2001, A3. A recent Japanese analysis of China's long-range potential in microelectronics particularly stressed the large number and high quality of electrical engineering graduates emerging from Chinese universities. Chinese "[u]niversities are narrowing their focus to the LSI/LCD industry and are pouring effort into cultivating and turning out talented engineers . . . . As for the problem of human resources, which are in short supply globally, China is supplied with an abundance of students from its huge population. The engineers within China steadily absorb the technologies that are adopted from overseas . . . . The image of China expanding into a great power in the field of LSIs and LCDs, right after Korea and Taiwan, has come into view." See FBIS, March 7, 2001, translation of "Special Projects – Part One – Industry, Government and Universities United in Enthusiasm and Talent for LSIs and LCDs," *Nikkei Microdevices*, March 2001 (JPP20010307000001).

[32]Semiconductor Industry Association, *1999 Annual Databook*.

[33]"Although the leading U.S. merchant semiconductor firms (such as Intel, TI, Micron, and AMD) spend 10 to 15 percent of revenues on R&D, the bulk of these expenditures focus on new product development. Intel has announced its intention to expand its long-term research program, but few

Intel, AT&T) historically conducted a significant amount of long-term basic R&D in microelectronics, but as competition has intensified "and time to market has become a more important determination of corporate success, even these companies have been forced to channel their research more closely to areas of strategic importance."[34]

While the U.S. industry's applications-oriented pattern of investment has been criticized as myopic, it reflects powerful, widely recognized economic and commercial imperatives:

> Long-term research is inherently uncertain, and the payback can be decades in the future. Furthermore, basic research can result in revolutionary technologies that undermine the existing product areas upon which the established firms rely on for their success. Economists have also described the so-called appropriability problem: because of the broad applicability of the results of fundamental research, it is difficult—if not impossible—for a company to capture all the economic returns from its research investments—and to prevent their competitors from doing so. Hence, individual firms tend to invest less in fundamental research than would be optimal for society.[35]

The Focus Center Research Program

Reflecting the growing belief that a greater research effort is required, the U.S. semiconductor industry in 1997 launched the Focus Center Research Program. The Microelectronics Advanced Research Corporation's (MARCO) Focus Center Research Program is a wholly owned and separately managed subsidiary of the Semiconductor Research Corporation, a not-for-profit research management organization that funds and operates university-based research centers in microelectronics.[36]

---

other semiconductor manufacturing firms conduct much R&D beyond development of next-generation products. None of the new leaders in digital communications perform any fundamental research or maintain much internal semiconductor R&D, instead focusing their efforts on the development and marketing of next-generation semiconductor products." Jeffrey T. Macher, David C. Mowery, and David C. Hodges, "Reversal of Fortune? The Recovery of the U.S. Semiconductor Industry," *California Management Review*, Fall 1998, p. 122.

[34]An IBM executive, Kathleen Kingscott, commented on this type of basic R&D in 2000: "They're not going to do that anymore. IBM used to do it; AT&T used to do it; but we can't do that anymore. In this capital-intensive industry companies are very hard pressed to find the resources to move from one generation to the next. It's particularly true in the smaller semiconductor supplier companies." Cited in Nance, *op. cit.* Erich Bloch and Jerry Sheehan, "Federal IT Funding: Creating an Infrastructure, Growing an Infrastructure, Growing an Industry," *IMP Magazine*, March 1999, http://www.cisp.org/imp/march_99/03__99bloch)sheehan.htm.

[35]*Ibid.*

[36]*About MARCO*, MARCO Web site at <http://marco.fcrp.org> (accessed January 26, 2001). MARCO has its own management personnel but uses the infrastructure and resources of the Semiconductor Research Corporation.

MARCO has been charged with the establishment of the Focus Center Research Center, which is designed to support pre-competitive, cooperative, long-range microelectronics R&D at U.S. universities. Each focus center consists of a team of U.S. universities tasked with conducting exploratory long-range (eight or more years) R&D on silicon-based integrated circuits in order to "address gaps and barriers anticipated in the development of certain technologies" outlined in the roadmap (ITRS).[37] The Focus Center Research Program is funded jointly by Semiconductor Industry Association member companies (50 percent), by suppliers (25 percent), and by the U.S. government (25 percent), with a 6-year budget of $300 million.[38]

The National Laboratories

The U.S. government administers a number of facilities with excellent microelectronics R&D capability. Most notable among these facilities are the national laboratories, which are supervised by the Department of Energy.[39] U.S. and foreign companies may enter into cooperative research and development agreements (CRADA) with U.S. government research organizations, including the national laboratories—in effect, drawing on the infrastructure and expertise of those institutions to overcome specific technological hurdles with commercial implications.[40]

A dramatic example of the CRADA programs' potential for driving technological advances in microelectronics is a current ongoing effort to develop ex-

---

[37]*Focus Center Research Program*, MARCO Web site at <http://marco.fcrp.org>.

[38]FCRP Strategic Plan, MARCO Web site at <http://marco.fcrp.org>.

[39]The U.S. National Laboratories were originally established under the Manhattan Project and the Atomic Energy Commission. They are owned by the U.S. government and are managed by the Department of Energy (DOE). They perform R&D for DOE programs and R&D work for other federal agencies and the private sector on a cost-reimbursable basis. Under the Federal Technology Transfer Act of 1986 laboratory directors can enter into cooperative R&D agreements with private companies and negotiate licensing agreements and contracts. See generally Manufacturing Studies Board and National Materials Advisory Board, Commission on Engineering and Technical Systems, National Research Council, *The Semiconductor Industry and the National Laboratories: Part of a National Strategy*, Washington, D.C.: National Academy Press, 1987, pp. 10-11.

[40]Under CRADA agreements companies and government laboratories pool resources and efforts on a particular technological problem. The government may provide personnel, equipment, and laboratory facilities. The private-sector partners typically contribute funds and personnel. The government usually holds the patents on technologies developed under these agreements, but it gives its private-sector partners the exclusive license to market the technology, retaining the right to buy any product developed pursuant to the CRADA. Measuring CRADA outcomes is not self-evident, as David Mowery documents in *Using Cooperative Research and Development Agreements as S&T Indicators: What Do We Have and What Would We Like?* Presentation before National Science Foundation conference, *Workshop on Strategic Research Partnerships*, October 13, 2000, publication of proceedings pending.

treme-ultraviolet lithography (EUVL) technology for manufacturing semiconductor devices with line widths as small as 20 nanometers (below that size silicon transistors cease to function normally).[41] The EUVL project has already won high praise for overcoming "technical hurdles that once looked insurmountable."[42] The apparent success of this effort, however, will make it possible to squeeze a higher level of miniaturization out of existing silicon-based technologies, not to make the radical leap to an alternative system that would surmount the brick wall.

The National Nanotechnology Initiative

Nanotechnology is the science and engineering of assembling materials and components atom by atom, or molecule by molecule, to create large structures with fundamentally new molecular organization.[43] It is highly likely that the technological solutions to the brick wall, when they are found, will reflect advances in nanotechnology.

In January 2000 the Clinton administration announced a National Nanotechnology Initiative pursuant to which 10 federal agencies would sponsor research and development into a broad range of nanotechnology themes. In FY 2000 $270 million in federal funds were allocated to this effort, and President Clinton's proposed 2001 budget raised this figure to $497 million. The largest block of federal money was allocated to the National Science Foundation ($97 million in 2000), with significant funding also going to the Department of Defense ($70 million) and Department of Energy ($58 million). Approximately 70 percent of the total funding will be directed toward university-based research.

---

[41]The EUVL CRADA draws together the Lawrence Livermore, Sandia, and Lawrence Berkeley national laboratories and five companies (Intel, AMD, Motorola, Micron Technologies, and Germany's Infineon) in a $350-million R&D effort. The three labs collaborate in what is termed a virtual national laboratory (VNL); the actual R&D is conducted on the premises of the labs. The private-sector participants have formed a limited-liability corporation to participate in this project, EUV LLC. This entity will transfer technology developed by the virtual national laboratory to semiconductor manufacturing equipment suppliers. EUV LLC members will be granted rights of first refusal on the equipment these toolmakers produce. They also will receive royalties on sales of equipment incorporating technology developed under this effort. The greatest benefit to participants "will be early access to VNL know-how, an advantage that could give member companies a two-year head start on the rest of the industry." Keith Dieffendorff, "Extreme Lithography," *Microprocessor Report*, June 2000, p. 1. For a review of the EUVL consortium and the special challenges it faces, see Greg Linden, David C. Mowery, and Rosemarie Ham Ziedonis, "National Technology Policy in Global Markets" in Maryann P. Feldman and Albert N. Link, (eds.), *Innovation Policy in the Knowledge-based Economy*, Boston: Kluwer Academic Publishers, 2001, pp. 311-36.

[42]Dieffendorff, *op. cit.*, p. 1.

[43]*National Nanotechnology Initiative: The Initiative and its Implementation Plan* (National Science Council, Committee on Technology, Subcommittee on Nanoscale Science, Engineering and Technology, July 2000), p. 19.

While a significant part of this project will involve either the development of nanoelectronic devices or research regarding materials and processing technologies with microelectronics applications, its scope is far broader and its resources must be spread across fields as diverse as medicine, space exploration, agriculture, environmental protection, and transportation networks.[44]

The Advanced Technology Program

The Advanced Technology Program was created by an act of Congress in 1988 and is administered by the National Institute of Standards and Technology, which has a large, technically skilled staff.[45] Through this program the institute supports industry R&D efforts through provision of startup funding; equipment, facilities, and personnel; organizational and technical support; cost sharing for periods up to five years; and in some cases government participation in joint ventures.[46] Financial assistance is provided in relatively small amounts (e.g., up to $2 million per project and normally up to $5 million for joint ventures).

Many U.S. industries, including the semiconductor industry, have benefited from this program.[47] The program has been a lightning rod for criticism by opponents of government support for industry, and a number of powerful members of Congress have periodically threatened to shut the program down.[48] Nevertheless, the program grew substantially under the first Bush administration. Its subsequent embrace and rapid expansion by the Clinton administration, and concomi-

---

[44]The National Nanotechnology Initiative implementation plan envisions that nanotechnology "will foster a revolution in information technology hardware rivaling the microelectronics revolution begun about 30 years ago" (*ibid.*, p. 51). The initiative foresees, for example, research that will develop new approaches to nanostructure synthesis to permit affordable fabrication of electronic nanodevices for commercial use (the current estimated cost of one fabrication plant for 70-nm microelectronics is estimated at over $10 billion), *ibid.*, p. 52.

[45]See generally National Research Council, Charles W. Wessner (ed.), *The Advanced Technology Program: Challenges and Opportunities*, Washington, D.C.: National Academy Press, 1999. For a comprehensive look at the program also see National Research Council, Charles W. Wessner (ed.) *The Advanced Technology Program: Assessing Outcomes*, Washington, D.C.: National Academy Press, 2001.

[46]Funding for the program peaked at $341 million in 1995 and has remained at $200 million or less in subsequent years. The program has achieved successes in areas such as medical devices.

[47]A project involving the Diamond Semiconductor Group developed technology for enabling faster processing of large semiconductor wafers with better process control; another project involving free research developed process improvements for growing large, silicon-carbide crystals, a semiconductor material used in making optoelectronics devices such as blue light-emitting diodes. National Institute of Standards and Technology, *Building Bigger and Better Semiconductor Wafers*, February 1999, at <http://www.nist.gov/public_affairs/factsheet/diamond.htm>.

[48]"Stark Warning Issued on Advanced Technology Program," *Bulletin of Science Policy News*, March 21, 1997.

tant congressional opposition, resulted in the curtailment of its budget rather than its elimination.[49]

Programs to support technology development are frequently misunderstood and not uniformly accepted, reflecting a strong bias for market-based solutions and perhaps a lack of familiarity with past U.S. practice. Although small in size, the Advanced Technology Program may be uniquely controversial partly because of its broad mission (to support enabling technologies with economy-wide benefits) and partly because it became involved in the political machinations of the mid-1990s. But, SEMATECH was also controversial, at both its inception and its renewal; opponents initially argued that it would not work and later argued that it had worked too well. The longstanding U.S. debate about the principle of government support for industry, rather than its efficacy, has few if any parallels overseas.

FORMS OF GOVERNMENT SUPPORT OUTSIDE THE UNITED STATES

As a general proposition government interventions in microelectronics are far larger in scale and broader in scope outside the United States. They are also more plainly directed at specific commercial objectives, yet they generate considerably less domestic controversy than do the U.S. programs, which are limited virtually entirely to pre-competitive R&D. Outside the United States, governments also fund pre-competitive research, particularly in Japan and Europe, but they are also providing large-scale financial support for R&D with direct and immediate commercial application. The table at the end of this paper provides a partial summary of significant government measures supporting microelectronics R&D outside the United States.[50] In addition, above and beyond assistance for basic and applied R&D, foreign governments are providing capital for the construction of wafer-fabrication facilities and in some countries, such as Taiwan and Singapore, they have taken direct equity positions in semiconductor enterprises.

Governments are also utilizing industrial policy measures to replicate, to the extent possible in a non-U.S. environment, what are seen as the strongest features of the U.S. system. Thus, governments are countering the strong attraction that

[49]Christopher T. Hill, "The Advanced Technology Program: Opportunities for Enhancement," in Lewis M. Branscomb and James H. Keller (eds.), *Investing in Innovation: Creating a Research and Innovation Policy That Works*, Cambridge, MA: MIT Press, 1998.

[50]The dollar figures depicted reflect conversion at exchange rates prevailing in April 2001. The dollar totals are not a good benchmark of the true level of government support, because in many cases such support takes the form of contribution of government facilities, transfer of technology from government laboratories, the services of government researchers, and other in-kind contributions that cannot readily be quantified in dollar terms. The chart does not depict numerous government-supported acquisition-and-diffusion R&D projects under way in newly industrializing countries.

the United States has traditionally exerted for foreign engineers and scientists by establishing incentives to draw such individuals into their own microelectronics programs. In Taiwan, Japan, Korea, and the European Union, governments and regional authorities are attempting to create new Silicon Valleys, utilizing various incentives to encourage the clustering of high-technology firms in designated zones. Some are also attempting to create parallels to the U.S. venture capital system by pumping government funds into new "venture" enterprises.

Policies Related to Research and Development

In Japan and Europe, by far the most important forum of government support for the microelectronics sector is in the area of research and development. Government funds are being channeled into a number of large industry-government R&D consortia with both pre-competitive and commercial themes. Typically these projects are sustained, long-term efforts involving the principal firms in each country, with very substantial government funding levels.[51] In both Europe and Japan, the main government-sponsored programs are designed to enhance national advantage with respect to what are seen as the fastest potential-growth end-markets, such as wireless and wired telecommunications and digital home appliances. The U.S. industry, which dominates PC-based end markets, is seen as vulnerable as non-PC end markets grow as a proportion of the total market for semiconductors.

Some of the large-scale R&D projects being undertaken in advanced countries outside the United States offer a vehicle for reducing some of the technological burdens confronting the U.S. semiconductor industry. U.S. firms participate in and benefit from some foreign government-supported R&D programs. IMEC, a highly regarded, partially government-funded European microelectronics R&D center, performs leading-edge R&D on a contract basis for non-European firms, including Motorola, Texas Instruments, and AMD.[52] Most major U.S. semiconductor producers have R&D centers abroad and/or R&D alliances and

---

[51]Japan's recently inaugurated Future Information Society Creation Laboratory, which will develop high-speed semiconductor devices, will reportedly receive $300 million in government funding (*Handotai Sangyo Shimbun*, September 6, 2000, p. 1; see FBIS, October 1, 2000, translation of "Development of 0.07 to 0.05 micron LSI Technology," *Nikkei Microdevices*, October 2000, pp. 67-70 (JPP20001018000105). The European Union's MEDEA Plus project, which begins in 2001, will reportedly receive about 200 million euros in government support. See FBIS, June 9, 2000, translation of "Billions for Chip Research," *VDI Nachrichten*, p. 31 (EUP200006226).

[52]The regional government of Flanders in Belgium provides slightly less than half of IMEC's annual budget and has pledged to do so through the end of 2001. About 25 percent of IMEC's contract research revenue is derived from contracts with government entities, primarily the European Union. See FBIS, December 8, 1995, translation of "For Research and Industry," *Markt und Technik*, December 8, 1995 (FTS19951208000472) and "For Research and Industry," *Markt und Technik*, December 8, 1995; *IMEC Annual Report* 1999, p. 1.

technology exchange arrangements with foreign firms. All these are channels through which they may obtain technology originally developed in foreign government-supported programs.

The historical pattern of transnational cooperative R&D in this strategic industry is, however, uneven and is likely to remain so, reflecting abiding economic nationalism and national security concerns.[53] An official from MEDEA,[54] Europe's most significant microelectronics R&D program, explained frankly why participation in that project was limited to European-based firms: "[A]fter all, we have no desire to fund technology transfer to other regions."[55] While Japan has permitted some foreign participation in its microelectronics R&D projects, the issue remains controversial, and foreign involvement is likely to remain closely regulated and monitored.[56]

Taiwan, Korea, and Singapore are not seeking to emulate the large-scale R&D consortia found in the most advanced countries. Instead they are pursuing R&D strategies characterized by

- Rapid acquisition of foreign leading-edge technology;
- Efficient diffusion within the national industry; and
- Efforts to foster the ability to manufacture leading-edge products.[57]

Governments adapting this strategy pursue policies to attract inward investment and joint ventures involving advanced foreign semiconductor firms. They

- Buy foreign technology outright and establish listening posts in advanced countries;

---

[53]This tension and the frequent need for cooperation were addressed in the 1996 National Research Council study, *Conflict and Cooperation.*

[54]MEDEA is the new industry-initiated pan-European program for advanced cooperative research and development in microelectronics. It has been set up and labeled within the framework of EUREKA to ensure Europe's continued technological and industrial competitiveness in this sector. The Ministerial Conference in Hanover granted the EUREKA label on June 23, 2000.

[55]MEDEA Office Director Gerard Matheson, FBIS, June 9, 2000, translation of "Billions for Chip Research," *VDI Nachrichten*, p. 31 (EUP200006226).

[56]In Japan's Association of Super-Advanced Technologies (ASET), foreign membership is permitted but foreign firms do not take part in some R&D areas. Foreign participation was controversial because the basic purpose of government funding of ASET was "to strengthen the Japanese electronic industry through technological development." After deliberation, ASET's members decided to allow foreign participation, but "only when it can promote research and development." *2000 Edition of Nihon Handotai Nenkan.* Tokyo Press Journal, March 25, 2000, pp. 196-97.

[57]Taiwan's ASTRO project, which is sometimes analogized to the European and Japanese projects and to SEMATECH, is smaller in scale and emphasizes technology acquisition and diffusion. ASTRO is reportedly suffering from defections of personnel and the non-participation of the leading Taiwanese firms.

• Establish diffusion-enhancing research institutes, consortia, and associations; and

• Offer incentives for repatriation of their nationals working for semiconductor firms abroad.[58]

Taiwan has been so successful in adapting technology that a TSMC official estimated in 1999 that his firm and Intel were equals in terms of manufacturing technology. "[N]ow our company and Intel are both at the 0.18-micron level, so there is no technological gap."[59]

Policies to Meet Burgeoning Capital Requirements

Since the beginning of the 1990s the sheer size of the investments required to establish state-of-the-art semiconductor manufacturing facilities has become an obstacle to growth. With the cost of a single fab rising above $1 billion and with markets volatile, the risks have become too great for many private companies to bear. In response, in the European Union national and regional governments have committed substantial investment resources to ensure the retention of at least some state-of-the-art manufacturing capacity within the European Union.[60] In Korea the government has traditionally channeled capital to the semiconductor industry through its influence over the banking system. While the implementation of International Monetary Fund-backed reforms has ostensibly ended this practice, there is substantial evidence that the industry in Korea continues to receive capital assistance from the government through a variety of channels.[61] Still, the most dramatic and comprehensive government support for capital in-

---

[58]For a comprehensive treatment of this subject see John A. Matthews and Dong-Sung Cho, *Tiger Technology: The Creation of a Semiconductor Industry in East Asia*, Cambridge: Cambridge University Press, 2000.

[59]See FBIS, August 25, 1999, translation of Motohiko Kitahara, "Interview with Hsuan Ming-chih of UMC Regarding Taiwan's Semiconductor Foundry Strategy," *Nikkei Sangyo Shimbun*, August 1999, p. 18 (FTS19991003000139).

[60]The Semiconductor 300 project established the world's first operational wafer-fabrication facility utilizing 300-mm wafer technology. This project was a collaboration among Siemens (Infineon), Motorola, Wacker Siltanic, the German Research Ministry, and the Free State of Saxony. Approximately 40 percent of the funding for this project has reportedly been derived from government sources. See "300-mm Fab in Dresden Makes First 256-Mbit DRAMs," *Frankfurter Zeitung/Blick Durch Die Wirtschaft*, January 12, 1997 and FBIS, February 4, 1998, translation of "Europe's Largest Electronics Project Starts in Dresden," *Frankfurter Zeitung/Blick Durch Die Wirtschaft*, January 12, 1997, p. 6. A similar project in Crolles, France, involves the joint construction by Philips and ST Microelectronics of a 300-mm fab. This project is reportedly being funded by the departmental government of Isere, with additional capital investment support being sought from the French government.

[61]In early 2001 the Korea Development Bank reportedly bailed out Hyundai Electronics by purchasing its bonds at concessional rates of interest. At his confirmation hearing U.S. Trade Representative-designate Robert Zoellick commented that the apparent preferential treatment of Hyundai was "part of a deeper problem in Korea, which is slipping from the restructuring that the government

vestment in microelectronics is now found in two significant emerging producing countries, Taiwan and Singapore.

Government Assumption of Capital Costs

Taiwan and Singapore have exploited the capital investment dilemma confronting the global semiconductor industry, in effect, by assuming the costs and risks of the high-investment levels themselves. In mid-2000 Taiwanese planners envisioned that Taiwanese firms would build a total of 21 new 300-mm fabs and 9 new 200-mm fabs by the year 2010. A Japanese survey taken in mid-2000 of known planned 300-mm wafer-fabrication facilities around the world concluded that half of the total facilities either were in Taiwan or were being wholly or partially financed by Taiwanese firms in other countries.[62] The government of Singapore has publicly set a goal of 20 fabs by the year 2005.[63] A substantial proportion of the fabs that are planned or being built will be dedicated to foundry operations—the manufacturing of foreign firms' designs on a contract basis.

Taiwanese Policy

The aggressive capital investment programs under way in Taiwan supports its acquisition-and-diffusion technology strategy and partially compensates for the relative absence of larger-scaled R&D consortia. The performance of foundry manufacturing services creates a close, interdependent relationship between the foundry and the designing firm, and has enabled Taiwanese firms in particular to surge to the forefront with respect to semiconductor manufacturing technology.

Taiwan's TSMC and UMC are considered the best foundry operations in the business. They have learned and perfected semiconductor-manufacturing techniques through the operation of foundries, emphasizing flexibility, speed, high yields, and low cost so successfully that they have eclipsed even the Japanese industry in terms of manufacturing skill. In a new joint venture between Taiwan's

---

promised to do." See "U.S. Questions Bailout of Hyundai Electronics," *Choson Ilbo*, January 31, 2001. A Korean editorial observed with respect to this transaction that "[i]ndustries strategic to national development such as semiconductors and construction are so important that they can't be left to suffer. . . ." See FBIS, February 2, 2001, translation of "Preferential Treatment for Hyundai," *Choson Ilho*, February 2, 2001, (KPP20010202000082). "Some use an expression that the government is putting a blanket over the whole market in order to assist Hyundai . . . . A researcher in a government-funded research institution predicted that the current government would keep Hyundai afloat until its term is over." See "The Government Above the Market, the Market Above Hyundai?" *Chugan Choson*, March 29, 2001.

[62]FBIS, July 1, 1997, translation of "Playing to Our Strengths," *Bild der Wissenschaft*, July 1997, pp. 88-89 (FTS19970730000530).

[63]The 20-fab figure may be overstated given that a fab is defined by Singapore planners as a facility capable of producing 20,000 wafers per month. Thus, a single facility with twice that capacity would count as two fabs.

UMC and Hitachi, UMC is transferring Taiwanese manufacturing know-how to Hitachi—a startling reversal of roles.[64]

Tax Holidays

Taiwan's planners estimate that the planned 30 new fabs will cost approximately $60 billion. This huge sum will not be financed directly by the government in whole or in part but by Taiwanese companies, which are expected to raise the money through a combination of retained earnings (50 percent) and funds raised on the private equity markets (50 percent). Even so, they are in a position to make these investments as a direct consequence of government policy. Because semiconductors are considered a strategic high-technology industry, the government provides substantial tax holidays to the semiconductor industry. For example, Taiwanese producers usually pay no income tax. TSMC has accumulated so many tax credits that in every recent year its after-tax earnings have actually been higher than its pre-tax earnings.[65] Raising capital in the equity markets has been further facilitated by the lack of a Taiwanese capital gains tax and the government's identification of the semiconductor industry as the number-one priority industry.[66]

The Foundry Phenomenon

The advent of the foundry model, which may revolutionize the semiconductor business, is a direct consequence of decisions made by Taiwanese government planners in the late 1980s. UMC and TSMC, the two Taiwanese firms that lead the world in foundry manufacturing, are creations of the Taiwanese govern-

---

[64]See FBIS, February 8, 2000, translation of Y. Nagahiro, "Hitachi-UMC Tie-Up to Generate Test Cases that Will Forecast the Future of LSI Technology and Business," *Nikkei Microdevices*, February 2000, pp. 149-50 (JPP200002080000029). "Taiwanese companies are unimaginably fast . . . . Taiwanese manufacturers' technological prowess grew so much that it surprises us [Japanese]."See FBIS, June 1, 2000, translation of A. Minamikawa, "Investment in 300 mm Plants Heating Up, 32 Heating Up, 32 New Lines to be Built," *Nikkei Microdevices*, June 2000 (JPP20000706000024).; FBIS, May 1, 2000, translation of Y. Nagahiro, "World's First 300 mm Production Line – Establishing World Standard 300 mm Technology" *Nikkei Microdevices*, May 2000, pp. 16-17 (JPP20000516000091).

[65]TSMC Annual Reports, 1995-99.

[66]The Taiwanese semiconductor industry was shaken in the latter half of 2000 by a sharp drop in stock prices and a slump in personal computer sales, which reduced the prospect that all of the investments envisioned in mid-2000 would actually proceed—particularly with respect to capacity for commodity products like DRAMs. However, a substantial proportion of the planned new fabs are earmarked for the provision of foundry services to foreign device markets, which spreads the risk otherwise associated with over-commitment to a particular device type. Taiwan's two major foundry producers, TSMC and UMC, made clear that they would still "push for vast expansion without changing the business model" and that they had the capital resources with which to do so. See FBIS, January 1, 2001, translation of Peter Chang, "Expanding Si Foundry Business with 300mm and 0.13 Micron Technology," *Nikkei Microdevices*, January 2001 (JPP20010126000005).

ment. TSMC was established as the world's first pure-play foundry, meaning that it only produces semiconductors on a contract basis for sale by other firms and does not market its own line. This concept was considered so radical and risky at the outset that sufficient private capital could not be raised for the TSMC venture. As a result, 44 percent of TSMC's capitalization was provided by the Development Fund of the Executive Yuan (Cabinet), a special fund utilized for high-risk industrial projects. TSMC has grown rapidly, in significant part by absorbing other Taiwanese semiconductor firms that have also received government financial assistance.[67]

Singapore and Malaysia

Singapore, like Taiwan, is making major investments in fabs, which it will produce on a foundry basis. Singapore's Chartered Semiconductor Manufacturing (CSM) is already the world's third-largest foundry producer. The government has played a major role in ensuring that the necessary capital is available to achieve its increasingly ambitious expansion plans. The Economic Development Board invests directly in semiconductor enterprises that establish wafer fabs in Singapore, and has helped Chartered Semiconductor Manufacturing to emerge as a major foundry operator. The Economic Development Board manages a number of funds through which it channels grants, investments, and loans to microelectronics enterprises.[68] Singapore's tax rules provide incentives to priority indus-

---

[67]The Worldwide Semiconductor Manufacturing Corporation and Acer, both of which have been absorbed by TSMC, received substantial financial support from the China Development Industrial Bank, a government bank with the mission of providing long-term credit and investment funds to strategically important industries. "Account Update—Taiwan," *The Asian Banker Journal* April 19, 2000; "Management Statement," *China Industrial Development Bank* at <http://www.cdcdpbnk.com/english/e-y-2.htm/>. Taiwan's attraction of high-technology industries is documented in National Research Council, *Conflict and Cooperation*, p. 33, Box A.

[68]The prototype investment was its 26 percent stake in TECH Singapore, established in 1991 with Texas Instruments, Canon, and Hewlett-Packard. The Cluster Development Fund, created in 1993 under the Economic Development Board's (EDB) direction, made its first investment in 1994 in Chartered Semiconductor Manufacturing's Fab II ($100 million Singapore). EDB Investments ("EDBI") funded a fab to produce DRAMs in conjunction with a Japanese-owned firm, Hitachi Nippon Steel Semiconductor Singapore. In 1997 EDBI joined CSM and Agilent Technologies to finance the construction of a fab to manufacture 8-inch wafers for the digital consumer and communications industries. In 1998 Philips and TSMC joined with EDBI to create a foundry operation in Singapore. The EDB reportedly took a 20 percent equity stake. In 1996 EDBI took a 30 percent equity stake in a joint venture between Hitachi and Nippon Steel to establish a wafer-fabrication facility for 64-megabit DRAMs. See "Government Makes First Major Investment from Cluster Development Fund in Wafer Plant," *The Straits Times,* February, 26, 1994; Singapore Investment News with Hitachi Nippon Steel Semiconductor Singapore Pte Ltd *Company Profile*, http://www.has.hitachi.com.sg/hns/profile/c-about.htm.; EDB, "New U.S. $1.2b Wafer Fab to Start Production in 2000," *Singapore Investment News*, November 1, 1998; See also FBIS, "Philips Semiconductor Plant in Singapore," *Algemeen Dagblad*, September 30, 1998, p. 14 (BR3009120698).

tries, including the semiconductor industry, on a sliding scale, with tax rates ranging from 13 percent to zero (normal tax rates are 25.5 percent). An enterprise operating a wafer fab in Singapore will be taxed at the zero rate if it utilizes the most advanced technology.

Malaysia appears to be following the Singapore model; its government investment arm, Khazanah Nasional Berhad, has taken or is planning to take large equity positions in several new semiconductor-manufacturing enterprises in Malaysia.[69]

Implications

The full long-run implications of the foundry phenomenon for the U.S. semiconductor industry are not clear. The ability of U.S. producers to outsource the production of their designs to some of the most efficient and skilled manufacturers in the world reduces the otherwise-enormous investment risks they would face by producing their own devices and helps to ensure a high-quality end product. Many observers in East Asia warn that if the U.S. industry does not make full use of the foundry opportunities available to it in Taiwan and Singapore, others will (e.g., their Japanese and European competitors). Such a development would work to the serious disadvantage of the U.S. industry. On the other hand, the foundry model requires designing firms to share their proprietary designs with the foundry producers, a relationship that has made European policy makers, at least, uneasy.[70] If the U.S. industry surrenders too much of the manufacturing

---

[69]Khazanah took a 40 percent equity stake in a project involving Atmel of the United States to build a wafer fab in Malaysia's Kulim Hi Tech Park. It made a 25 percent equity investment in a joint venture with Malaysia's AIC Semiconductor to establish a semiconductor packaging enterprise. It reportedly will take an equity share of up to 32 percent in Silterra, a new semiconductor foundry; the government reportedly also plans to guarantee $750 million of Silterra's debt, waive taxes on $1 billion in future profit, and exempt it from import and export duties. See AIC Corporation Berhad, January 18, 2002, http://www.klse.com.my/website/listing/lc/aic.htm; FBIS, October 15, 1997, reprint of Shamsul Akmar and Lee Yuk Peng, "U.S. Firms Investment Arm Tie Up for Wafer Fab Project," *The Star*, October 15, 1996, (FTS19971015000152); *AIC Corporation Berhad*, Press Release, September 10, 1998.

[70]In Germany the state government of Bavaria has directed the Fraunhofer Institute of Solid State Technology (IFT), an institute of applied research, to undertake the local commercial manufacture of application specific integrated circuits (ASICs) designed in Europe. The reason for this move was to give local ASIC designers a European alternative to Asian foundries. "The European market for this type of component was worth around DM 4 billion in 1995. However, only half the ASICs sold were produced in Europe. 'This means that many medium sized firms have to reveal their know-how to foreign manufacturers,' said [IFT] Deputy Director Konrad Hieber." See FBIS, August 15, 1996, translation of Andreas Beuthner, "Fraunhofer Tech Transfer to Industry Viewed," *Computer Zeitung*, August 15, 1996, p. 4 (FTS19960815000685).

function to offshore foundries, it is not clear that it can retain its position of world leadership, which is based increasingly on design functions alone.

Taiwan, for example, is promoting an indigenous semiconductor design capability the way it previously encouraged semiconductor manufacturing. Taiwan has over 100 design houses, which benefit from a number of government programs. In January 2000 ERSO (Electronics Research and Service Organization) organized a club to provide local design houses with a platform to exchange intellectual property, the Silicon Intellectual Property Consortium. The objective of the club is to promote the intra- and inter-company circulation of IP and design reuse. Taiwan's Chip Implementation Center, administered by the National Science Council, is currently designing over 500 devices in its research programs and is turning out excellent design engineers.[71]

While Taiwanese design houses today do not constitute a significant challenge to the U.S. industry, it is possible to envision future scenarios in which it is common for an OEM to commission a design from a firm in Taiwan or Europe and contract with a foundry like TSMC to manufacture it, effectively bypassing the integrated device manufacturers, which today account for the preponderance of U.S. semiconductor sales. Some transactions of this kind are occurring already.[72]

Qualified Workforce

The rapid growth of semiconductor industries around the world has created a massive demand for skilled engineers, scientists, and technicians. At the same time the number of individuals graduating from U.S. universities with electrical and electronic engineering degrees is declining.[73] The U.S. has been able to offset this trend to a considerable degree because of its abiding appeal to talented immi-

---

[71]In 1999 the Industrial Technology and Research Institute established an Integrated Design Service Center with a staff of 46 engineers and assistants. This center builds standard cell libraries for advanced foundry devices and is expected to expand design capability with respect to system-on-a-chip technology. "Taiwan's Semiconductor Industry's Sales Jump in 1999; After 2000 Taiwan to Take the Offensive in Assembly, Testing and Design Fields," *Semiconductor FDP World*, June 2000; Industrial Technology and Research Institute, *Annual Report*, 1999, p. 43.

[72]The French telecommunications firm Alcatel has commissioned TSML to design and (apparently) manufacture systems-on-a-chip for use in so-called Bluetooth products, a potentially revolutionary system that will establish short-wave radio links between mobile phones. "PCs and home appliances," Taipei Central News Agency, Taipei, wire service report, November 1, 2000.

[73]In 1988 approximately 24,000 people graduated from U.S. universities with bachelor's degrees in electrical and electronic engineering. By 1997 this total had fallen below 14,000, and it is not forecast to increase significantly in the foreseeable future. Engineering Workforce Commission statistics and Semiconductor Research Corporation projections presented by Michael Polcari, National Research Council *Symposium on National Programs to Support the Semiconductor Industry*.

grants and foreign scientists and engineers who become U.S. residents.[74] A recent article by an official in Japan's Ministry of Foreign Affairs made the following observation:

> Many of the top students at Stanford University are Asian Americans, and 20 percent of the engineers in Silicon Valley are Asian Americans as well. . . . The source of the "power" of the United States, which was founded by immigrants to begin with, probably lies in the fact that the society is built on its basic openness, social dynamism, and "competition predicated on change." . . . For a Japan that does not have a history as a nation of immigrants, it is rather difficult to transform itself into this kind of society.[75]

Although the United States continues to exert a powerful attraction to foreign talent as a place to live, work, and pursue opportunity, significant countervailing forces are inducing an increasing number of foreign graduates of U.S. universities to return home and U.S.-resident foreign scientists and engineers to relocate abroad.

Personnel shortages in the semiconductor industry are reported throughout the developed and newly industrializing world and are forecast to become more acute in the next decade. Planners in Taiwan and Singapore view prospective workforce shortages as the single most important constraint on the industry's expansion plans. In Europe an estimated 500,000 IT-related jobs are unfilled, and an annual shortfall of 3,500 new graduates in electronic engineering currently exists and is expected to widen.[76]

All major semiconductor-producing countries in East Asia are implementing substantial programs to train more specialists and engineers for the semiconductor industry, but these are generally acknowledged to be inadequate for the actual need. One result of this global shortage has been a furious competition to attract

---

[74]An Israeli assessment of the attraction of Silicon Valley: "Why, in fact, do foreign high-tech experts, especially Israelis, flock to Silicon Valley? The answer is simple. The Valley is the Mecca of the high-tech industry. Whether or not they are entrepreneurs at heart, many seek to gain professional experience on the industry's business and technological front. Others want a change of atmosphere, are looking to take a breather for a few years. The major decision of whether or not to remain in the Valley will come later, depending on circumstances. Dan Avida of EFI defines the émigré's stay in the Silicon Valley as 'a period of time that gets protracted on an annual basis.' Opinions are divided as to the influence of the financial considerations on the lengthening of one's stay in the Valley. Some feel money is of no account. It's a matter of career, they claim. Others believe that those who have become accustomed to the high living standard of the United States will be in no hurry to give it up." See FBIS, May 9, 1997, reprint of Nir Nathan, "We've Taken Silicon Valley," *Tel Aviv Globes*, May 6, 1997 (FTS19970509000829).

[75]See FBIS, November 1, 2000, translation of Hitoishi Tanaka, "New Development in Japan's Economic Diplomacy: Free Trade Agreements," *Chuo Koron*, November 1, 2000, pp. 50-65 (JPP20001019000039).

[76]See FBIS, October 6, 2000, translation of D. Wendeln, "Europe on the Way to World Class in Microtechnology and Nanotechnology," *VDI Nachrichten*, October 6, 2000 (EUP20001011000369).

skilled specialists from other countries or to establish design centers in other countries to attract local talent.

Taiwanese planners concede that despite numerous training programs, there is no realistic possibility Taiwan can produce enough engineers on its own—and that "we will find them in the U.S., in Southeast Asia, and in Mainland China."[77] Korea has expressed alarm that foreign countries, particularly Taiwan, have been scouting out and recruiting semiconductor researchers and engineers from Korean companies—a concern that is ironic, given Korea's own aggressive overseas recruiting practices.[78] Singapore has implemented an extremely liberal immigration policy, and its Economic Development Board operates an International Manpower Program, which scours foreign universities for talent.[79] In China, where local governments once competed with each other to attract inward foreign investment, they are now engaged in a fierce rivalry to attract foreign talent, particularly ethnic Chinese who "bring with them their S&T accomplishments."[80]

---

[77]Taiwan has established an extensive network of government organizations to attract scientists and engineers to Taiwan, particularly Taiwanese expatriates and ethnic Chinese. The National Youth Commission maintains a database tracking Taiwanese graduates abroad, including over 11,000 engineers, who are periodically contacted and offered incentives to return to Taiwan. Over 100 of the companies in Hsinchu Science Park were founded by returning expatriates, and by the mid-1990s over 6,000 expatriates were returning annually, reflecting the government's concerted recruiting efforts. *National Science Council Yearbook,* 1998; Hsinchu Science Park Web site at <http://www.sipa.gov.tw/en/seconde/induse/induse.html>.

[78]In 1993 Korea's Ministry of Science and Technology (MOST) established a program to "attract and make good use of" the approximately 400 ethnic Koreans working overseas in science and technology fields. This effort was intended to encourage these individuals to help Korea "acquire at an early date the newest science, technology and know-how in the R&D stages in advanced countries." Under the MOST program foreign scientists with five years' postdoctoral experience abroad are given airfare and high salaries in return for a six-month commitment to work at a South Korean facility. During the first year of this program offers were made to 57 such individuals, the majority of whom had Korean surnames and were living in the United States. One-third of the 31 acceptances were in the electronics field. (*Maeil Kjongje Sinmun*, January 25, 1994; *Hanguk Kjongje Sinmun*, December 27, 1994). Korea's semiconductor makers have an estimated 4,000 professional researchers, of which 400 are said to be targeted by foreign recruiters. An international incident of sorts erupted in 1998 when a number of Samsung semiconductor engineers defected en masse to Taiwan's Nan Ya Technology Corp., triggering Korean outrage. Korea nonetheless maintains its own aggressive recruiting efforts abroad. Korea's MOST maintains a database of ethnic Koreans with science and technology skills living abroad who are targeted by recruiters with generous incentives. The Korean Institute for Science and Technology seeks to recruit "outstanding Korean scholars living in Japan." *(Hanguk Kjongje Sinmun*, December 27, 1994; *Maeil Kjongje Sinmun*, March 15, 1993).

[79]Singapore's Chartered Semiconductor, Ltd., has recruited 750 people through this program since 1995. See "Singapore's Transition into a Knowledge-Base Economy–Investing in the Future Today," *Singapore Investment News* (press release from the Economic Development Board, no pagination), April 1, 2000.

[80]Between June 1999 and June 2000 the Beijing municipal government enacted at least six sets of incentives to attract ethnic Chinese to its Zhongguancun Science Park. See "China Actively Partici-

At present the single most powerful policy-related employment inducement worldwide is the compensation packages made possible by Taiwan's tax laws—a source of much grumbling by competitors throughout East Asia. One of the most commonly cited sources of the startling competitive advantage achieved by Taiwanese firms like UMC and TSMC is that they "can attract and keep the best people" with compensation arrangements that foreign firms cannot match.

The market value of UMC and TSMC stock has risen dramatically since the date of original issuance, but employees given stock as compensation are taxed on the face value of the shares, not the market value—which is sometimes over 50 times as high. Moreover, when the shares are sold, whether acquired directly or through the exercise of stock options, no capital gains tax is levied, because Taiwan eliminated its capital gains tax in 1990.[81] Thus, engineers signing on with these firms face the prospect of becoming very wealthy in a relatively short time, "and that becomes a big reason why Taiwan can attract the best talent in the high-tech industry from at home and abroad."[82]

Government-backed Venture Programs

Citing the enormous success of the U.S. free-market model for developing and commercializing new technologies, a number of foreign governments have begun to emphasize what they characterize as entrepreneurism and venture capitalism in their high-technology promotion policies. In contrast to the 1980s, when governments established a series of semiconductor mega-projects involving the major producers, each aimed at producing a very specific milestone device such as a 1-gigabit DRAM, governments now have begun to encourage smaller players to explore a wider range of new technologies, some of which may grow into internationally competitive products.[83]

China, for example, is establishing microelectronics incubators, organizations intended to generate small, startup semiconductor enterprises systemati-

pates in Fight for Overseas Talent," *SinaNet*, August 14, 2000, online article, no page citation, http://www.zgc.gov.cn/news/epnews/0008114-4.htm and "Beijing Issued Implementation Rules on Recruiting Talents from Outside City," *Beijing Morning Post*, August 14, 2000, online article, no page citation, http://www.zgc.gov.cn/news/epnews/000814-5.htm.

[81] *Income Tax Law*, Article 4-1.

[82] *Nikkei Microdevices*, "Investment in 300mm Plants Heating Up; 32 New Lines to be Built."

[83] Some significant successes have already been achieved in this regard. The German ASIC design house SICAN was established with 100 percent public funding in 1990. In 1997 SICAN's founder claimed he had achieved a competitive edge over U.S. rivals in designs for Asynchronous Transfer Mode and Motion Picture Expert Group technologies. At that point SICAN's percentage of public funding had been reduced to 20 percent. See FBIS, July 1, 1997, translation of "Playing to Our Strengths," *Bild der Wissenschaft*, July 1997, pp. 88-89 (FTS19970730000530).

cally.[84] Japan's Ministry of Education is funding a Venture Business Laboratory Program at Japanese universities to promote R&D, foster "talented people brimming with entrepreneurial spirit," and generate "student ventures" arising out of original research.[85]

The terms employed in the implementation of these policies—"venture" is the overwhelming favorite, although Singapore has coined the word "technopreneurship"—convey the notion that these countries are emulating the free market policies that have helped foster U.S. venture capitalism, such as an open and transparent securities market and minimal government intervention in choosing the technologies to be financed. A closer look at the actual venture capital programs in microelectronics outside the United States reveals that while they seek to replicate the success of the U.S. system in promoting innovation, the role of government is much more pervasive, and many of the fundamental aspects of traditional industrial policy remain. For example,

- *Governments choose the technologies that are to be supported for development.*

Bureaucrats continue to consult with industry and academia to identify promising products: the conventional industrial policy practice of providing financial incentives to firms that they hope will be winners, with recognition that some will be losers. In a change from the past, however, the decision-making tends to be more decentralized and flexible, with broad sectors specified by long-range policies but individual funding decisions made by industry-government committees and specific government venture funds. In Korea, for example, the venture companies supplying most of the capital are creations of various government ministries. In Taiwan a number of startups have arisen from government-initiated projects at ERSO, the government microelectronics research institute.[86]

---

[84]China began establishing "hi-tech business incubators" in 1987 as one element of the Torch Program, a sweeping government plan to promote high-technology enterprises. The Beijing Microelectronics Enterprise Incubator was established in 2000 under the joint administration of the Aerospace Bureau and the Beijing Automatic Measuring and Testing Research Office. It features a series of "microelectronics professional incubating workshops" within a floor space of 6,000 square meters. The incubator will provide up to 30 incubating microelectronics enterprises with financing, design platforms, analysis, testing, evaluation, and intermediary services. The municipal government of Beijing is contributing land, loans, and the tax relief. See FBIS, November 26, 2000, translation of interview by Zhou Wei of Zhu Lilan, "PRC Official Discusses Nurturing High-Tech Enterprises," *Jingji Ribao*, September 22, 2000, p. 5 (CPP20000926000052).

[85]Students receive grants of 5 to 10 million yen to pursue independent research themes. In the Tohoku University Venture Business Laboratory, the main research themes are micro-machining, integrated microsystems and microelectronics, micro-optics and optoelectronic mechanical systems. "Toward the Creation of New Industries," *Tokyo Trigger*, December 1998.

[86]ERSO (Electronics Research and Service Organization) is a government-funded, nonprofit applied research institute. It develops microelectronics technologies that are transferred to Taiwanese

• *Governments provide the funds for development and commercialization of new technologies through these government-supported ventures.*

Japan, which has come to view venture businesses as the driving force behind U.S. economic growth in the 1990s, has deployed a broad array of government financial support measures to fund startups in microelectronics. This includes MITI's Support Program for System LSI Development.[87] There are as well benefits available under the New Business Creation Law[88] and the Creative Activity Law[89], and funding by the Japan Development Bank.[90]

In Korea the government has established dozens of venture capital companies to provide direct equity injections, equity, loans, and managerial advice to small- and medium-size enterprises; although some of these venture companies have been nominally privatized and have received some private funds, they continue to obtain funds from ministry "promotion funds" for industry.[91]

---

companies, and many of its employees leave after five to six years to work at Taiwanese electronics firms. ERSO is best known for its practice of spinning off pieces of itself to form successful microelectronics enterprises, such as UMC, Vanguard, and the Taiwan Mask Corp.

[87]This program, established in 1998, is designed to provide government financial support to small semiconductor design companies and "contribute to strengthening the design capabilities of our country's semiconductor industry." MITI Machinery and Information Industries Bureau. See *Denshi Kogyo Nenkan 1998*, Tokyo: Dempa Publications, Inc., May 10, 1998, p. 212.

[88]This law provides for R&D subsidies to small- and medium-size ventures to develop new technologies. One current program funded under this law supports development of system-on-a-chip advanced design technology. *Tsushansho Koho*, May 16, 2000, pp. 1-4.

[89]Under this law a small- or medium-sized venture whose business plan is approved by a prefectural governor can receive low-interest loans, tax incentives, and assistance in lining up investors in order to develop new technologies. Venture businesses benefiting from this law include the Nomura Electronics Technology Research Institute (researching VLSI composed of ceramic-type condensers); Tekunosemu (development of test contracts for the next generation of semiconductors); and Futo Electronics Industry Co. (development of high-quality glass packaging for semiconductor sensors). *Nikkan Kogyo Shimbun*, p. 3, April 12, 1997; *Nihon Keizai Shimbun*, March 29, 1996, p. 1; MITI SMEA Web site at <http://www.chusho.miti.go.jp>.

[90]In April 1995 the Japan Development Bank established the New Business Support Program to fund venture businesses. This program allows small firms to pledge their intellectual property as collateral, and most of the initial loans went to ventures in the semiconductor and computer industries. One of these laws went to the multimedia venture, Dome, to develop a "semiconductor for animation compression." In 1997, however, Dome filed for bankruptcy. *Yomiuri Shimbun,* July 8, 1995, p. 17; *Nikkei Sangyo Shimbun,* November 21, 1996, p. 1; July 7, 1997, p. 24.

[91]Korea Development Bank Web site at <http://www.kdb.co.kr/web>; KTB Venture Capital Web site at <http://ktbuc.com/main.htm/>; Korea Technology and Banking Network Web site at <http://ktb.co.kr/eng/investment.htm>; "ROKG, Business to Raise W150 Billion for IT Venture Fund," *Korea Times*, January 12, 2000; Press release, Ministry of Science and Technology, *KAIST's High Technology Complex (HTC) Now Open*, March 12, 1999, http://www.most.go.kr/press-e/51.htm.

• *Governments continue to support the large semiconductor producers through the venture programs.*

Although the beneficiaries of government venture backing in the first instance are small- and medium-size enterprises, the ultimate beneficiaries are often the large, internationally competitive firms. In Taiwan several of the more successful smaller companies arising from ERSO activities have been absorbed into TSMC and UMC, a pattern that relieves the large private producers of the costs borne by ERSO for semiconductor projects that fail.

In Korea the explicit aim of supporting small- and medium-size enterprise ventures in the semiconductor area is to provide the semiconductor chaebol with less expensive local inputs and equipment, and R&D projects that are designed to support chaebol product lines are given priority in receiving government research support.[92] The government has also partnered with the chaebol in investing venture capital in U.S. firms that can provide Korean industry with promising technologies.

New Silicon Valleys?

Every country outside the United States that has made a major effort to promote an indigenous semiconductor industry has implemented programs to replicate what it views as the best features of Silicon Valley. Such features include the clustering of mutually supportive enterprises and research organizations within a limited geographic area.[93] Typically the government sets aside land for high-technology industrial parks and offers investors incentives for locating there. The sites are chosen in areas that highly educated scientists and engineers would find attractive places to live and work or, at least, preferable to other alternatives in the region.

The government usually ensures onsite technical and infrastructural support in the form of applied research institutes, nearby universities with good electronics programs, national laboratories, secure supplies of electricity and pure water,

[92]For example, in 1996 the Korean government, academia, and industry established a program to assist the transfer of technology to small- and medium-size enterprises, the University Industry Technology Force (UNITEF). A small- or medium-size enterprise facing a technological hurdle can apply to UNITEF for assistance and, if approved, a research task can be assigned to one of 500 professors taking part in the program. The chaebol help govern and manage this program, and small- and medium-size enterprise subcontractors of the chaebol are given priority in receiving technical support through UNITEF. See "Korean Engineering Professors from a National Group," Asian Technology Information Program, May 9, 1997 (Report AT1P97.042), no page citation, http://www.atip.or.jp/public/atip.reports.97/atip97.042html.

[93]Annalee Saxenien's review of the growth of Silicon Valley provides a recent example of the cluster phenomenon. Annalee Saxenian, *Regional Advantage: Culture and Competition in Silicon Valley and Route 128*, Cambridge, Mass.: Harvard University Press, 1994. For a multi-faceted view, see Martin Kenney, (ed.), *Understanding Silicon Valley: The Anatomy of an Entrepreneurial Region*, Stanford: Stanford University Press, 2000.

and even pilot integrated-circuit production lines to assist local companies in refining process skills. Some of these initiatives have failed or face prospects that are at best uncertain. However, a number of "new Silicon Valleys" have succeeded in attracting a critical mass of semiconductor device makers and various supporting enterprises in concentrated geographic zones that offer significant operational and cost advantages.

- *Hsinchu Park*

Taiwan's Hsinchu Science-Based Industrial Park is universally acknowledged as a dramatic success, reflecting not only the generous incentives available to investors but also the presence near the park of (what is probably the foremost applied research organization in the world) the Industrial Technology and Research Institute (ITRI) and its microelectronics affiliates, ERSO.[94]

The production of integrated circuits accounts for nearly 60 percent of the total revenues generated by firms located in the park.[95] Hsinchu and Taiwan's other science-based industrial parks foster close physical proximity of OEMs, device makers, design houses, and other enterprises engaged in the production of IT equipment. This greatly enhances efficiency and shortens cycle times. U.S. companies that operate in the parks comment that "everything we need is right here," and that transactions and interactions that require several days in the United States (because of physical distances) can be achieved in a matter of hours in Hsinchu.

- *Kumamoto Prefecture*

Japan's Kumamoto Prefecture on Kyushu (Japan's "Silicon Island") was one of 26 sites designated in 1984 for development as a "technopolis"; while many technopolis sites failed to thrive, Kumamoto has become a complex of semiconductor businesses, including companies such as NEC, Fujitsu, NTT, Omron, and Tokyo Electron, and the U.S. testing firm Teradyne.[96] In addition to incentives offered at the national level[97] the prefectural government offered investors exemptions from fixed property taxes, reduced enterprise taxes, low-interest loans, and direct subsidies. The prefecture has also sponsored a collaborative semiconductor R&D project to develop sub-0.1-micron manufacturing technology.[98]

---

[94]A newly established domestic or foreign company in Taiwan's science-based industrial parks is exempt from corporate income tax for the first 5 years. The rate of corporate and commodity tax on products or services exported by enterprises in the parks is zero. Taiwan Ministry of Economic Affairs Web site at <http://it.moeaidb.gov.tw/committee/english/b-4.html>.

[95]Hsinchu Science-Based Industrial Park Administration.

[96]*Handotai Sangyo Shimbun*, June 14, 2000, p. 2.

[97]Semiconductor firms locating in technopolis sites qualify for accelerated depreciation of buildings and equipment located in the sites. MITI, *Handbook on Specific Facilities Subject to Special Depreciation Allowances*, August 10, 2000, p. 155.

[98]Kumamoto Technopolis Web site at < http://www.kmt-technopolis.or.jp >.

- *Taedok Valley*

Korea's Taedok Science Town ("Taedok Valley") was established in the 1980s with the objective of geographically concentrating the country's R&D resources. Roughly 37 percent of the 566 venture firms in Taedok are engaged in some aspect of the semiconductor business, leading Korean observers to characterize Taedok as "Venture Valley." Taedok has 70 national, public, and private research institutions and over 14,000 employees holding at least a master's degree.[99] The government is establishing an advanced-technology commercialization center in Taedok to provide support for local venture firms.[100]

- *New Sites in Singapore*

The government of Singapore has earmarked about 160 hectares of land at three sites for the location of up to 25 wafer fabs and various supporting enterprises.[101] These sites (Woodlands, Tampines, and Pasis Ris) are being developed by Singapore's Jurong Town Corporation, a statutory board created to take over the task of developing industrial estates from the Economic Development Board.[102] Since 1995 Jurong Town has "pulled out all the stops to develop facilities—ahead of demand—to attract the big players in the high-value-added wafer-fabrication industry."[103]

- *"Chinese Hsinchus"*

China is "putting into practice what was successful in Taiwan," establishing high-technology development zones with the industrial infrastructure for semiconductor manufacturing and financial and tax incentives for enterprises locating in the zones. The scale of the Chinese effort is ambitious: "It is the kind of setting in which ten science and industrial parks of the order of Hsinchu Science and Industrial Park . . . can spring up all at once." One semiconductor equipment firm predicts that Chinese versions of Hsinchu "will extend over the entire breadth of China in five to ten years."[104]

---

[99]See FBIS, May 4, 2000, translation of Yi Che, "Reportage: Taedok Valley–Over 14,000 Researchers with a Master's Degree or Better," *Chugan Choson*, May 4, 2000 (KPP20000510000022).

[100]FBIS, April 10, 1998, translation of Chong Ch'ang-hun, "Taedok to Become Mecca for Venture Firms," *Chonja Sinmun*, April 10, 1998, p. 1 (FTS19980705000165).

[101]See FBIS, July 29, 1997, reprint of Jennifer Lien, "Germany to Put up First Silicon Wafer Manufacturing Plant," *Singapore Business Times*, July 29 1997, no page citation (FTS19970729000242).

[102]The chairman of Jurong Town Corporation's board, Major General Lim Neo Chian, is also the deputy chairman of the Economic Development Board. Jurong Town implements development plans for high technology at the direction of the government. EDB Singapore, *1998 EDB Board*, EDB website, www.sedb.com/edbcorp/an_1998_02.jsp.

[103]Jurong Town Corporation, *Annual Report 1997/98*.

[104]See FBIS, March 7, 2001, translation of M. Kimura, "Industry, Government and Universities United in Enthusiasm and Talent for LSIs and then LCDs," *Nikkei Microdevices*, March 2001, p. 62 (JPP20010307000001).

EVENTS ABROAD: GOVERNMENT POLICIES AND SIGNIFICANT RECENT TRENDS

The question of whether government interventions in microelectronics are effective or counterproductive and whether they are to be condemned, emulated, or simply ignored will no doubt continue as the subject of active controversy in this country for many years. For purposes of this overview it will suffice to note that government policies outside the United States are contributing to a number of significant trends in the global industry which have implications for the future competitive position of the U.S. industry. These include Japan's attempt to bring about a national revival in microelectronics; Taiwan's continued rapid expansion and its potential influence on the emergence of China as a major competitor; and the resurgence of the European semiconductor industry, particularly in the rapidly growing telecommunications field.

Japan Seeks an Industry Revival

At the beginning of the 1990s the U.S. and Japanese semiconductor industries stood approximately in a position of parity as the dominant players in the global industry. Since then Japan's position has eroded rapidly, with its market share dropping from about 50 percent of the worldwide total in 1990 to 26 percent in 1998. The Japanese industry concentrated heavily on DRAMs, which became a low-priced commodity with the entry into the market of Korean and (more recently) Taiwanese DRAM producers. The U.S. industry, with leadership in microprocessors, rapidly gained market share (expressed in terms of share of total revenue) as the explosion in demand for personal computers drove demand for higher-value logic devices. At present Japanese firms are burdened with high-costs and large levels of debt as a result of over-investments in DRAMs.[105]

Planning a Comeback

The precipitous decline of the Japanese industry's competitive position has led to a wrenching reappraisal within Japan of what has gone so badly wrong and the emergence in 2000 of a coherent strategy for achieving a national revival of Japanese leadership. The government exercises a more important role in this strategy than at any time since the 1970s.

---

[105]"[T]hat which felt the greatest brunt of the collapse in DRAM prices and the wave of the silicon cycle is Japan's semiconductor industry." Japanese industry analysts speak of "the ten years that were lost in the 1990s" in a result of "reorganization and withdrawal from the DRAM business." See FBIS, August 1, 2001, translation of Michio Mizogami, "Prescription for Japan's Revival," *Nikkei Microdevices*, August 2000, pp. 207-09 (JPP20000817000061).

Leaders of the Japanese semiconductor industry have been unsparing of themselves in analyzing the missteps of the past decade. The have publicly acknowledged strategic blunders (like the over-commitment to DRAMs), complacency about the U.S. resurgence and the rise of Korea and Taiwan, and the erosion of their leadership in manufacturing efficiency.[106] The implications of the foundry business model pioneered by Taiwan were not appreciated by the Japanese industry, it is now acknowledged, until the very end of the decade of the 1990s.[107]

Japanese industry leaders also note that the United States undertook a comprehensive and successful collaborative industry-government effort "to incorporate scientific methods into semiconductor production technology, production management technology, on-site maintenance, and the like," while Japan by contrast "lost the sharing of roles among government, industries and universities" and thus "lost our national strategy for the LSI industry." Similarly, the strong backing of governments in Korea and Taiwan is seen by these leaders as a major reason underlying why "Japan was usurped of its position" by industries based in these countries, "just as the U.S. was by Japan 20 years earlier."[108]

Japanese Consortia

In April 1994, 10 major Japanese semiconductor manufacturers established the Semiconductor Industry Research Institute Japan (SIRIJ), which was to serve as a think tank for the Japanese semiconductor industry, analyzing the overall industrial environment, formulating strategy, and drafting industrial policies.[109]

---

[106]One Japanese observer recalls that at the time U.S. firms began emphasizing semiconductor manufacturing science, the Japanese industry took a disparaging attitude. Speaking of biannual symposia on semiconductor manufacturing science that were held in the United States, he notes: "Guest speakers from Japan were invited to the symposia. One Japanese speaker jokes that the level of papers from the U.S. was low. In Japan was there such a view, or was it [rather] that the level of U.S. papers was ignored completely?" *Ibid.*

[107]The first significant foundry relationships between Japanese device makers and Taiwanese foundries were established in 1999. See FBIS, March 30, 1999, translation of "NGC and Hitachi Join Other Three Big Makers in Commissioning Taiwanese Companies with their LSI Production," *Kagaku Kogyo Nippo,* p. 11 (FTS19990427000356).

[108]See FBIS, November 2, 2001, translation of Hajime Sasaki, "Lack of Awareness of LSIs as a Leading Industry, Foundation of Industry Weakening," *Nikkei Microdevices*, November 2000, pp. 160-65 (JPP20001108000046) and FBIS, September 1, 2000, translation of Michio Mizogami, "Japanese LSI, LCD Makers Take Note, Part 2: Management Holding Down Production Technology is Reason for Japan's Decline," *Nikkei Microdevices*, September 2000, pp. 180-82 (JPP20000911000014).

[109]SIRIJ, *Outline of the Semiconductor Industry Research Institute Japan*. The intention of the member companies was to create a semiconductor-specific policy body similar to the Semiconductor Industry Association in the United States. The information was obtained through interviews by Maki Hishikawa with government officials and industry executives in Japan in July 2000.

"This move can be described as an attempt by the Japanese manufacturers, whose market share has stagnated, to stage a comeback."[110] SIRIJ has developed a succession of policy recommendations for government promotional measures that have subsequently been adopted by the Japanese government.

In March 1995 SIRIJ released a report that called for the establishment of a new generation of R&D consortia.[111] Japanese industry leaders noted with irony that Japan ended its subsidized large-scale joint R&D projects in microelectronics in the 1980s in response to criticism by the United States, while the U.S. launched SEMATECH and seized global leadership from Japan.[112] By the mid-1990s the Japanese government had concluded that the eroding national position in microelectronics was a more serious problem than any that might arise from friction with the United States.

A new generation of government-supported joint R&D projects was initiated, bringing to an end the so-called "15 years blank period," which began at the end of MITI's first VLSI project in 1980. The most important project was the Association of Super-Advanced Electronics Technologies (ASET), a MITI-sponsored consortium engaging 21 Japanese companies to pursue many microelectronics themes, including new forms of lithography (X-ray, laser, electron-beam, extreme-UV). ASET received about $50 billion yen (about $430 million) in government funds between 1995 and 2000. The government's Japan Key Technology Center founded a seven-company consortium to develop 400-mm silicon wafers, the Super Silicon Crystal Research Institute (SSi).[113] In addition to these government-sponsored programs the Japanese industry established two major privately funded research consortia, STARC, which is patterned on the SRC in the

---

[110]FBIS, December 1,1994, Taro Okabe, "Semiconductor Industry Research Institute Resolves Policy Framework," *Nikkei Microdevices*, December 1994, pp. 120-21 (FTS19941201000063).

[111]"Cooperative Research and Standards in the Japanese Electronics Industry—Hideo Setoya, Executive Director of ASET," *Technology Standards & Standardization Processes—Their Impact on Innovation and International Competitiveness*, Stanford: Stanford University/U.S.-Japan Technology Management Center, June 1999, p. 105; and SIRIJ, *Research Institute Profile,* undated.

[112]Japan's government-subsidized joint R&D projects, such as MITI's 1975-80 VLSI Project, "came under fire from the United States during the Japan-U.S. trade friction in the late 1980s; thereafter Japan gave up on using such a system. Ironically, after Japan gave up the cooperation between the government and private sector, the U.S. organization called SEMATECH, which is modeled after the Japanese [VLSI] institute, was inaugurated." See FBIS, October 31, 1995, translation of Emi Yokata, "Japanese Manufacturers to Launch Joint Research on New Chip Technology to Counterattack U.S. Manufacturers," *Ekonomisuto*, October 31, 1995, p. 42 (FTS19951031000245) and FBIS, November 2, 2001, translation of Hajime Sasaki, "Lack of Awareness of LSIs as a Leading Industry, Foundation of Industry Weakening," *Nikkei Microdevices*, November 2000, pp. 160-65 (JPP20001108000046).

[113]*1998 Edition of Nihon Handotai Nenkan*, Tokyo: Press Journal, March 25, 1998, p. 119.

United States, sponsors university-based R&D,[114] and Selete, which concentrates on technology for manufacturing 400-mm wafers.[115]

Revival of the Japanese Semiconductor Industry

In March 1999, amid a sense of crisis arising out of declining semiconductor sales, SIRIJ established a "Semiconductors in the New Century Committee" (SNCC) to draft proposals for a new generation of government-industry R&D projects that were to revive the Japanese semiconductor industry. In March 2000 SNCC delivered a final report, *Proposal: Revival of the Japanese Semiconductor Industry,* to MITI. The report recommended that industry, government, and universities join together in a new generation of cooperative R&D projects. These would build on recently established consortia like Selete and STARC and would emphasize system-on-a-chip technologies. SNCC's chairman declared that "no matter what is said overseas, Japanese [semiconductor producers] who have been defeated in a landslide in the 1990s have no option but to form an all-Japan alliance in order to launch a counterattack."[116]

The Ministry of the Economy, Trade, and Industry (METI)—previously called MITI—appears to be implementing the basic elements of the SIRIJ program. This program calls for a developmental effort aimed at "system on chip, which is said to be the brains of the industry."[117] Japanese industry leaders calculate that the biggest growth markets for semiconductors in the twenty-first century will not be in PC equipment, where the U.S. industry leads, but in cell phones and digital consumer equipment, where Japan leads. The new-generation R&D projects are designed to establish further Japanese dominance in these areas.[118]

---

[114]The Semiconductor Technology Academic Research Center (STARC) is co-funded by 11 Japanese firms. It funds university R&D based on three criteria: (1) R&D for new technologies that will become industry standards or mainstream products; (2) pre-competitive R&D that can be transferred to industry for application in 5-10 years; and (3) R&D that fosters young researchers likely to make future contributions to the industry. (*Denshi*, p. 34, November 1995; *Nikkei Sangyo Shimbun*, November 12, 1998, p. 19; STARC Web site at <http://www.starc.or.jp>.

[115]Semiconductor Leading Edge Technologies (Selete) is a privately funded joint venture company established by 10 Japanese semiconductor producers in 1995 to evaluate next-generation manufacturing equipment and materials. This organization parallels some of the functions of the U.S. SEMATECH consortium and was in fact established as a Japanese alternative to a proposal by SEMATECH for an international joint venture for evaluating next-generation semiconductor manufacturing equipment. *Denshi*, November 1995, p. 34; *Nikkei Sangyo Shimbun,* February 14, 1996, p. 11; *Tokyo Semiconductor,* June 2000, pp. 36-40; Selete *Annual Report Fiscal 1998* and *Fiscal 1999.*

[116]*Nikkei Sangyo Shimbun*, December 6, 1999, p. 8.

[117]*Nikkei Microdevices*, December 1999, pp. 98-103.

[118]See FBIS, January 2, 2001, translation of "From Stagnation to Growth, the Push to Strengthen Design," *Nikkei Microdevices*, pp. 106-124 (JPP20010125000012).

The head of its Machinery and Information Industries Bureau, Akira Kubota, summarized the plan:

> [T]his time the government will take the lead in semiconductor projects that involve joint efforts by industry, government, and universities. Leading-edge technology development will be the government's role. First, we considered building clean rooms at national laboratory sites. That is because nothing will come of silicon technology research themes unless there are clean rooms. For the clean-room construction we appropriated 16.5 billion yen in the FY 1999 supplementary budget. Next, we will launch two semiconductor projects. One is the development of basic next-generation LSI technology using the government's clean rooms. In that effort we will work with industry in the development of new materials, measurement technology, and so forth. The other semiconductor project has to do with the development of equipment, including a small-scale production line.[119]

The specifics of the new government-industry effort were clarified in late 2000 and early 2001, indicating a dramatic resurgence in government support for this sector:

- *Super Clean Room*

METI will build a 4,500-square-meter "super clean room" in Tsukuba Science City to foster industry/government/academia cooperation in the development of system-on-a-chip technology.[120] The project aims primarily at improving the performance of transistors belonging to the 70-nm technology generation through research on high-k gate dielectric process technologies and other relevant technologies. The research targets also include low-k interlayer dielectric process technologies and lithography and mask technologies for the 70- to 50-nm technology generation. METI reportedly sought 8.1 billion yen (about $70 million) in the FY 2001 budget for this project.[121] In all, the government is expected to invest 30 billion yen (about $260 million) in the 7-year project.[122]

- *Future Information Society Creation Laboratory*

The Japanese government plans to invest 30 billion yen (about $260 million) in this 5-year project, which will create a "new, small-scale and very short-term

[119] Akira Kubota, "Semiconductors Support the IT Revolution; Silicon Technology the Key in the Next Decade," *Nikkei Microdevices*, October 2000.

[120] Japan Information Processing Development Corporation, *Information Whitepaper 2000*, (Tokyo: Computer Age, June 16, 2000), p. 406.

[121] See FBIS, October 3, 2000, translation of "Asuka Joint Industry Project for Micro-Lithography System LSI Outlined," *Kagaku Kogyo Nippo*, October 3, 2000, p. 9 (JPP20001004000002).

[122] *Handotai Sangyo Shimbun*, September 6, 2000, p. 1. To construct the clean room 16.5 billion yen (about $139 million) was appropriated from the FY 1999 supplemental budget. See FBIS, October 1, 2000, translation of "Development of 0.07 to 0.05 micron LSI Technology," *Nikkei Microdevices*, October 2000, pp. 67-70 (JPP20001018000105).

microelectronics production line" and develop super-high-speed LSI devices (SOILSI) and high-performance system LSI devices, "rather than in the area of memory, as in the past."[123] The small-scale line is seen as an advantage in producing devices for digital home electronics, an area where consumer preferences change quickly and short delivery time is required. The laboratory is headed by Tohoku University Professor Tadahiro Ohmi, who has been serving on various government advisory councils, including the Industrial Technology Council, an advisory body to METI for industrial technology policy. He is regarded as one of the most influential scholars with respect to government promotion of the semiconductor industry in Japan.[124]

- *"Asuka" Sub 0.10-Micron Project*

In September 2000 it was reported that 11 Japanese semiconductor producers had agreed to invest 76 billion yen (about $750 million) in a 5-year project to develop manufacturing technology for circuit widths of 0.10 to 0.07 micron and less by 2005. STARC will oversee Asuka's R&D with respect to design technology, and Selete will manage development of device and process technology.[125] A staff of 250 researchers from the 11 core companies will be assigned to Selete and 90 to STARC.[126] The Electronic Industries Association of Japan has asked the government for assistance for the Asuka project in four areas.

1. Joint research in METI's Super Clean Room;
2. Research on materials and measurement-related component technologies;
3. Support for semiconductor manufacturing equipment makers' development of leading-edge technologies; and
4. Support for the development of system LSI design technology.[127]

METI reportedly will assist this effort with its own parallel R&D effort and is requesting $60 million for FY 2001 for the first year of an anticipated 7-year effort.[128] (See Table 1.)

---

[123]The small-scale production line is intended to conserve energy and shorten production time. The Japanese team is aiming for a line for which the scale of production can be varied flexibly in units of 2,000 to 3,000 wafers per month, a factor of 10 less than the production units used in a conventional line. *Ibid.*

[124]*Handotai Sangyo Shimbun*, September 6, 2000, p. 1.

[125]See FBIS, November 1, 2001, translation of "Can Next-Generation Semiconductor Project Save Japan?" *Nikkei Microdevices*, November 2000, pp. 12-14 (JPP20001103000001).

[126]See FBIS, March 16, 2001, translation of Yatsuka Yoshida, "Asuka Project Targets Semiconductor Industry Revival," *Tokyo Trigger*, March 2001, pp. 14-15 (JPP20010316000013).

[127]See FBIS, October 4, 2000, translation of "Asuka Joint Industry Project for Micro-Lithography System LSI Outlined," *Kagaku Kogyo Nippo*, October 3, 2000, p. 9 (JPP20001004000002).

[128]*Nihon Keizai Shimbun*, September 20, 2000, p. 3.

TABLE 1 Government-Supported Microelectronics R&D Initiatives Outside the United States

Country	Project	Research Period	Government Contribution	Themes
Japan	Next-Generation Semiconductor R&D Center (super clean room)	2001-08	$300 million ($60 million in 2001)[a]	Process and device technology for 70-mm generation
Japan	Future Information Society Creation Laboratory	2001-06	$300 million	Create small-scale, very short-term semiconductor production line
Japan	Asuka	2001-06	Will use METI super clean room	Develop design technologies for 0.10- to 0.07-micron system-on-a-chip and device process technologies
Japan	NEDO projects	2001-	Budget for one NEDO project, development of a gas-cluster ion-beam system is reportedly 2 billion yen (about $20 million)	Cluster ion-beam process technology, system-on-a-chip design technology, advanced parallel-compiler technology
Japan	ASET	1995-	$500 million	Lithography, semiconductor manufacturing technology
Japan	Selete	1996-	[b]	Manufacturing technology for 300-mm wafers
Japan	STARC	1995-	[c]	Basic research
European Union	MEDEA	1997-2000	$720 million (est.)	Process technology, design, applications
European Union	MEDEA Plus	2001-09	$1,350 million (est.)	Systems-on-a-chip, UV lithography
European Union	PIDEA	1998-2002	$135 million (est.)	Packaging and interconnection
Germany	Semiconductor 300	1996-2000	$680 million	300-mm wafer technology
France	Crolles I and II	1998-	$136 million (est.)[d]	Pilot 300-mm fab
Belgium	IMEC	Permanent research institute	$40 million/year (est.)	System-on-a-chip designs; next-generation (sub 0.10-micron) production technology; packaging
France	LETI (GRESSI, PLATO and PREVUE)	Permanent government laboratory	Performs R&D for transfer to industry	CMOS technologies, alternatives to CMOS, UV silicon semiconductor technology
Germany	BMFT programs	Permanent	Direct funding of R&D and contributions to FHG institutes of applied research	Flexible manufacturing, microsystems, non-silicon semiconductor technology
Taiwan	ASTRO	2000-	Government will fund half	Technology induction, upgrading of local industry

[a]METI requested $60 million in FY2001 budget for first year of a 7-year project.
[b]Privately funded but received NEDO contract to develop technology to cut PFC use.
[c]Mostly private funding; Key Technology Center provided subsidy for CAD software development.
[d]Crolles I reportedly received subsidies of FF 900 million to FF 1 billion. Additional subsidies have been requested for Crolles II.

- *NEDO Initiatives*

METI's satellite R&D organization, the New Energy and Industrial Technology Development Organization (NEDO), will contract with domestic organizations to foster "cluster ion-beam process technology," "system-on-a-chip advanced design technology," and "advanced parallel-compiler technology."[129] One of these projects got under way in October 2000: a 5-year effort to develop a full-scale gas cluster ion-beam system for ultra-low-energy ion implantation, lateral surface sputtering, and formation of high-quality thin films.

The project, based in Kyoto University, involves a variety of private companies and government and university research institutes. The budget is 2 billion yen (about $20 million). The director of the project, Isao Yamada, explained that given the relative scarcity of venture capital in Japan to support R&D projects of this scale, "the development of key technologies such as semiconductor processes can only be carried out in Japan with financial assistance from the national government."[130]

- *Nanotechnology Initiatives*

The Japanese government has since 1985 been promoting nanotechnology R&D; Japanese scientists believe their country is dominant in this field.[131] However, the Clinton administration's announcement of the U.S. National Nanotechnology Initiative caused a "nanotech shock" in Japan, which saw a U.S. threat to its perceived leadership. METI announced plans to form a Japanese Nanotechnology Consortium, and plans were announced to set up a Nanomaterial Research Center linked to industry and academia. METI's Electrotechnical Laboratory is "pushing forward with the front line" of nanotechnology research, examining themes such as three-dimensional optical device structures and devices in which memory can be read or written "by means of a single photon."[132] The government's fiscal 2001 budget for nanotechnology research was increased by 25 percent, to 38.2 billion yen.[133]

---

[129]*Tsusansho Koho*, March 24, 2000, p. 19.

[130]See FBIS, March 1, 2001, translation of "Japan Pioneers Ultra-Low-Energy, High-Density Ion Implantation," *Tokyo Trigger*, March 2001, pp. 31-33 (JPP20010327000004).

[131]See FBIS, December 1, 2000, translation of Kazuhiro Yoshihura, "World Expanding With Nano Technology," *Kagaku Gijutsu Janaru*, December 2000, pp. 20-21 (JPP20010118000005); FBIS, March 1, 2001, translation of "The Full Story on Nanotechnology; the U.S. Aims for it after IT! Japan Bets Recovery on it!" *Tokyo Trigger*, March 1, 2001, pp. 6-10 (JPP20001031600001l). For a description of early Japanese R&D efforts in this field see JTECH Panel Report on *The Japanese Exploratory Research for Advanced Technology (ERATO) Program*, December 1988, pp. 4.17-4.19.

[132]See FBIS, March 1, 2001, translation of K. Amagasa, "Nano-Intelligent Materials That Utilize Self-Organization," *Tokyo Trigger*, March 2001, pp. 28-30 (JPP20010327000012).

[133]See FBIS, February 23, 2001, translation of "Nanontechnology Challenge: Hidden Power in Nanometer Size, Key to Japan's Revival," *Nikkei Sangyo Shimbun*, February 19, 2001, p. 1 (Internet

• *Promoting Design Capability*

METI secured $68 million in its FY 2000 budget for "strengthening semiconductor design capabilities" and "semiconductor microprocessing basic technology."[134] It requested about $100 million in the FY 2001 budget for R&D on microelectronics-related subjects, "system-on-a-chip advanced design technologies," and semiconductor device process technologies."[135]

Telecommunications Policy

The Japanese effort to challenge U.S. leadership in microelectronics emphasizing non-PC-based applications has been assisted both directly and indirectly by Japanese telecommunications policy. Nippon Telephone and Telegraph (NTT) holds a dominant position in Japan's telecommunications market. Although NTT has styled itself a private entity since 1985, the government retains a 53.15 percent equity stake; NTT remains in many respects a public organization. NTT—the most profitable enterprise in Japan—utilizes revenues garnered from its near-monopoly of telephone service to fund R&D in telecommunications-related fields, including microelectronics.[136] In recent years NTT has pursued research themes emphasizing applications with respect to cellular phones and other forms of wireless communications.[137]

The deregulation of telecommunications in the United States is widely cited as a key factor in U.S. dominance of the Internet in the 1990s, and NTT's regulatory policies—which have restricted competition and prevented low-cost Internet access in Japan—are widely criticized in Japan as the main reason Japan has lagged in incorporating Internet technology throughout its economy.[138]

---

version of *Nikkei Sangyo Shimbun* monitored via Nikkei Telecom 21 website) (JPP20010223000047) and FBIS, January 1, 2001, translation of "United States, Japan Compete for Nanotechnology Research," *Nikkon Kogyo Shimbun* (Nikkei Telecom Database Version), January 1, 2001, p. 3 (JPP20010125000031).

[134]Japan Information Processing Development Corporation, *op. cit.,* p. 408.

[135]*Handotai Sangyo Shimbun*, September 13, 2000, p. 3.

[136]*Nikkei Weekly,* May 29, 2000, p. 1.

[137]In August 2000 the NTT Science and Core Technology Laboratory Group announced that it had developed LSIs in which the structure of the circuit could be freely modified in accordance with the type of data handled for image processing and voice recognition. The new device will be cost effective because it is versatile and can be used in products with different standards, such as cell phones. NTT's semiconductor R&D has emphasized low-energy wireless circuits, low-energy analog/digital circuits, and low-energy LSI manufacturing processes. The objective of these efforts is to reduce the power consumption of semiconductors used in communications equipment. *Nihon Keizai Shimbun*, August 26, 2000, p. 15; NTT, *R&D in Review for 1998*, pp. 29-33.

[138]See FBIS, May 19, 2000, translation of "Sakakibara on NTT Connection Fee Cuts," *Mainichi Shimbun*, May 14, 2000, Morning Edition (Nikkei Telecom Database Version) (JPP20000519000006) and FBIS, November 1, 2001, translation of "IT Strategy Head Interviewed on Obstacles to IT Revolution," *Chuo Koron*, November 1, 2000, pp. 64-71 (JPP20001019000040).

However, with respect to wireless communications, U.S. regulatory policy has fostered the emergence of five incompatible wireless standards, making it difficult to establish a uniform subscriber base. By contrast, Japan utilizes a single standard and partly as a result is far ahead of the United States in wireless communications.[139] Most Japanese semiconductor makers are placing a high priority on developing devices that build this growing leadership, which they see as a high-volume technology driver enhancing their overall competitiveness.[140]

IT Strategy

In the latter half of 2000 the Japanese government launched a "Basic IT Strategy," a sweeping array of promotional measures and legal reforms with the stated objective of overtaking the United States as a "high-speed Internet superpower" within five years.[141] One element of this plan is an ambitious public works spending program designed to link most of Japan's households with high-speed fiber-optic broadband connections by 2005. This would give Japan far more pervasive broadband coverage than the United States.[142]

Japanese electronics hardware manufacturers are developing products alongside which will take advantage of Japan's near universal broadband coverage. These include interactive televisions and home appliances. These may, at least

---

[139]"[I]n Japan wireless technology is two years ahead of Europe, and Europe is two years ahead of the United States," in Ray Tsuchiyama, "Deconstructing Phone Culture: How Japan Became a Leader in Mobile Internet," *ACCJ Journal*, July 2000, p. 14. NTT's 67 percent-owned wireless subsidiary, NTT DoCoMo, holds a near-monopoly on domestic wireless communications. NTT DoCoMo developed a wireless Japan-only digital standard, PDC, and subsequently opened it to other carriers, who adopted it. NTT DoCoMo pioneered so-called third-generation (3G) wireless service, introducing it ahead of Europe and the United States, with corresponding advantages for the suppliers of 3G products. NTT DoCoMo introduced wireless Internet access service in February 1999, and this has proven so extraordinarily popular that wireless access is likely to leapfrog PC-based access as Japan's primary medium for connecting with the Internet.

[140]The CEO of Rohm Corporation, J. Hitaka, commented in January 2001: "It's safe to say that the cell phone is now the semiconductor technology driver . . . . No matter what, we will stick it out in the production of system LSIs for cell phones . . . [T]he percentage of semiconductors carried in a cell phone is high. We can say that cell phones are semiconductor canned goods. And they are produced in volumes of hundreds of millions of units. That is why we absolutely want to capture the market [for system LSIs for] cell phones." "The Game—System LSIs for Cell Phones; Making the Most of Equipment Know-How in Design," *Nikkei Microdevices*, January 2001.

[141]This objective was fixed by the IT Strategy Council, a blue-ribbon industry-government committee chaired by Sony Chairman Naboyuki Idei, established at the direction of Prime Minister Yoshiro Mori, *Nikkei Weekly*, September 4, 2000.

[142]One Japanese observer commented that "the day that NTT announces the completion of turning all Japan into an 'optical empire' will be the 'X day' that the United States fears most," in FBIS, August 1, 2000, translation of Kazunori Ishiguro, "IT War: US Strategy to Contain the NTT Optical Empire," *Shokun*, pp. 40-51 (JPP20000821000008). See also FBIS, November 15, 2000, "Japan Ministry Contracts 5 Firms for Development of High Speed Optical Network," *Nikkan Kogyo Shimbun*, November 9, 2001, p. 1 (JPP20011115000004).

partially, displace the PC and by so doing "blast a hole in U.S. dominance of the Internet."[143] MITI's investment in a small-scale production line is expressly intended to enable Japanese semiconductor manufacturers to produce devices for digital home appliances running on broadband connections.

The Japanese national microelectronics effort has taken several years to launch and has been characterized by no small amount of confusion and disagreement.[144] But the outline of a coherent plan has now been established and very substantial public and private resources have been committed to implementing that plan. Professor Tadahiro Omi, who runs the project tasked with developing a small-scale production line, says of this effort that "Japan is pursuing the way to victory."[145]

European Resurgence

The European semiconductor industry—long the butt of criticism from both inside and outside of Europe—was thought at the beginning of the 1990 to be in a state of irretrievable decline.[146] The billions of dollars of government subsidies poured into attempts to reverse the industry's fortunes appeared to have no effect. Various strategies based on consortia, national champions, and bureaucratic community-level programs seemed wasteful and futile.

This European decline has reversed through the 1990s.[147] All three of Europe's semiconductor firms are in the top 10 worldwide and are growing faster than some of the other top 10 firms.[148] The European technology base is world class, and Europe is said to be on the leading edge in developing next-generation

---

[143] See FBIS, December 1, 2000, translation of Osamu Katayama, "With the Start of Digital TV Broadcasts, Japanese Home Appliance Makers Now Engaged in a Fierce Struggle for Dominance," *Chuo Koron*, December 1, 2000, pp. 216-23 (JPP20001270000051).

[144] "'What the heck is going on?' For awhile that expression was the most apt way to describe attitudes toward the semiconductor projects," in FBIS, October 1, 2000, translation of "Development of 0.07 to 0.05 micron LSI Technology," *Nikkei Microdevices*, October 2000, pp. 67-70 (JPP20001018000105).

[145] *Ibid.*

[146] EC Commission, DGXIII, Telecommunications, Information and Industries and Innovations, Communication from the Commission to the Council, *The European Electronics and Information Technology Industry: State of Play, Issues at Stake, and Proposals for Action,* SEC (19) 565, April 3, 1991; European Report No. 1608 (August 4, 1990).

[147] A Japanese observer commented in September 2000: "Ten years ago, European [semiconductor] manufacturers continued to lose profits and there were rumors of withdrawals from the semiconductor business, but recent developments have been remarkable and all three [European] companies are in good shape," in Mizogami, *op. cit.*

[148] The European firms are Infineon Technologies AG (formerly Siemens), Philips, and STMicroelectronics.

manufacturing technologies.[149] A recent study by the European Electrical Engineering, Electronics, Information Technology Association (VDE) predicted that, based on current observable trends, Germany along with the rest of Europe would lead the world in microelectronics and nanotechnology by the year 2010, "followed by the United States and Japan/Asia." The VDE chairman, while endorsing these findings, remarked, "I caution against arrogance."[150]

One of the major factors in Europe's remaining success within the microelectronics field has been the region's strength in non-computer-related OEM markets. These relate primarily to telecommunications but also involve automotive electronics, smart cards, and multimedia consumer products. European producers are far less heavily invested in the production of standard memory chips or major PC microprocessors than are producers in the rest of the world. As a result they have been less affected by the volatility experienced in those product areas.[151] Most European fabs combine analog with digital circuitry on the same chip to make semi-custom parts for telecom and embedded system applications.

European fabs are have also learned to efficiently place increasing amounts of interconnectivity onto a complex integrated circuit instead of a circuit board, as well as to mix intellectual property cores on the same chip.[152] In the mid-1990s the telecommunications sector began to grow rapidly, "and it became clear that the communications market would [eventually] stand side-by-side with the PC and public service sectors in providing traction for the LSI market."[153] A number

---

[149]"An Executive Report on the MEDEA Programme: Mid-Term Assessment," *Future Horizons*, online report, http://www.minez.nl/kamerbrieven1999/execut.pdf, *accessed on* July 19, 1999. ASML's proposed acquisition of SVGL, the principal remaining U.S. lithography firm, supports this perspective, although at this writing the acquisition has not yet been approved.

[150]See FBIS, October 11, 2000, translation of D. Wendeln, "Europe on the Way to World Class in Microtechnology and Nanotechnology," *VDI Nachrichten*, October 6, 2000, (UEP20001011000369).

[151]See FBIS, October 6, 1995, translation of "Largest Eureka Project a Resounding Success," *Economische Zaken*, October 6, 1995, pp. 8-9 (FTS19951006000613); FBIS, May 24, 1996, translation of "Remarkable Results of JESSI 'Phase Shift' Project," *Computable*, May 24 1996, p. 3 (FTS19960524001698); "European Market Roundup: Continued Growth Predicted Through 2001". *Solid State Technology*, March 1, 2000, p. S10.

[152]At the STMicroelectronics fab in Crolles, France, chip designers now work at workstations right in the fab, and SOITEC has built a fab manufacturing silicon-on-insulator wafers directly adjacent to the STMicroelectronics fab. "Fab Trends: Are Europe and Asia Leading the Way?" *Solid-State Technology*, March 1, 2000, p. 8.

[153]"European LSI manufacturers are riding on a growth track for 2001 using the sudden growth of the LSI market for communications as the tailwind. Until now U.S. forces have primarily controlled the personal computer (PC) field, which has borne the role of traction for the LSI market, and Japanese manufacturers have controlled the public sector. European LSI manufacturers, on the other hand, have focused heavily on the communications field. The communications sector began growing rapidly in the latter half of the 1990s, thus indicating that the strategy of European LSI manufacturers was right on target. Rapid growth is expected for the communications sector after 2001, and European manufacturers intend to ride this wave and are making necessary preparations to achieve high growth,"

of industry observers believe that the European semiconductor producers' strategy of concentrating their efforts in this area is now paying off.[154]

Europeans freely concede that their often derided, sometimes cumbersome government promotional programs have made an important difference in the rebound of their semiconductor industry. These include:

- *"Top down" pre-competitive research and development*

"Top down" pre-competitive research and development, sponsored, planned, and implemented pursuant to the European Commission's Framework Programs, is funded in part from the European Community budget. This category embraced the ESPRIT project (concluded in 1998) and now involves a substantial work program in microelectronics pursuant to the European Union's Fifth Framework R&D program. The Essential Technologies and Infrastructures division of the Fifth Framework, of which microelectronics and optoelectronics comprise a significant subcategory, is being funded at the level of 1.1 billion euros between 1999 and 2002.[155] The microelectronics effort is concentrated on system-on-a-chip technology and telecommunications devices, where the European industry sees its greatest future advantage.[156]

- *"Bottom up" joint R&D by European companies*

"Bottom up" joint R&D initiatives by European companies to develop marketable products under the rubric of the EUREKA program are often funded by national governments. The first major microelectronics project of this kind was JESSI (1988-96), a $3.6 billion effort widely credited with a dramatic improvement in the competitiveness of European semiconductor makers.[157] Among JESSI

in FBIS, January 1, 2001, translation of "3 major European LSI Makers Show Stable Growth Through Large Investments," *Nikkei Microdevices*, January 1, 2001, pp. 88-93 (JPP20010131000003).

[154]Ulrich Schumacher, CEO of Germany's Infineon Technologies AG (formerly Siemens), made the following comments in January 2001: "[I]n the communications sector, we are already the world's leading company in wireless communications. We will strive to become the world's leader in the wired communication sector in 2001. Concretely, we are the world's leader in the fields of RF, ISDN, VDSL (very high-bit-rate digital subscriber line), and fiber technology, and we are maintaining the number two position in the base band processing field. Next, we intend to reach the number one position in the remaining fields, and then realize a 'full system solution' for which all of these fields are combined with all functions realized in one chip." FBIS, January 2001, translation of Ulrich Schumacher, "2001 to be a Good Year, Expand DRAM in Areas other than PC," *Nikkei Microdevices*, January 2001 (JPP20010131000003).

[155]*IST Workprogramme 2000* at <http://www.cordis.lu/ist/present.htm>.

[156]IST Workprogramme 2000 URL is <http://222.cordis.lu/ist/bwp_en10.htm>.

[157]"European Market Roundup: Continued Growth Predicted Through 2001," *Solid-State Technology*, March 1, 2000, S10. In a 1996 French analysis of the rebound of Europe's semiconductor industry three factors were cited: (1) a strong market in Europe, (2) restructuring by the European Union producers, and (3) JESSI. "And the third element playing a role in the European industry's recovery was JESSI. Launched in 1989, this research program enabled the closing of the European industry's

accomplishments were "[contributing] substantially to IMEC's ascent as the leading center for microelectronics R&D in Europe,"[158] the development of two devices that "buttressed . . . Europe's leading role in the sector of cellular telephones," and the fostering of much closer cooperation between European universities and the semiconductor industry.[159]

JESSI was succeeded by the MEDEA microelectronics project, which began in 1997. It will conclude at the end of 2000 and be succeeded by MEDEAPlus, a 9-year follow-up program with an annual budget of 500 million euros, of which about 40 percent will be government funded.[160] MEDEA has reportedly yielded good results in developing technologies to support chips in the telecommunications, multimedia, and automobile industries.[161] A related EUREKA project, PIDEA, has been established to address packaging and interconnection technologies, subjects that were not covered by MEDEA.[162]

---

technological gap. 'We are now on a par with the best,' is the word repeatedly heard at SGS-Thomson. JESSI also contributed to a strengthening of the ties upstream between the European manufacturers of semiconductors and their suppliers of machines and materials and, above all, downstream with their clients. This has produced virtual vertical integration that has enabled a closing in on the Japanese position. The growth of the substantial role being played by customized IC's at SGS-Thomson illustrates the importance of the part that JESSI has played in this regard." See FBIS, January 30, 1996, translation of Philippe Le Coeur, "European Semiconductors Industry in Economic Turnaround," *Le Monde*, January 30, 1996, p. 15 (FTS19960130001915).

[158]See FBIS, January 26, 1996, translation of Richard Sietmann, "With Research and Development in First Place; Belgian Microelectronics Institute IMEC in Demand as Partner for Cooperation in International Research and Development," *VDI Nachrichten*, January 26, 1996, p. 13 (FTS1996012600087).

[159]See FBIS, October 5, 1995, translation of "JESSI Has Developed Single-Chip Solutions for GSM Telephones," *Mikroelektronik und Mikrosystemtechnik* No. 4, 1995, p. 52 (FTS19951005000666).

[160]See FBIS, June 9, 2000, translation of "Billions for Chip Research," *VDI Nachrichten*, p. 31 (EUP200006226).

[161]See FBIS, October 13, 1999, translation of "No End to Global Competition," *Elektronik*, September 21, 1999, pp. 16, 18 (FTS19991013001537). "The GSM telephone, DAB (Digital Audio Broadcasting) digital radio, the MPEG format for compressing video images, and even the DVB (Digital Video Broadcasting) are all European inventions and originated within JESSI and MEDEA." See FBIS, May 24, 2000, translation of "The Strengths of Research Europe," *La Tribune*, May 24, 2000 (EUP20000524000474).

[162]Asked in retrospect what should have been included in MEDEA that was left out, MEDEA Chairman Jürgen Knorr responded, "Without a doubt the packaging and testing of these chips! It does no good for the chip to have the desired high operating speed if problems in the wiring of the chip and bottlenecks in packaging technology make it impossible to take advantage of that speed!" see FBIS, October 13, 1999 translation of "No End to Global Competition," *Elecktronik,* September 21, 1999, pp. 16, 18 (FTS 19991013001537). PIDEA's budget is 400 million euros, of which various governments will provide 30 to 50 percent. See FBIS, December 11, 1997, translation of Frederic Fassot, "A European Plan for Passive Components," *Electronique International Hebdo*, December 11, 1007, pp. 1, 12 (FTS19980122000510).

• *National and regional government financial support for semiconductor R&D and state-of-the-art production facilities*

The German federal government and the state government of Saxony heavily subsidized a project to build a pilot 300-mm wafer fabrication facility in Dresden. This became the first such facility in the world to become operational.[163] A second 300-mm fab is also being established at the Dresden site by a new entity—Infineon Technologies SC300 GmbH & Co. The government of Saxony will hold an equity position of 115 million euros in this and the German Federal Ministry of Education and Technology is expected to contribute another 75 million euros to this project.[164]

A similar effort is being sponsored with government assistance at a site in Crolles, France. In 1998 ST Microelectronics established a research facility and a pilot 300-mm fab with government subsidies estimated at Fr 900 million to Fr 1 billion.[165] In April 2000 ST Microelectronics and Philips announced plans for a second 300-mm fab at Crolles that would serve as a research facility and a commercial fab whose output the two firms would share equally. They reportedly planned to ask the French government for contributions to the capital and research costs of this project.[166]

• *National and regional government support for major microelectronics research laboratories and organizations*

IMEC, founded by the regional government of Flanders in Belgium, is a large, independent, extremely highly regarded center that contracts with European governments and European and foreign companies to conduct microelectronics R&D. IMEC has spun off numerous semiconductor enterprises and has won global acclaim for a succession of technological achievements.[167] The gov-

[163]See John Baligu, "Top Fabs of 2000," *Cahners Semiconductor International*, September 7, 2000, pp. 1-4, http://www.semiconductor.net/semiconductor/issues/2000/200005/six0005fab.asp. This plant was achieving yields of 90 percent for 64-Mbit DRAMs by March 2000. "Infineon 300-mm Processes Achieve 90 percent yields on 64-Mbit DRAMs," *Semiconductor Business News*, April 3, 2000.

[164]"Infineon Plans $1 Billion Move to 300-mm Production in Dresden," Semiconductor Business News, March 31, 2000, <http://www.siliconstrategies.com/story/OEG20000331S0023>.

[165]Of this total, 200 million was contributed by Department of Isere and 200 million by local administrations. See FBIS, May 13, 1998, translation of Gilles Musi, "Strategy – SGS-Thomson Investing Almost 3 Billion Francs in Grenoble," *La Tribune* (Internet Version), May 12, 1998, (FTS19980513000697) and FBIS, May 12, 1998, translation of Anne Chatel-Demerge, "The SGS Thomson Group Invests 3 Billion Francs in Grenoble," *Les Echos*, May 12, 1998, p. 13 (FTS19980512000875).

[166]See "Chip Giants Prep. Fabs as Wafer Suppliers Gear Up – Europe, Japan Tool 3mm," *Electronic Engineering Times*, May 1, 2000. (Nexis reprint).

[167]IMEC researchers were the first in the world to make a CMOS image-recording semiconductor that performs as well as a charge-coupled device (CCD), a key component in cameras and camcorders. The CMOS device is far less expensive than a CCD and, as a result, "within a few years video and

ernment of Flanders has guaranteed to fund nearly half of IMEC's $90 million annual budget through the end of 2001.[168]

In France the government Laboratory for Electronics, Technology, and Instrumentation (LETI) is conducting microelectronics R&D that is transferred to private companies, is supporting the creation of startups, and is sponsoring consortia addressing specific microelectronics themes, such as GRESSI (CMOS technology and non-volatile memories), PLATO (sub-1.10-micron CMOS and alternatives to CMOS), and PREVUE (ultraviolet lithography).[169]

Finally, the German Research Ministry (BMFT) funds microelectronics R&D both directly[170] and through Germany's excellent system of applied research institutes, the Fraunhofer Gesellschaft (FHG).[171]

---

electronic cameras will be so cheap that you will find them virtually anywhere." See FBIS, September 16, 1997, translation of Stephen Stroeytkens, "Flemish Technology will Make Video Cameras Very Cheap," *Groot-Bijgaarden De Standaard,* September 15, 1997, p. 12 (FTS19970916000718).

[168]See FBIS, August 1, 1997, "Research for the Next Millenium," *Electronik,* January 21, 1997, pp. 14-15 (FTS19970801001209) and FBIS, January 26, 1996, translation of Richard Sietmann, "With Research and Development in First Place; Belgian Microelectronics Institute IMEC in Demand as Partner for Cooperation in International Research and Development," *VDI Nachrichten,* January 26, 1996, p. 13 (FTS1996012600087).

[169]"Upbeat Europe Aims to Upgrade Its Chip Technology," *Solid State Technology*, December 1998, p. 32. See also FBIS, June 27, 1996, translation of "LETI Joins Forces with Equipment Manufacturers in 0.25-Micron Technologies," *Electronique International Hebdo*, June 27, 1996, p. 12 (FTS19960919001047).

[170]The BMFT directly funds R&D in microsystems technology (the integration of semiconductors with precision mechanical systems. "Technology Networking Will Speed European Research," *Solid State Technology*, March 1, 1999, p. S26; "Upbeat Europe Aims to Upgrade Its Chip Technology," *Solid State Technology*, December 1998, p. 32.

[171]The Fraunhofer Microelectronics Alliance, with roughly 1,000 employees, combines seven FHG research institutes "to turn product ideas into serial products." During the past five years the FHGs have helped German semiconductor manufacturers reduce costs through automation and flexible manufacturing techniques; produced ASICs on a contract basis for European firms; pursued alternative-to-silicon technologies such as GaAs, iridium phosphide, arsenic, and antimony; developed very-high-frequency transistors and monolithic microwave integrated circuits (MMICs) for use in wireless communications; developed semiconductors lasers; and conducted R&D with respect to optical components. See FBIS, November 19, 1997, translation of Guenter Weimann, "Compound Semiconductors for Mobile Communications," *Fraunhofer Magazin*, No. 3, 1997, pp. 18-19 (FTS19971119000552); FBIS, August 14, 1996, Franz Miller, "Diamond Coatings for Electronics," *Der Fraunhofer*, November 1, 1996, pp. 20-23 (FTS1996081Y000702); FBIS, August 15, 1996, translation of Andreas Beuthner, "Fraunhofer Tech Transfer to Industry Viewed," *Computer Zeitung*, August 15, 1996, p. 4 (FTS19960815000685); FBIS, May 16, 1996, translation of Elisabeth Feder, "GaAs Technology is Becoming an Individual Priority in Germany," *Electronique International Hebdo*, May 16, 1996, pp. 18-19; FBIS, October 4, 1996, translation of Leo Ploner, "Compute Electronically, Compete Optically. Photonics: Focus on Research Attention Not the Optical Computer but on the Synergy of Microelectronics and Optics," *VDI Nachrichten*, October 4, 1996, p. 23 (FTS1996100400963).

Euraccess

The Euraccess initiative is a program organized under European Union auspices and run jointly by IMEC and LETI. It has the objective of "identifying platforms where academic and industrial institutions can jointly study new ideas [with respect to deep sub-0.01-micron microelectronics] and their industrial feasibility." It establishes a network of hubs and labs—institutions active in leading-edge microelectronics R&D. Hubs examine themes such as the limits of MOSFET functionality, alternative gate dielectrics, optical interconnections, and alternatives to silicon device structures. Research results within hubs are shared within common teams, and industrial partners assume responsibility for transferring the technology developed within the hubs to pilot fabs. "Labs focus on specific technologies, with limited industrial support. Researchers are expected to move between the hubs and labs."[172]

Telecommunications Policy

The European resurgence in microelectronics has been based in substantial part on European strength in wired and wireless telecommunications markets.[173] As in Japan, European regulatory policy has resulted in a single wireless standard, rather than the five incompatible standards that characterize the U.S. market.[174] The European telecommunications industry has itself benefited from a series of long-term, large-scale European Union-sponsored R&D projects that began in the 1980s and continue today.

The first project, RACE (1985-95), successfully achieved the introduction of integrated broadband communication (IBC) systems throughout Europe, either in specialized scientific networks or limited public services.[175] RACE was followed by ACTS (1994-98), which pursued a variety of themes related to the implementation of advanced communications systems.[176] The Wireless Strategic Initiative

[172]"Technology Networking Will Speed European Research," *Solid State Technology*, March 1, 1999, p. 12; IMEC, "What is Euraccess?" at <http://www.imec.be/EURACCESS/‡‡summary>; "Deep-Submicron R&D Program Hopes to Open Doors to Advances for Europe's Chipmakers," *MicroMagazine.com*, March 1999, at <www.micromayazine.com/archive/99/03/break_1.htm/>.

[173]European Union Commissioner Erkki Liikanesi commented in April 2000 that in a competitive comparison between the European Union and the United States, "The positive side is of course mobile communications, because here Europe is the leader," in FBIS, April 17, 2000, translation of interview by Stefan Krempl with Erkki Likanne, "EU Commissioner: We Need a Sense of Urgency," *Munich Telepolis*, April 17, 2000, no page citation given (EUP20000417000266).

[174]The European telecommunications standards authority required the European Union member states to develop a single digital network standard, GSM (Global System for Mobile Telecommunications).

[175]ACTS Web site at <http://www.ukinfowin.org/ACTS/ANALYSYS/INTRO/chap1.htm>.

[176]*Official Journal of the European Communities*, L.126/1, May 18, 1994.

(WSI), part of the European Union's Fifty Framework Program, will evolve into the so-called Wireless World-Net (next generation of wireless systems).[177]

In addition to these programs a number of stand-alone projects administered by national governments within the European Union are developing microelectronics technologies with wired and wireless applications.[178]

The Potential Emergence of China

The government of China has been promoting an indigenous semiconductor industry since the 1980s. It has been hampered in this effort by resource shortages and by national security restrictions imposed by more developed nations, including the United States and Japan, on the export of advanced semiconductor manufacturing equipment to China. Existing Chinese wafer fabs can only produce semiconductors on 5- and 6-inch wafers with line widths of 0.5 to 1.6 microns.

A poor infrastructure, low productivity, and the small scale of all indigenous manufacturers disadvantage the industry. China must import almost all integrated circuits needed for the production of color televisions, air conditioners, refrigerators, computers, and communications devices.[179] At present China does not account for even 1 percent of the world semiconductor market.[180]

Despite the difficulties confronting the Chinese industry at present, many industry observers predict that China will emerge as a major competitor in 10

---

[177]WSI Web site at <http://www.ist-wsi.org/project.htm>.

[178]For example, the RF Front End Project (1998-2001), 50 percent funded by the German government, is intended to improve and speed up radio frequency (RF) circuit design and help integrate that design into the wireless system design flow. "Joint European Electronics Consortium—Cadence Targets RF Design," *Electronic Engineering Times*, March 16, 1998. Since the early 1990s the German Ministry of Research (BMFT) has provided major research funding for optical communications systems. See FBIS, July 31, 1997, translation of *Photonik-Foerderschwerpunkt-Photonik in Rahmen des Foerderkonzepts Informationstechnik, Cologne, August 5, 1997*, no author given, no pagination (FTS19970731000454 97G16107).

[179]See FBIS, November 1, 2000, translation of Chu Dechao, "Overview of the Semiconductor Industry in China," *Tokyo Semiconductor FDP World*, November 2000, pp. 176-79 (JPP20001127000111). "[O]ur country's integrated circuit industry has remained rather weak and small in terms of its overall scale and has lagged relatively far behind in terms of its production technology development capability, product design and development capability and standard, and so on, with its product sales volume taking up only 8 per thousand of the world integrated circuit sales volume and less than 20 percent of the domestic market demand. As the majority of our products are of intermediate or low standard, we have had to import core key products from abroad (such as CPU, DSP for mobile telecommunications, and so on)." FBIS, May 15, 2000, translation of Qu Weizhi, vice-minister of Information Industry, "How to Develop Integrated Circuits Industry," *Renmin Ribao*, May 15, 2000.

[180]See Grant Johnson, "Is China's Semiconductor Market Worth the Risk for Multinationals? Definitely!" *Electronic News*, March 1999, p. 10, <http://www.instat.com/insights/semi/1999/china42999.htm>.

years. (Some Japanese observers, noting that Chinese engineers "absorb [semiconductor] technologies at an unimaginable speed," believe that it will happen sooner.[181]) The government has frequently indicated the importance that it places on fostering this industry.[182] China already manufactures or assembles a broad array of electronic equipment incorporating semiconductors, accounting for nearly 6 percent of world semiconductor consumption.

By 2010 China is forecast to be the world's second-largest semiconductor market, after the United States.[183] The Chinese government is using the prospect of access to this growing market as well as financial and operational incentives to attract world-class foreign semiconductor device, equipment, and materials producers to invest in China.[184] The U.S. producer Motorola is building a facility in Tianjin that, when completed, "will become one of the largest semiconductor, manufacturing facilities in the world."[185] NEC is investing heavily in semiconductor production facilities through joint ventures in Shanghai and Beijing,[186] and other Japanese producers have begun to map out ambitious investment initiatives in China.[187]

Much of the Chinese government assistance is being provided pursuant to Project 909, a $1.2 billion program initiated in 1995 to establish five device manufacturing companies and at least 20 design and development centers in a Pudong

---

[181]See FBIS, March 7, 2001, translation of M. Kimura, "Industry, Government and Universities United in Enthusiasm and Talent for LSIs," *Nikkei Microdevices*, March 2001, p. 62 (JPP20010307 000001).

[182]"As integrated circuits concern economic development and national security, we should promote integrated circuits development by integrating the state will with a market mechanism . . . . [T]he state should map out preferential policies to support its development, including preferential revenue, investment, capital coordination, and qualified personnel policies, as well as other incentive policies, in order to speed up its development." See FBIS, May 15, 2000, translation of Qu Weizhi, "How to Develop Integrated Circuits Industry," *Renmin Ribao*, May 15, 2000, p. 11 (CPP20000515000066).

[183]Michael Pecht, Weifeng Liu, and David Hodges, *China's Semiconductor Industry* (Office of Naval Research, 2000) at <http://intri2.org/ttec/aemo/report/index.htm>.

[184]See FBIS, July 28, 1997, translation of "State Encourages Foreign-Founded Microelectronics, Basic Electronic Product Items," *Zhongquo Dianzi Bao*, January 14, 1997, p. 9 (FTS19970728 001781); FBIS, April 7, 1999, translation of Hu Angang, "How Does China Attract Foreign Direct Investment," *Guangzhou Gangao Jingji*, February 15, 1999, pp. 33-37 (FTS19990407001093); FBIS, July 12, 2000, translation of Suo Yan, "PRC's State Council Drafts New Policies to Develop IT Industry," *Renmin Ribao*, July 12, 2000,(CPP20000712000106).

[185]The Motorola plant will make devices for incorporation into Motorola handsets, also produced in Tianjin. See FBIS, August 22, 2000, reprint of Matthew Miller, "Beijing Approves Motorola Base, To Work With Taiwan Tycoon, Jiang Zemin's Sun," *South China Morning Post (Internet Version)*, August 22, 2000, no page citation (CPP20008220000287).

[186]See FBIS, February 2, 2001, translation of "Japan's NEC to Expand Semiconductor Production in China," *Kagnaku Kogyo Nippo* (Nikkei Telecom Database Version), January 12, 2001, p. 9 (JPP20010202000013).

[187]See FBIS, March 1, 2001, translation of "LSI Makers Eye China for Developing New Business Strategies," *Nikkei Microdevices*, March 2001, p. 72 (JPP20010307000003).

New Area of Shanghai. Most of these new enterprises are being created in partnership with foreign firms.

U.S. Export Control Policy

U.S. policy toward China at present is to deny approval for export licenses for semiconductor-manufacturing equipment capable of producing devices using design rules less than 0.35 microns. This policy has impeded the development of China's semiconductor industry.[188] However, governments in Japan and Europe are becoming less restrictive, with the result that Chinese fabs are procuring from vendors in those countries equipment that cannot be obtained from U.S. sources.[189] NEC, for example, is supplying DRAM manufacturing technology to Shanghai's Hua Hong NEC Electronics Co., which will use 0.25-micron design rules, more advanced than permitted under the U.S. export control regime.[190]

Taiwan in China

Just as Taiwan has transformed the global competitive picture during the past five years, Taiwanese initiatives may accelerate China's emergence as a first-rank global competitor. At present, encouraged by both the Chinese and Taiwanese governments and driven by a growing workforce shortage in Taiwan, the Taiwanese information industry is relocating much of its manufacturing operations to the mainland.[191] Taiwanese Minister of Economic Affairs Lin Hsin-yi said in November 2000 that Taiwanese firms controlled 50 to 60 percent of China's total production of information technology hardware.[192] The Taiwanese semiconductor industry is widely expected to relocate most of its low-end (200-mm wafer, 0.25 micron and above) operations to China by 2005. Taiwan's desktop computers are now largely produced in Taiwanese-owned facilities on the mainland, as well as 56 percent of Taiwan's motherboards, 88 percent of its scanners, 74 percent of its CD drives, and 58 percent of its monitors.[193]

---

[188]The United States, Japan, Korea, and the European Union participate in the Wassenaar Arrangement, a multilateral export control regime that obligates participants to administer export control licensing requirements for advanced semiconductor manufacturing equipment to all locations.

[189]"Compared with the United States, both Japan and Europe have greater freedom to approve the export of advanced semiconductor equipment [to China]." Pecht et al., *op. cit.*, p. 41.

[190]See FBIS, February 2, 2001, translation of "Japan's NEC to Expand Semiconductor Production in China," *Kagaku Kogyo Nippo* (Nikkei Telecom Database), January 12, 2001, p. 9 (JPP2001020 2000013).

[191]"Taiwan Electronics Industry Migrates to China," *Taiwan Central News Agency*, August 24, 2000.

[192]See FBIS, November 30, 2000 "Taiwan's High-Tech Advantage over China Eroding," *Taipei Times* (Internet version), November 30, 2000, (CPP20001130000168).

[193]"Taiwan's Electronics Industry Migrates to Mainland China," Taiwan *Central News Agency*, Wire Service Report, August 24, 2000.

The Chinese government, "which has carefully observed how Taiwan succeeds in the LSI industry, hopes to as far as possible put into practice measures that were successful in Taiwan."[194] It is putting in place special incentives to lure Taiwanese semiconductor investment. Meanwhile, the government of Taiwan—although officially banning direct investment in China—is encouraging, through its tax policy, domestic semiconductor producers to relocate their less sophisticated operations to the mainland.[195] Some significant moves are already in progress:

- Chang Ju-ching, former president of Taiwan's Winbond Semiconductor Manufacturing Company (SMC), has announced plans to use Toshiba's technology to build two 200-mm wafer fabs in Shanghai, one of which will use 0.25-micron design rules.[196]
- Winston Wang, son of Formosa Group Chairman Wang Yung-Chang, has formed a joint venture—the Hongli Semiconductor Company—with Jiang Mianheng, son of China's President Jiang Zemin. The venture is to build six fabs in Shanghai, three of which will eventually utilize 300-mm wafer technology. Japan's Oki Electric will reportedly provide technical assistance, and the Chinese government has reportedly provided preferential financing and a tax holiday.[197] Wang reportedly will invest $1.63 billion in this enterprise. Ground was broken on the first 8-inch fab on November 18, 2000, with the startup of mass production (50,000 wafers/month) set for the second quarter of 2002. Wang reportedly stated at the ceremony that "as long as the market needs, Hongli can start to produce chips of 0.18 microns instead of chips of 0.2 microns as it is currently planned."[198]

---

[194]See FBIS, March 7, 2001, translation of M. Kimura, "Industry, Government and Universities United in Enthusiasm and Talent for LSIs," *Nikkei Microdevices*, March 2001, p. 62 (JPP20010307 000001).

[195]"Investment in Mainland China by Taiwan Enterprises: Present Status, Problems and ROC Government Assistance," Taiwan Industrial Development and Investment Center, internal memorandum; "Beijing Welcomes Taiwan Semiconductor Firm," *Taiwan Economic News*," June 28, 2000, at <http://www.taiwanheadlines.com>.

[196]"Since it is a joint venture between the most powerful young master and the richest one across the Taiwan Strait, government officials on the Mainland naturally give all the green lights. The Shanghai Municipal Government guarantees full support. And the state-owned bank on the Mainland has agreed a generous loan of 2.5 billion US dollars. Moreover, the government will provide the most favored treatment to this special project, including tax exemption for five years, etc." See FBIS, May 11, 2000, reprint of Wen-Hung Fung, "Taiwan's IT Production Continues Moving Into China," Taipei *Central News Agency*, May 11, 2000, (CPP20000511000146) and FBIS, December 5, 2000, translation of Xia Wensi, "The Eldest Young Master Jiang Sailing Smoothly Through Business World – China's Telecom King Jiang Mianheng," *Kai Fang*, December 5, 2000, No. 168, pp. 11-13 (CPP20001211300004O).

[197]See FBIS, May 11, 2000, reprint of "Jiang Zemin's Son, Taiwan Company Team Up on Chip Venture," *Taipei Times* (Internet Version) May 11, 2000, no page citation (CPP20000511000120).

[198]FBIS, November 20, 2000, translation of Pai Te-hua, Wang Chung-ning and Wang Shih-Chi, "Construction of Taiwan-Funded Microchip Plant Begins in Shanghai," *Chung-Kuo Shih-Puo*

• Taiwan's Advance Device Technology has already built a 6-inch wafer fab in southeastern China.[199]

• In October 2000 it was reported that Taiwan's Wafer Works Corporation would set up a foundry in connection with China's Beijing Oriental Electronics Group, using second-hand equipment.[200]

In December 2000 TSMC Chairman Morris Chang announced plans to visit Beijing, where he was to meet with several delegations organized by the Chinese government—fueling speculation that TSMC was planning to build a foundry on the mainland. A company spokeswoman said the real purpose of his trip was to participate in a bridge tournament. "Our chairman is an avid bridge player," she said.[201]

Chinese engineers and specialists will staff the Taiwanese plants in China and may also help to operate semiconductor facilities in Taiwan, where multiple institutional structures exist to diffuse advanced technology.[202] Although the Taiwanese are determined never to surrender "the high end" to China, it is unclear how they can prevent diffusion of advanced technology to China—their own as well as that absorbed from their foreign partners and customers. This diffusion occurs through the mobility of personnel, particularly as the boundaries between the two industries become increasingly blurred.

One U.S. Taipei-based semiconductor executive notes that a decade ago, many Taiwanese managers were working for U.S. companies. Many have since migrated to Taiwan to pursue their own business opportunities. The results of this are by now well known. In a similar manner Chinese engineers working for Taiwanese and other foreign-invested firms on the mainland or for Taiwanese firms in Taiwan will migrate just like the Taiwanese managers. The PRC government is trying to accelerate this process.

Taiwan has always prohibited direct trade and investment with the mainland, and its stated policy toward the relaxation of restrictions on direct investment in

---

(Internet Version), November 20, 2000, p. 1 (CPP20001120000035). In January 2001 rumors were circulating in Taiwan that this enterprise was encountering financial difficulties and that "further preparation of the site has stalled," with the result that Taiwanese mid-level executives who had been lured away from TSMC to work on the project were returning to Taiwan. See FBIS, January 13, 2001, "China-Taiwan Princelings Joint Venture Stalled," *Taipei Times* (Internet Version), January 13, 2001 (CPP20010116000160).

[199]"Taiwan's IC Makers Eye China—May Contract Low Cost Plants There, but the Island's Government is Balking," *CMP Media Inc.* at <http://www.ebonline.com>.

[200]See FBIS, October 3, 2000, "Taiwan Company to Set Up 6-Inch Chip Foundry in China," *Taipei Times* (Internet Version), October 3, 2000, no page citation (CPP20001003000135)

[201]See FBIS, December 11, 2000, reprint of "Taiwan Semiconductor Chair Going to China for 'Bridge Tournament,'" *Taipei Times*, December 9, 2000, no page citation (CPP20001211000132).

[202]"Taiwan Legislator Warns of High Tech Exodus," *South China Morning Post*, July 6, 2000, online report, <Nikkei BP AsiaBizTech http://www.nikkeibp.asiabiztech.com/ archive/onnet/ 200007onnet.html>.

the mainland remains one summarized by the slogan "no haste, be patient."[203] However, despite official prohibitions the government has supported domestic firms' investments in China on a controlled basis, providing legal advice and tax benefits to firms making desired investments, but imposing fines and other penalties on firms making "undesirable" investments. Divisions exist within the government and the Taiwanese information technology sector over the appropriate scale and pace of future Taiwanese investments on the mainland.[204] Despite these constraints the migration of Taiwanese information technology manufacturing, including the semiconductor industry, to the mainland may already be unstoppable regardless of the policy Taiwan's government chooses to adopt.

> The fact is the new government is powerless to stop a new wave of hi-tech firms investing on the mainland . . . about 30,000 Taiwan firms have invested more than US$40 billion on the mainland in the past decade, the vast majority of them routed through Hong Kong [and third countries] to skirt a ban on direct investment.[205]

DIRECTIONS FOR U.S. POLICY

Government intervention in the global semiconductor industry poses complex challenges for the United States. U.S. economic doctrine opposes government intervention intended to produce specific commercial outcomes, but with a few exceptions during the past several decades the United States has not been able to persuade other countries of the wisdom of leaving the evolution of strategic industries like semiconductors to the vagaries of the market.

Instead, beginning in the mid-1980s the U.S. government was drawn into a series of limited market interventions to counter the adverse effects of foreign government measures. In semiconductors these include the Semiconductor Trade Agreement (STA),[206] the formation of SEMATECH, and the development of the

[203]Because of the prohibition on direct investment, all Taiwanese mainland investments are indirect, utilizing offshore corporations to preserve the legal fiction that Taiwanese investments are not being made in China.

[204]In January 2001 the chairman of Acer, Inc., Taiwan's largest PC manufacturer, published an opinion piece blasting Taiwan's restrictions on high-tech investment on the mainland. But the same month Morris Chang, the chairman of TSMC, said that his firm would not invest in China for at least three years, citing cronyism, corruption, and U.S. export control regulations. See FBIS, January 18, 2001, PRC: Taiwan Computer Firm CEO Cited on Importance of Mainland Expansion," *China Daily*, (Hong Kong Edition, Internet Version), January 18, 2001 (CAP20010118000038); See also FBIS, January 13, 2001, "Taiwan Business Leader's Advice on High Tech (Hong Kong Edition, Internet Version), Investment in China: 'Donts'" *Taipei Times* (Internet Version), January 13, 2001, (CPP20010116000158).

[205]See FBIS, January 18, 2001, "PRC: Taiwan Computer Firm CEO Cited On Importance of Mainland Expansion," *China Daily* (Hong Kong Edition, Internet Version), January 18, 2001, (CAP20010118000038)

[206]The objective and the effect of the STA were to open the Japanese semiconductor market and to

long-range industry-government-academia partnership embodied in the Semiconductor Roadmap.

Ironically, as these and other manifestations of government support for the U.S. industry are being phased out, foreign countries are emulating them. These countries see the enhanced role of government as an important element in a strategy to challenge U.S. leadership in information technology in general and microelectronics in particular.

Should We or Shouldn't We?

It is not within the scope of this paper to address the question of whether the U.S. government should provide support to the U.S. commercial semiconductor industry—such as through financing for capital investment or support for applied research aimed at producing a specific products. Any major-scale proposal of such an initiative would be so controversial that it is unclear whether it could be implemented. Rather—in light of the major technical challenges facing the industry and the scale and increases in interventions in this industry by governments abroad—a more practical question relates to what positive role, if any, the U.S. government can play. Such a role might be a part of a broader U.S. response—one that does not contravene conventional U.S. views about the appropriate place of government in the economy.

The role of the state in the economy is perhaps a less controversial subject today than it was in the era of Hamilton and Jefferson, but it is not a settled question. A rough consensus has emerged over the past century—articulated first by Theodore Roosevelt and reiterated by successive generations of U.S. leaders—that in economic affairs the government should serve as a sort of neutral referee. As such, it would act where necessary to ensure that competition is "fair" and that the public is not victimized by unscrupulous commercial practices. At the same time, this consensus holds that the government should not intervene to promote or protect the interests of individual competitors or sectors.[207]

---

halt Japanese dumping. The STA was criticized for allegedly causing Japanese DRAM producers to form a cartel, which limited DRAM supply and raised DRAM prices. However, the historical record demonstrates that the Japanese DRAM producers were jointly restraining output for market stabilization purposes a year before the STA was implemented. See Kenneth Flamm, *Mismanaged Trade? Strategic Policy and the Semiconductor Industry*. Washington, D.C.: Brookings Institution Press, 1996), pp. 168-69, and Tyson, *op. cit.*, pp. 117-22. Collective output restraints for price stabilization purposes are common in Japanese industry and are generally a response to domestic price erosion, not pressure from the United States. In 1996 and 1997, for example, Japanese and Korean DRAM makers reportedly reached an agreement to curtail their output of 16M DRAMs, a move that led to the doubling of prices for this product in early 1997. See FBIS, June 14, 1997, "Media Report ROK, Japanese Strategy to Control DRAM Prices," summarizing articles in *Digital Choson Ilbo*, January 29, 1997; *Choson Ilbo*, April 17, 1997; *Chugang Maegyong*, April 30, 1997; and *Maeil Kjongje Sinmun*, April 8, 1997, (FTS1997062Y002280).

[207]While Roosevelt was the first major U.S. leader to express the concept of the state as neutral

In addition, it is widely accepted that the government should take certain affirmative steps to promote general economic welfare, such as: the sponsorship of roads, bridges, and other elements of the transportation and communications infrastructure; promotion of scientific advances; and measures to improve the quality and availability of education and training.[208] In the latter half of the twentieth century it also became generally accepted that the government must take certain steps necessary to ensure that industries essential to the national defense exist and remain strong enough to meet the needs of U.S. military forces.

But beyond these limited areas in which a role for the state is generally acknowledged, the consensus unravels. While beleaguered U.S. companies and industries sometimes succeed in securing government assistance in the form of bailouts, import protection, special tax relief, and the like, such measures are almost always controversial and for that reason frequently short lived. An imperative of the global economy, however, is that U.S. preferences and practices be measured against the policies and practices of major and emerging competitors, not necessarily for emulation but for a careful assessment of impact and value.

Meeting Challenges

Arguably, the most serious challenges confronting the U.S. semiconductor industry today are in areas where the government can play an important and positive role without contravening generally accepted U.S. notions of the proper limits on the intrusion of the state in the economy. The brick wall confronting the

---

arbiter, the idea itself had already gained widespread acceptance in U.S. society. Advocated by influential economic thinkers during the nineteenth century, it was "reinforced by the frontier process and eventually became embedded in U.S. folklore. Economists taught it to their students; politicians paid homage to it; businessmen gave it lip service when they engaged in oratory for public consumption . . . ." Ellis W. Hawley, *The New Deal and the Problem of Monopoly*. Princeton: Princeton University Press, 1966, p. 47. While Herbert Hoover will never enjoy stature among his countrymen for his achievements in the economic realm, he was able to articulate these widely shared U.S. values succinctly in a speech delivered during the election campaign of 1928: "It is as if we set a race. We, through free and universal education, provide the training for the runners; we give to them an equal start; we provide in the government the umpire of the fairness of the race. The winner is he who shows the most conscientious training, the greatest ability, and the greatest character." See Richard Hofstadter, *The American Political Tradition*, New York: A.A. Knopf, 1948.

[208]In the nineteenth century the federal and state governments substantially underwrote construction of railroads and in the twentieth century the construction of airports, highways, canals, and ports. Federal funding of research and development has led to such advances as atomic energy, the Internet, the Global Positioning System, lasers, solar-electric cells, storm windows, Teflon, communications satellites, jet aircraft, microwave ovens, genetic medicine, and a wide array of advanced materials and composites. Office of Science and Technology Policy, *Fact Sheet on How Federal R&D Investments Drive the U.S. Economy*, June 15, 2000, at <http://www-es.ucsd.edu/stpp/whouse(rp).htm.

industry is the type of large technological hurdle that the U.S. government has previously helped industry surmount through support for basic science and pre-competitive R&D.

- An increase in the volume of federal funding of programs such as MARCO and the Nanotechnology Initiative—*basic research initiatives* that have enjoyed strong bipartisan support—would be an important first step in attacking emerging challenges in microelectronics. Ensuring the *availability of a trained, educated workforce* is a core government responsibility. The workforce shortage confronting the industry can be partially addressed by providing additional resources to U.S. universities and incentives for students and faculty to enter and remain active in fields that are critical to the challenges confronting this industry—electrical engineering, physics, and chemistry.
- The post-Cold War changes in defense and other *research budgets* must be clarified and redressed. Without the necessary additional funds the students will not be trained, and in any event, cannot be trained quickly.
- Consequently, even in the face of reduced retention rates, U.S. immigration policy should be administered in a manner that facilitates the *attraction of foreign talent* to the United States.

The potential problems posed by foreign industrial policies are more complex. The United States possesses an array of *trade remedies* that can be invoked against certain defined "unfair" trade practices, but these tools have frequently proven crude and/or ineffective in the complex realm of global competition in technology-intensive industries. However, a very substantial proportion—perhaps most—of the foreign programs summarized in this survey do not constitute "unfair" trade practices as defined by U.S. law and are not proscribed by any existing multilateral rules. They simply exceed abiding U.S. notions of the appropriate role of the state in the economy.

Learning From Past Success

The success of the U.S. semiconductor industry during the past 15 years reflects, in substantial part, a series of improvisations by the government and the industry working together to hammer out solutions to challenges arising out of foreign industrial policies without a fundamental departure from U.S. economic values.

- The Semiconductor Trade Agreement, while controversial and in some respects unique, was nevertheless consistent with a long line of comparable limited actions by the U.S. government designed to open markets and promote competition—and it was phased out when market-based competition in the industry was restored.

• SEMATECH was created to address specific national defense-related concerns arising out of the growing dependency of U.S. weapons systems on critical components for which a secure domestic production base was regarded as essential.

The flowering of the joint industry-government-university effort, reflected in the Semiconductor Roadmap, was in part a response to the strategic promotional efforts under way in Japan and Europe. More broadly they reflected a recognition that a cooperative approach is required to sustain the tremendous benefits offered by this rapidly evolving industry.

More improvisation and greater cooperation will be required in the coming decade.

POSTSCRIPT

Since this paper was written in early 2001, a number of developments have occurred which deserve to be noted here. Most dramatically, China is beginning to emerge rapidly as a major production base for semiconductors. A massive influx of foreign investment and skilled manpower, predominantly Taiwanese, is resulting in the establishment of new semiconductor foundries in the Yangtze Delta region and in Beijing. At this writing (at the end of 2002) two of these new foundries are operational, six more are under construction or will enter the construction phase by early 2003, and at least 11 more are planned.[209] This expansion reflects a new Chinese government promotional effort designed to replicate Taiwan's success in microelectronics on a much larger scale in China, drawing heavily on Taiwanese and other foreign capital, management and technology.[210] China's new policy measures closely resemble those utilized by Taiwan,

---

[209]In September 2002, Shanghai-based Semiconductor Manufacturing Corp. (SMIC) had two fabs operational and planned at least two more, and Grace Semiconductor Manufacturing International, also based in Shanghai, had two fabs under construction and two more planned (interviews with senior executives at SMIC and Grace, Shanghai Zhangjiung Science & Technology Park, September 2002). In Suzhou, He Jian Technology Corporation, widely reported to be affiliated with Taiwan's UMC, had one fab under construction and five more planned (interview with officials of the Suzhou Industrial Park, Suzhou, September 2002). Taiwan's TSMC had announced plans to build at least one fab in Songjiang and has acquired sufficient land for additional facilities (interview with official of Shanghai Songjiang Industrial Zone, September 2002). In Beijing, the Beijing Semiconductor Manufacturing Corp. had two fabs in the early stages of construction and at least three more planned (interview with officials of the Beijing Economic Development Area, September 2002).

[210]The principal Chinese policies are spelled out in the *Tenth Five Year Plan (2001-2005) — Information Industry*, http://www.trp.hku.hk/infofile/china/2002/10-5-yr-plan.pdf, and Circular 18 of June 24, 2000, *Several Policies for Encouraging the Development of Software Industry and Integrated Circuit Industry*, published in Beijing Xinhua Domestic Service, 04:49 GMT, July 1, 2000. The

including the establishment of science-based industrial parks, tax-free treatment of semiconductor enterprises, passive government equity investments in majority privately held semiconductor companies, and preferential financing by government banks. In addition, China is providing a protected market for semiconductors manufactured domestically, levying an effective value-added tax of 3 percent for locally made semiconductors versus 17 percent for imported devices.[211]

The abrupt rise of China as a significant site for semiconductor manufacturing reflects the erosion of several longstanding impediments to China's development in this sector. With the end of the Cold War, the export control regime administered by Western countries restricting the flow of semiconductor manufacturing equipment and technology to China has loosened substantially, and the new mainland foundries have experienced little difficulty in acquiring equipment and process technology to produce 8-inch wafers using 0.18- to 0.25-micron design rules.[212] Taiwan's legal constraints on investment and technology transfer to the mainland have been relaxed, and many Taiwanese are circumventing such restrictions in any case.[213] Finally, in the wake of its entry into the WTO, the Chinese government has abandoned or is phasing out a number of longstanding policies which have deterred inward foreign investment in microelectronics, such

---

municipal governments of Shanghai and Beijing have issued their own circulars articulating promotional policies to be implemented within their jurisdictions to augment the national-level measures. These are, respectively, Shanghai Circular 54 of December 1, 2000, *Some Policy Guidelines of This Municipality for Encouraging the Development of the Software Industry and the Integrated Circuit Industry*, Shanghai Gazette, January 2001; and Beijing Circular 2001-4, *Measures for Implementing 'Policies for Encouraging the Development of Software and Integrated Circuit Industries' Issued by the State Council*, Jing Zhen Fa No. 2001-4 (February 6, 2001).

[211]Pursuant to State Council Circular 18, qualifying integrated circuit (IC) manufacturing enterprises are eligible for rebates of the VAT on indigenously manufactured ICs which result in an effective VAT rate of 6 percent. Qualifying IC design firms are eligible for rebates which result in an effective VAT rate of 3 percent (Circular 18, Article 52). The effective VAT rate for domestically manufactured ICs has reportedly been lowered to 3 percent. Investors currently establishing semiconductor manufacturing facilities in China are operating on the assumption that the eventual effective VAT rate will be 14 percent. FBIS, February 15, 2002, translation of "Interview with SMIC President Richard Chang," *Nikkei Microdevices* (February 2002), pp. 16-17 (JPP20020215000003); FBIS, March 28, 2002, Interview with Tsuyoshi Kanawishi, outside director of SMIC, *Nikkei Telecom* (02:02 GMT, March 28, 2002) (JPP20020328000017).

[212]See generally General Accounting Office, *Export Controls: Rapid Advances in China's Semiconductor Industry Underscore Need for Fundamental U.S. Policy Review* (April 2002) GAO-02-620; "China Finds Way to Beat Chip Limits," *New York Times* (May 6, 2002), p. 4.

[213]FBIS, March 7, 2002, translation of Hsu Yu-chun, "Tsai Ying-wen Says Punishment Will Be Meted Out to Enterprises Which Have Moved to the Mainland," *Ching-Chi Jih-Pao* (March 2, 2002), p. 1 (CPP20020307000018); FBIS, August 9, 2002, "MOEA Announces Regulations on Wafer Fab Transfers to Mainland China," *Taipei Central News Agency* (08:55 GMT, August 9, 2002) (CPP20020809000092).

as the general prohibition on 100 percent foreign-owned enterprises and stringent restrictions on "trading rights."[214]

Taiwan's government planners are seeking to adjust to the growing migration of the island's semiconductor manufacturing operations to China by implementing a new "roots in Taiwan" strategy in microelectronics. This approach accepts the loss of much of Taiwan's commodity manufacturing functions and production-line jobs to China but attempts to retain in Taiwan the most advanced manufacturing, design, and distribution functions.[215] Specifically, the government hopes to sustain a high concentration of 12-inch wafer fabs on the island, to enhance Taiwan's capabilities with respect to systems-on-a-chip, and to improve Taiwan's position in upstream (materials and semiconductor manufacturing equipment) and downstream (assembly, test, packaging) functions. Reflecting this new emphasis, government promotional initiatives are curtailing (although not eliminating) financing for wafer fabs and increasing aid for IC design and upstream and downstream microelectronics functions. Revised tax incentives place greater emphasis on R&D, training, and maintaining "operational headquarters" in Taiwan.

Japan has launched a number of government-supported industry-government R&D projects in 2001-2002. The Millennium Research for Advanced Information Technology (MIRAI) was initiated by METI in 2001 to develop next-generation semiconductor materials and process technologies, such as measuring and mask technology for 50 nm-generation devices.[216] In 2002, METI launched a 5-year industry-government R&D project to develop extreme-ultraviolet (EUV) lithography for 50-nm device manufacturing in conjunction with an association of 10 Japanese device and lithography equipment purchasers.[217] In July 2000, 11 Japanese semiconductor manufacturers established a new R&D company, Ad-

---

[214]World Trade Organization, *Accession of the People's Republic of China* (Decision of November 2001), WT/L/432 (November 23, 2001), Parts I.5, I.7.

[215]Taiwan Ministry of Economic Affairs, *Promotional Strategies and Measures* (2002); Government Information Office Release, *Liberalization of Mainland-Bound Investment of Silicon Wafer Plants* (2002).

[216]Government funding for this seven-year project was set at 3.8 billion yen for the first year. The project is being operated jointly by ASET and METI's new semiconductor R&D organization, Advanced Semiconductor Research Center ("ASRC") in the Tsukuba Super Clean Room. MIRAI website, <http://unit.asit.go.jp/asrc/mirai/index.htm>; *Handotai Kojo Handobukku* (December 5, 2001), pp. 4-5.

[217]The producers have formed the Extreme Ultraviolet Lithography System Development Association ("EUVA") to undertake the project. First-year government funding was set at 1.09 billion yen. Japan Patent Office General Affairs Department Technology Research Division, *Handotai Rokogijutsu Ni Kansaru Shutsugan Gijutsu Douko Chosa* (May 10, 2001), p. 17; METI, *Heisei Yonnendo Jisshi Hoshin* (March 8k, 2002), p. 1; *Handotai Sangyo Shimbun* (January 16, 2002), p. 3.

vanced SoC Platform Corporation ("ASPLA") to standardize design and process technologies for systems-on-a-chip utilizing 90-nm design rules. METI reportedly will provide 31.5 billion yen for this effort, which will feature partnership with STARC and Selete. METI's motivation for supporting this project is to create an "All Japan Foundry"—a standardized production line that can be used by all of Japan's device makers.[218] METI is prodding Japanese device makers — with some success—to consolidate their manufacturing divisions and specialize in design.[219]

[218]ASPLA website, <http://www.aspla.com/jp>; *Ekonomisuto* (July 2, 2002), p. 20.

[219]"Next Generation Semiconductor Project — METI Tells Firms to Discard Own Plants," *Nikkei Sangyo Shimbun* (June 12, 2002); FBIS, November 1, 2002, "Final Phase of LSI Industry Restructuring, Some Non-Winning Scenarios," *Nikkei Microdevices* (November 2002) (JPP20021120000009).

SEMATECH Revisited: Assessing Consortium Impacts on Semiconductor Industry R&D

Kenneth Flamm and Qifei Wang
University of Texas at Austin

Over 1987-1988, in the midst of a perceived crisis in the U.S. semiconductor industry, 14 U.S. semiconductor manufacturers formed the SEMATECH (for **se**miconductor **ma**nufacturing **tech**nology) R&D consortium with the support of the U.S. government.[1] In terms of its size, visibility, and public policy impact SEMATECH has perhaps been the most significant private R&D consortium formed in the almost two decades that have passed since the passage by the U.S. Congress of the National Cooperative Research Act of 1984, which granted partial antitrust exemption to registered U.S. R&D consortia.

SEMATECH's cooperative structure to some extent was stimulated by Japan's VLSI projects of the late 1970s, which were perceived in the United States (and in Japan, for that matter) as having greatly advanced the technological competence of Japanese semiconductor producers.[2] A 1987 Defense Science Board report pointing to deterioration in the relative position of U.S. semiconductor manufacturers as a possible national security issue played an important role in a U.S. government decision to have the Defense Department pay half of the cost of a consortium budgeted at $200 million annually. While the objective of improving U.S. semiconductor manufacturing technology was fairly clear, the spe-

[1]We thank Jongwoo Kim for his outstanding research assistance.

[2]See K. Flamm, *Mismanaged Trade? Strategic Policy and the Semiconductor Industry*, Washington, D.C.: Brookings Institution, 1996, chapter 2, pp. 39-126, and J. Sigurdson, *Industry and State Partnership in Japan: The Very Large Scale Integrated Circuits (VLSI) Project*, Lund, Sweden: Research Policy Institute, 1986, for detailed discussions of the Japanese VLSI projects and their impact. A revisionist assessment can be found in M. Fransman, *The Market and Beyond: Cooperation and Competition in Information Technology Development in the Japanese System*, Cambridge: Cambridge University Press, 1992.

cific means by which SEMATECH was to do so were the subject of considerable debate, and SEMATECH's focus zigged and zagged in its first few years of existence. It was restricted to U.S. companies; Japanese producer NEC, which had a U.S. production plant, was turned away when it sought to join in 1988.[3]

A number of U.S. semiconductor materials and equipment producers formed a complementary organization, SEMI/SEMATECH, in 1987 specifically for the purpose of cooperating with SEMATECH. SEMI/SEMATECH was granted SEMATECH membership and a seat on the SEMATECH board of directors, and became the official vehicle for the organization of SEMATECH development projects and the teaming of materials and equipment producers. In 2000 SEMI/SEMATECH renamed itself the Semiconductor Industry Suppliers Association (SISA).

Given the relatively large theoretical literature on R&D cooperation that sprang up in the late 1980s and thereafter,[4] there has been surprisingly little empirical analysis of the impacts of R&D cooperation on industrial R&D.[5] The highest profile R&D consortium in the United States, SEMATECH, appears to have been the subject of precisely three studies with any claim to rigor. One study, undertaken by Link, Teece, and Finan, calculated an internal private rate of return for SEMATECH member companies on a sample of SEMATECH projects that

---

[3]Good resources on the history of SEMATECH are SEMATECH's own Web site at <www.SEMATECH.org> and the corporate chronology contained within; W.I. Spencer and P. Grindley, "SEMATECH After Five Years: High Technology Consortia and U.S. Competitiveness," *California Management Review*, vol. 35, 1993; P. Grindley, D.C. Mowery, and B. Silverman, "SEMATECH and Collaborative Research: Lessons in the Design of a High-Technology Consortia," *Journal of Policy Analysis and Management*, vol. 13, 1994; L.D. Browning and J.C. Shetler, *SEMATECH, Saving the U.S. Semiconductor Industry*, College Station: Texas A&M Press, 2000; John Brendan Horrigan, "Cooperation Among Competitors in Research Consortia," unpublished doctoral dissertation, University of Texas at Austin, December 1996. For an overview of the consortium's contributions and its relationship to other policies, see National Research Council, *Conflict and Cooperation in National Competition for High-Technology Industry*, Washington, D.C.: National Academy Press, 1996, especially p. 48 and supplements A and B.

[4]For example, M.L. Katz, "An Analysis of Cooperative Research and Development," *RAND Journal of Economics*, vol. 17, 1986; C. D'Aspremont and A. Jacquemin, "Cooperative and Noncooperative R&D in Duopoly with Spillovers," *American Economic Review*, vol. 78, 1988; M.L. Katz and J.A. Ordover, "R&D Cooperation and Competition," *Brookings Papers on Economic Activity: Microeconomics* 1990; M.I. Kamien, E. Muller, and I. Zang, "Research Joint Ventures and R&D Cartels," *American Economic Review*, vol. 82, 1992; R.D. Simpson and N.S. Vonortas, "Cournot Equilibrium with Imperfectly Appropriable R&D," *Journal of Industrial Economics*, vol. 42, 1994.

[5]In addition to the SEMATECH studies mentioned here, see M. Sakakibara, "Evaluating Government-sponsored R&D Consortia in Japan: Who Benefits and How?" *Research Policy*, vol. 26, No. 4-5, December, 1997; L.G. Branstetter and M. Sakakibara, "Japanese Research Consortia: A Microeconometric Analysis of Industrial Policy," *Journal of Industrial Economics*, vol. 46, June 1998; L.G. Branstetter and M. Sakakibara, "When Do Research Consortia Work Well and Why? Evidence from Japanese Panel Data," NBER Working Paper No. W7972, October 2000, for empirical studies of Japanese R&D consortia.

had been completed as of April 1993.[6] No attempt was made in this study to specifically measure the impact of SEMATECH participation on internal company R&D spending or to estimate a social return to investments in these projects.[7]

A second study, by Irwin and Klenow,[8] analyzes firm-specific R&D expenditures in a sample of semiconductor firms over the years 1971-1993 and attempts to estimate the impact of SEMATECH membership on the R&D-to-sales ratio of member firms. They find a small and marginally statistically significant negative impact on R&D spending by members, and interpret this to mean that SEMATECH's impact on member R&D was primarily to reduce costs for R&D that would have been duplicated by individual member companies absent SEMATECH (as opposed to funding poorly appropriable R&D that would not have been undertaken at all without SEMATECH). Irwin and Klenow parenthetically assert that reduction of duplication through R&D cooperation "does not provide a rationale for government funding." There are a number of conceptual, empirical, and econometric problems in the Irwin and Klenow study that we address below.

A third study, by Horrigan[9], did not directly evaluate the impact of SEMATECH on member or industry R&D but did undertake so-called event studies of the impact of technical announcements from SEMATECH on the stock market valuation of consortium members and others. Horrigan found that technical announcements had a strong, positive impact on SEMATECH members who were both semiconductor producers and users (system manufacturers like IBM, DEC, Rockwell, Hewlett-Packard, and AT&T) but little impact on "pure play" semiconductor manufacturers (like Intel, TI, AMD). Interestingly, there was some evidence that firms that were not members of SEMATECH seemed to benefit somewhat from SEMATECH technical announcements, consistent with the notion that spillovers from SEMATECH to non-members were significant.

SEMATECH underwent significant changes in structure and research direction in the period shortly after the first two of these studies ended. Even in the early years there had been a growing emphasis on projects designed to improve the equipment and materials used by U.S. semiconductor makers but purchased

[6]Because member companies had never systematically or consistently estimated or tracked the economic value of benefits from SEMATECH membership, the perceived value of benefits reported by those willing to give a numerical answer to a survey is somewhat suspect, and unquantified benefits to companies unwilling to guess were assigned a value of zero. The authors note that their calculated rates of return are likely to underestimate true private returns.

[7]The cost structure of SEMATECH membership does not appear to be publicly reported. It is known that the originally approved cost structure consisted of 1 percent of a company's semiconductor revenues, with a minimum of $1 million, up to a maximum of 15 percent of the privately contributed portion of the SEMATECH budget ($100 million through the mid-1990s). See Browning and Shetler, *op. cit.* p. 35.

[8]D.A. Irwin and P.J. Klenow, "High-tech R&D Subsidies: Estimating the Effects of SEMATECH," *Journal of International Economics*, vol. 40, 1996.

[9]See Horrigan, *op. cit.*

from upstream equipment and materials producers. In 1992, after a new CEO had been brought on board and an internal reorganization undertaken, a new long-range plan (SEMATECH II) was adopted.[10] The new emphases were on a significant reduction in the elapsed time between introductions of new technologies, use of modeling and simulation in design of manufacturing processes and equipment, and greater systems integration of tools, processes, and operating systems within manufacturing plants.

Organizational changes also occurred. In 1992 for a variety of reasons three members withdrew, leaving 11 (of 13 original 1987 charter members, of 14 members in 1988). In 1995 a decision was made to partner with foreign companies in a project designed to accelerate the development of technology designed for use with 300-mm (12-inch) silicon wafers. In fiscal 1996, by mutual agreement, U.S. government funding for SEMATECH ended. In 1998 a separate organization, International SEMATECH, was formed as the umbrella for an increasing number of projects in which foreign chip producers were involved, and in 1999 the original SEMATECH restructured itself into International SEMATECH. Interestingly, International SEMATECH currently has 13 corporate members (8 U.S., 5 foreign), the same number as its parent SEMATECH when founded. The share of world semiconductor sales accounted for by the consortium's membership is now substantially greater than was the case in 1987.[11]

With a few vocal exceptions SEMATECH is widely credited within the U.S. industry with some role in stimulating a resurgence among U.S. semiconductor producers in the 1990s. The revealed willingness of its corporate members to continue funding SEMATECH at levels exceeding earlier private contributions after public subsidies ended in 1996 suggests that it is viewed as a privately productive and worthwhile activity. It was also certainly perceived as a major force in Japan, where the SEMATECH model greatly influenced the formation of a new generation of semiconductor industry R&D consortia in the mid-1990s.[12]

Despite these major changes there has been no new research on the impact of SEMATECH since these two early studies and little discussion of the limitations in the Irwin and Klenow study, the only one to date to seek to measure the impact of SEMATECH on semiconductor industry R&D. This paper attempts to update this early analysis and remedy some of its problems. We first review the literature on R&D cooperation and its predictions about firm R&D decisions. We next describe the simple analytical framework that justifies our (and, for that matter, the Irwin and Klenow) approach. We point out how some conceptual problems in Irwin and Klenow are addressed within this framework. We then review empirical and econometric issues in the original Irwin and Klenow study, and discuss a

[10]See Browning and Shetler, *op. cit.*, chapter 8.

[11]It is claimed that prior to its internationalization the SEMATECH membership never accounted for less than 75 percent of U.S. semiconductor industry sales, Browning and Shetler, *op. cit.* p. 197.

[12]See K. Flamm, "Japan's New Semiconductor Technology Programs," Asia Technology Information Program Report No. ATIP 96.091, Tokyo, November 1996.

strategy for addressing them. Finally, we produce new econometric results based on an expanded sample of firms extended over time and discuss the implications of our results.

R&D COOPERATION

The economics literature on R&D cooperation makes a distinction between two distinct forms of cooperation. The first is coordination. R&D undertaken by a firm may have spillovers to other firms, externalities that benefit others. If firms take these benefits into account and coordinate their R&D spending with other firms, they can generally receive greater profits than would be the case had they not coordinated their spending levels with other firms. When firms jointly commit to undertaking specific amounts of R&D taking into account spillover effects, we have coordination.

The second motive for cooperation is information sharing. Firms can actually share the results of their R&D investments, thus eliminating the need for duplicative investments. R&D can be completely appropriable (no undesired spillovers), yet firms may still benefit by pooling some of their R&D effort and reducing costs.

One form of cooperation explored in the literature is the case of pure coordination. If we call parameter β a "spillover coefficient," and assume R&D efforts among different firms are perfect substitutes, then Figure 1 shows an R&D cartel as the institutional realization of pure coordination. Without coordination firms 1

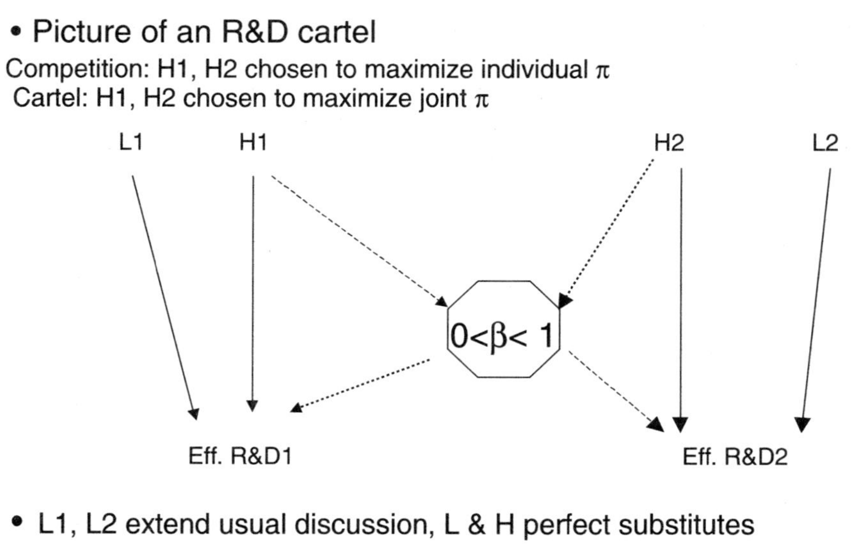

FIGURE 1 Pure coordination.

and 2 undertake levels H1 and H2 of "high spillover" R&D, with β dollars of R&D added on to firm 1's "effective R&D" level for every dollar of R&D undertaken by firm 2, and vice versa. Spillover coefficient β must clearly range in value from 0 to 1, with 0 for no spillover and 1 for complete spillover (your R&D investment benefits me as much as yourself). In making their R&D investment decisions absent coordination, it is natural to characterize a competitive equilibrium as one in which each firm takes the other firm's R&D level as a given. With the cartel the two firms commit to jointly fixing their R&D levels in order to maximize the sum of their joint profits.

The other form of cooperation that is well explored in the literature is an R&D joint venture. With this idealized story (shown in Figure 2) firms agree to perform all R&D through a joint venture that gives both firms access to all research results. Such an R&D joint venture could be organized in a competitive manner: Both firms could individually decide how much funding to give the joint venture in order to maximize their individual profits. Alternatively, the R&D joint venture could be organized as a cartel, with the two firms getting together and committing to R&D investments in order to maximize the sum of their individual profits jointly. In this latter case the story includes both coordination and information sharing, and the spillover coefficient β for another firm's R&D investment is always 1, since it is performed by the joint venture and given to both firms. If the joint venture is somehow less efficient than individual-firm R&D efforts, then β can be less than 1 and the possible values for the spillover coefficient are not materially different than in the case of the pure R&D cartel.

Without loss of generality the impact of R&D is assumed to be a lowering of production cost. In both of these stories firms are assumed to take into account the effect of their R&D investments at some first stage on their competitors' production costs at a second stage because of spillovers after all R&D is com-

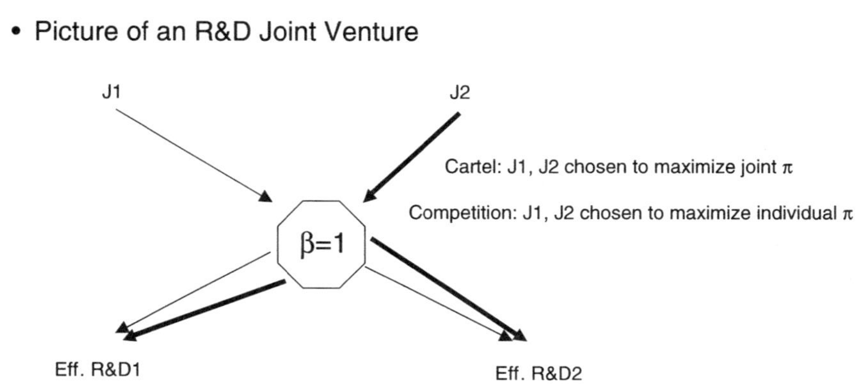

FIGURE 2 Coordination and information sharing.

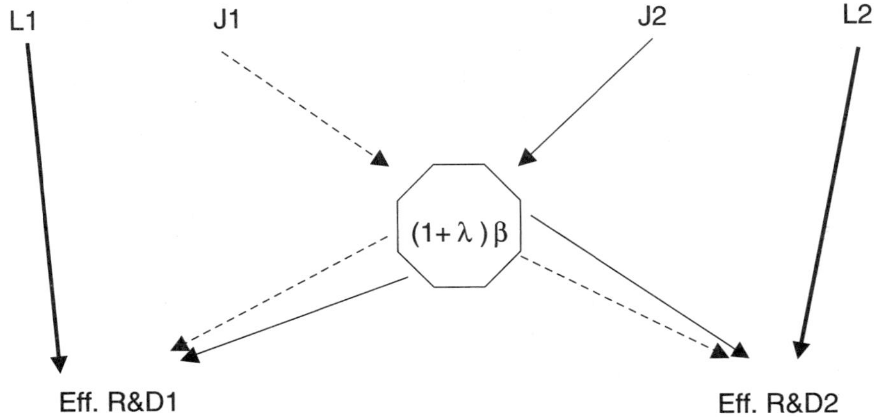

FIGURE 3 Picture of SEMATECH.

pleted and production is undertaken. Firms behave as Cournot competitors in the production stage, all firms are assumed to join a cartel when one is organized, and diminishing returns to R&D in reducing production costs are also assumed.

What then is the impact of cooperation? Although the precise details vary, the general result is that it depends on the magnitude of spillover parameter β. Without a joint venture (individual R&D efforts only) and very high spillovers (β large), R&D increases with a cartel. Without a joint venture and very low spillovers (β small), R&D decreases with a cartel. For any given degree of spillovers an R&D joint venture with uncoordinated funding (no commitment) gives the least "effective R&D" in any of the four possible cases, while the R&D joint venture with coordinated funding (commitment) yields the most "effective R&D."[13]

In short, if we argue that SEMATECH is like an R&D joint venture in which firms sit down together and jointly figure out which R&D investments are worth making in order to maximize the entire industry's profitability, then agree to fund those investments, then SEMATECH must increase the level of "effective R&D" being performed in the industry. Conversely, if SEMATECH were like an R&D joint venture in which members voluntarily decide how much to contribute, if they wish, with the ability to "free ride" on the contributions of others a real option, then SEMATECH would decrease the overall level of "effective R&D" performed in the industry.

Unfortunately, neither of these "pure" stories that have been explored in the theoretical literature is a good description of the real SEMATECH (see Figure 3).

[13]See, e.g., Kamien, Muller, and Zang, *op. cit.*, or Simpson and Vonortas, *op. cit.*, for models yielding these results.

SEMATECH members commit to R&D spending levels on some set of common projects deemed to benefit "the industry as a whole," while competing on other R&D projects they undertake internally in an environment designed to minimize spillovers to others. The theoretical literature generally assumes everyone joins the cartel or the joint venture. Partial industry membership—as in the real-world SEMATECH—is generally not considered. In addition, the real-world SEMATECH received a government subsidy for the first eight years of its life, and none of these models explicitly considers the impact of a subsidy—or the way that a subsidy might increase β in an R&D joint venture (matching federal funds could even transform a low spillover coefficient to a post-subsidy β greater than 1, with effective R&D exceeding total private R&D investment in the joint venture). Finally, it may be that one of the possible impacts of an organization like SEMATECH (even without additional federal funding) is to increase the spillover rate for privately conducted R&D, in effect altering something these models all take as exogenously set.

CONCEPTUAL ISSUES IN THE IRWIN AND KLENOW ANALYSIS

The models do suggest, however, that one must be cautious in issuing policy advice. Irwin and Klenow argue, for example, that if firms are motivated to participate in SEMATECH by a desire to fund R&D with significant spillovers, one would expect SEMATECH participation to result in a net increase in R&D undertaken by members after this "commitment" to such projects. Their econometric estimate of a negative impact on R&D is then interpreted as suggesting that it is "sharing" of information and not commitment that must have been the main motivation for SEMATECH's formation.[14] But as we have just noted, even if one were to quickly make the heroic leap from a highly stylized theoretical model to the real-world SEMATECH, firms getting together and jointly committing to funding of R&D with spillovers can actually *lower* R&D rather than increase it if spillovers are not sufficiently great. Even heuristic reasoning by analogy does not permit drawing the conclusion Irwin and Klenow propose.

Even more tenuous is the Irwin and Klenow claim that a government subsidy is clearly justified in order to stimulate industry commitment to R&D if externalities are present and completely unjustified if information sharing is the exclusive motivation for cooperation.[15] Kenneth Arrow in his seminal 1962 contribution

[14]"Recall the predicted impact of SEMATECH ranged from $-\theta^*$ of sales under the sharing hypothesis to +1 percent of sales under the commitment hypothesis. Although we did not explicitly test the sharing and the commitment hypotheses, our results are clearly more consistent with the sharing hypothesis." Irwin and Klenow, *op. cit.* p. 335.

[15]"Under the first motivation, SEMATECH induces members jointly to spend more on high spillover types of R&D. We call this the 'commitment' hypothesis. One problem with this hypothesis is that firms need not join SEMATECH, and those that do can leave after a 2-year notice. Firms

pointed out that even without any spillovers, if an invention were to be completely appropriable by the inventor (due, say, to a perfect patent system), the private return to invention could fall short of the social return (and therefore be a candidate for subsidy) if the inventor were unable to perfectly price-discriminate among users of the invention.[16] Later investigators established that excessive R&D investment could be an alternative outcome if, for example, patent races resulted as would-be monopolists raced to be first to patent and reap monopoly rents. The key point is that even with no spillovers from R&D it is possible (though not at all inevitable) for social return to exceed private return and a subsidy to be justifiable. The same holds true in the models reviewed above.[17] Theory seems only to suggest that it all depends on the specifics of the case and sweeping claims are best avoided.

Further, as Irwin and Klenow themselves acknowledge, their argument assumes that the projects undertaken by SEMATECH are perfect substitutes for the projects that companies fund internally. If this is not the case—if they are imperfect substitutes or even complements—the logic falls apart.

For example, Irwin and Klenow argue that if eliminating duplication in R&D—information sharing—is the sole motivation for cooperation, then a dollar of SEMATECH R&D simply replaces inframarginal investments in R&D within all member companies. Because the return to a marginal dollar invested in R&D in a company has not been changed, no additional R&D projects are

---

should be tempted to let others fund high-spillover R&D. Under this hypothesis, then, the 50 percent government subsidy and the equipment holdback provisions are crucial for SEMATECH's existence. The 'commitment' hypothesis both justifies these features and requires them to explain SEMATECH's membership." *Ibid.*, p. 329. "Unlike the commitment hypothesis, the sharing hypothesis does not provide a rationale for government funding. Firms should have every private incentive to form joint ventures to raise their R&D efficiency." *Ibid.*, p. 344.

[16]See K.J. Arrow, "Economic Welfare and the Allocation of Resources for Invention," in National Bureau of Economic Research, *The Rate and Direction of Inventive Activity: Economic and Social Factors*, Princeton: Princeton University Press, 1962. If there are no spillovers, of course, the problem of investment in R&D is not different conceptually from the case of a conventional fixed investment with the associated economies of scale created by this cost structure. The problem of understanding when a subsidy is justifiable in order to stimulate socially worthwhile investment in an industry subject to conventional increasing returns to scale has been well studied. See Larry E. Westphal, "Planning with Economies of Scale," in Charles R. Blitzer, Peter B. Clark, and Lance Taylor (eds.), *Economy-Wide Models and Development Planning*, Oxford: Oxford University Press, 1975, for a lucid discussion.

[17]See for example, Simpson and Vonortas, *op. cit.*, p. 86. In the context of the simple models we are discussing here imagine the following scenario: Assume SEMATECH effort is a perfect substitute for internal company effort, and that all R&D is performed through the consortium (no internal company effort, the case of a pure R&D joint venture—competitive or cartel—discussed in the theoretical literature). It remains entirely possible that the social return to an incremental R&D dollar could exceed the private return to an additional company dollar invested in the R&D joint venture and, therefore, justify a subsidy.

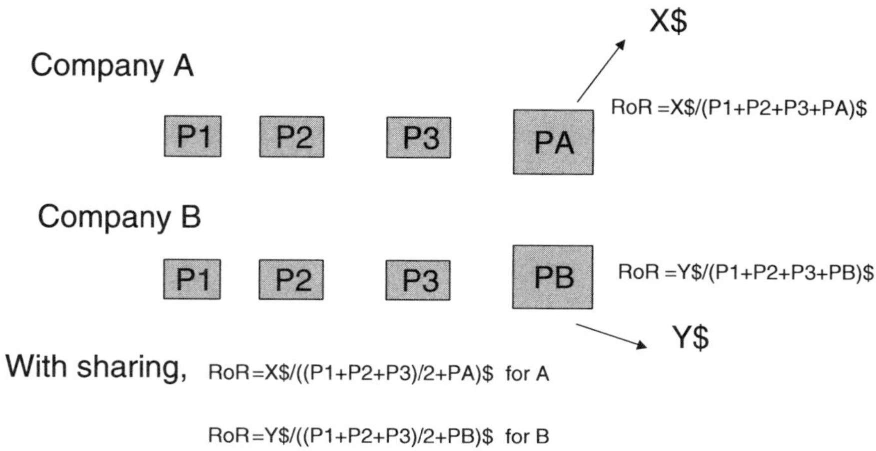

FIGURE 4 How information sharing through a consortium can stimulate R&D without spillovers.

started internally and the net impact of sharing of R&D is to reduce overall R&D spending.[18]

Suppose instead that the SEMATECH projects are complementary to internal company projects (and we maintain the assumption of no spillovers). It is then possible to construct entirely plausible scenarios where "sharing" of some projects is sufficient to increase the return on projects that would not otherwise have been undertaken to the point where they are. Figure 4 depicts such an example. Both companies have to step through three preliminary projects, costing P1, P2, and P3 dollars, before company A can spend PA in its own unique way, creating net revenues of X$, and company B can spend PB in its own unique way, with a subsequent revenue flow of Y$. Initially, doing all projects on their own, neither company can meet its hurdle rate of return. By combining their efforts on preliminary projects, however, each can halve its preliminary project spending to (P1 + P2 + P3)/2, perhaps increasing rates of return for the company-unique R&D projects to the point where they are now attractive investments.

In short, the assumption that internal-company and SEMATECH R&D are perfect substitutes is critical to claims that pure "sharing" must reduce overall industry R&D. There is no empirical evidence that this is the case, and one might even argue that the simple story told in Figure 4 is as plausible as any in describing the nature of R&D undertaken within SEMATECH.

Finally, Irwin and Klenow argue that if pure information sharing is what SEMATECH is about and SEMATECH R&D is a perfect substitute for internal R&D, the net impact of a SEMATECH effort of size S would be to cut every

[18]See Irwin and Klenow, *op. cit.*, p. 330, for this argument.

company's internal effort by S, since the many internal efforts of this size have now been replaced by a single common external effort.[19] In this case, with a dollar of SEMATECH effort a perfect substitute for a dollar of internal R&D within N member companies, aggregate industry R&D declines N dollars for every one dollar of SEMATECH effort, for a net decline of N-1 in R&D spending for the industry as a whole with every dollar of SEMATECH R&D. Note moreover that even though aggregate spending on R&D might have declined by N-1 dollars, *effective* R&D would remain unchanged—it would simply be done more efficiently as firms "passed around" the results of the common SEMATECH R&D dollar rather than duplicate it N times internally. SEMATECH would be socially beneficial even if it led to a sharp decline in overall industry R&D.[20]

But, do matters really stop here, where the Irwin and Klenow argument stops? Firms would now be earning an additional profit, since the same revenues would now be generated with lower total costs for all firms. The benefits of information sharing would thus create a rent for incumbent members, and entry by new firms into the industry—and SEMATECH—would therefore be stimulated. In the long-run, therefore, a SEMATECH R&D effort of size S, even if it were a perfect substitute for internal efforts by N member companies, would not reduce aggregate industry R&D by (N-1) × S, since declines in incumbent member spending on R&D would be offset to some extent by R&D spending by new entrants into the industry. "Effective" R&D summed up over the entire industry could actually rise as new entrants added their efforts to the unchanged totals for incumbent firms.

Our conclusion, as before, is that sweeping claims that estimated coefficients draw sharp lines in the sand between competing hypotheses about SEMATECH and the nature of its R&D projects are ill-founded.

ANOTHER CONCEPTUAL FRAMEWORK

To motivate the empirical analysis described by Irwin and Klenow we can construct a simple model that leads in an analytical way to the model to be estimated econometrically. Our point of departure is a model of industry R&D first described by Dasgupta and Stiglitz extended to include a shared R&D effort.[21] Appendix A to this paper gives details.

[19] *Ibid.*, p. 330. I am assuming that a dollar of SEMATECH R&D is as useful as a dollar of internal R&D and am ignoring federal subsidies for illustrative purposes.

[20] Which it would have, if this simple parable had been correct. Contributions to SEMATECH are basically treated as externally funded contract R&D by member companies and therefore show up as a component in company R&D spending reported in financial and statistical reports. Interview with Dan Damon, SEMATECH, October 2001.

[21] See P. Dasgupta and J. Stiglitz, "Industrial Structure and the Nature of Innovative Activity," *Economic Journal*, June 1980.

We take away from this model an analytical justification (see equation [8] in Appendix A) for a model with firm R&D share of sales a function of parameters related to the elasticity of cost reduction with respect to R&D (call this technological opportunity), industry concentration, and existence of and membership in an R&D consortium. Based purely on information-sharing motivations (assuming away spillovers between internal-firm R&D efforts) and assuming perfect substitution between consortium R&D and internal-firm R&D, we have seen that we would expect the formation of a "small" SEMATECH-like consortium to have a negative impact on R&D to sales ratios among members, but that would be a good thing, not a bad thing. "Effective R&D" within member firms could actually rise, costs and prices would decline, and society would be better off.

Heuristically we can further argue that if we allow spillovers between firms from internal R&D or consortium R&D to be a complement to rather than a substitute for internal efforts, the expected decline in firm R&D to sales ratios associated with pure information sharing might actually be reversed and a net positive effect on R&D to sales ratios observed. A positive impact of SEMATECH would show that the "coordination" effects more than counterbalanced an expected negative "sharing" effect. However, a negative impact would not disprove the existence of spillovers or imperfect substitution between SEMATECH and member company R&D but merely show that these impacts were not large enough to offset the expected negative impact of information sharing on R&D ratios.

MODELING ISSUES

We therefore will follow Irwin and Klenow in analyzing the impact of SEMATECH by estimating an equation where firm R&D ratios are taken to be a function of industry-wide factors (such as technological opportunity, industry concentration, factor prices) that are constant across firms but vary over time, and company-specific factors (product orientation, company technical capacity relative to the industry average, factor price differentials relative to others) that are assumed to be constant over time. We have motivated this analysis strategy somewhat differently than Irwin and Klenow, and propose different interpretations of the coefficients.

The basic equation to be estimated is

$$\text{R\&D/sales}_{it} = c_i + c_t + \text{rsl} * \text{R\&D/sales}_{it-1} + \text{ss}_t * \text{SEMATECH}_{it}$$

Where the c's are company and time dummy variables and SEMATECH is a set of dummy variables for consortium membership that varies from year to year by companies, depending on whether they were members, and whose coefficients also will be allowed to vary over time. Coefficient "rsl" measures the impact of the previous year R&D/sales ratio on the current ratio, reflecting a partial adjustment of actual to desired R&D levels. We expand on this point below.

We also take a different approach on other points.

Data

Close scrutiny of the original Irwin and Klenow sample shows that many firms classified in the code corresponding to semiconductors are not actually semiconductor device producers at all. Accordingly we have divided up the original Irwin and Klenow sample into what we call "broad" (B) semiconductor makers, producers with a broad portfolio of products they manufacture; "narrow" (N) integrated circuit (IC) firms, with a more limited product portfolio of relatively specialized ICs; fabless (F) design firms that do not actually manufacture semiconductors; producers of discrete devices (not ICs) (D); equipment and materials (EM) companies that sell specialized inputs to semiconductor device producers; and other firms (O) that do none of the above. Additionally, we have sought to identify which equipment and materials firms joined SEMI/SEMATECH over some period and which did not, a distinction not drawn by Irwin and Klenow.

Appendix B shows the composition of the original Irwin and Klenow sample as distributed within these industry groups and which firms were members of SEMATECH for at least part of the 1971-1993 period their sample covered. Of the 71 firms in their sample 11 belonged in the B group. All 6 of the SEMATECH members in their sample belonged to this same group. The N group contained 27 firms and the F group 10 firms. Eleven firms only produced discrete semiconductors (like individual transistors and diodes, not integrated circuits), 2 firms were materials producers, 5 produced other electronic components, and 5 still other products and services. In short, of the 71 companies in the original Irwin and Klenow sample, 11 were broad-based IC manufacturers (the B group, of which 6 were at some point SEMATECH members), another 27 were more narrowly focused IC manufacturers, 10 firms designed but did not manufacture ICs, 11 firms manufactured simple discrete semiconductor devices, and the remaining 12 firms were not directly involved in semiconductor production.

We have also extended the original sample of firms used by Irwin and Klenow from 1993 through 1998 and have added to the sample additional firms classified by Compustat as semiconductor producers, increasing sample size significantly. Some 46 additional firms were added: 3 in B (1 of which was to become a SEMATECH member), 15 in N, 19 in F, 5 in D, and 4 in M (see Appendix C).

Econometric Methods

We include lagged R&D-to-sales ratio as an explanatory variable to capture a partial adjustment process. Irwin and Klenow examine only the short-run instantaneous impact of SEMATECH on R&D by looking only at the coefficient of the SEMATECH variables; we also examine the long-run impact by examining the long-run multiplier ss/(1-rsl).

It is well known that with a lagged dependent variable, coefficient estimates are not consistent when a fixed effects estimator is calculated with panel data, as is the case in the original Irwin and Klenow paper. This is because consistency requires that the time dimension of the panel also become large, and typical time series of cross sections are quite short (often the number of observations per subject is a single digit). Irwin and Klenow have a relatively long data series (from 1971 to 1993) but only six years in which any of the SEMATECH membership variables are nonzero. By increasing our sample length to 1998, we almost double the time dimension of the SEMATECH membership variables and, we hope, also reduce bias issues coming from small sample size. In addition we have dealt with this problem by first differencing the model and then instrumenting the lagged R&D-to-sales ratio. R&D sales ratios lagged twice, and the value of company assets—current, lagged, and lagged twice—were used as instruments. These estimators, though not particularly efficient, are at least consistent.

Other Issues

Although we obtained some of the actual data from the authors, we were unable to reproduce many of their reported results when we ran their data and code in our version of the TSP econometric software package. Some of the results seemed to be clearly incorrect as published, or at least highly implausible; for example, their variable investment/sales has the same mean (to two significant reported digits) for a sample of 822 observations spanning 1971-1993 as for a subsample of 558 observations spanning 1980-1993.[22]

Our inability to reproduce their published results was somewhat less severe for the results we were concerned with (i.e., their analysis of R&D-to-sales ratios). (See Table 1, which shows both unweighted regression estimates and coefficients from regression weighting observations using a measure of real assets as weights.) Even there, however, there was a serious problem in their reported results. To begin, what they show as the regressions for R&D/Sales as the depen-

TABLE 1 Unweighted Regression Estimate and Coefficients for Regression Weights Observations Using a Measure of Real Asset As Weights

	IK Original Results Reported		IK Original Re-Estimated	
	Unweighted OLS	IKAssets-WLS	Unweighted OLS	IKAssets-WLS
RSL	0.26(0.06)	0.36(0.06)	0.26(0.08)	0.36(0.07)
Age2	7.8(1.5)	6.4(1.9)	7.8(1.9)	6.4(2.2)
Age35	9.3(1.3)	5.0(1.2)	9.3(1.6)	8.0(1.4)
Age6	9.7(1.3)	8.2(1.1)	9.7(1.7)	8.2(1.3)
S88	-1.04(0.56)	-1.53(0.45)	-0.01(0.56)	-1.53(0.52)

[22]See Irwin and Klenow, *op. cit.*, Table 7.

dent variable are actually the regressions for R&D/Assets and vice-versa.[23] Even when we make the appropriate substitution, there are further issues. Irwin and Klenow report their coefficients for the SEMATECH dummy as a percentage (i.e., multiplied by 100) to simplify interpretation. What they report in the published article as a coefficient of –1.04 for their unweighted least squares regression should actually be a coefficient of –.0104. Rather than being a third less than the –1.53 weighted least squares estimate, it is an order of magnitude smaller and well outside a 95 percent confidence interval for the latter! We were unable to reproduce most of their reported standard errors, using either normal or heteroskedasticity-consistent estimators, even though we were using the same data and econometrics software (TSP, though we used a later version).

Other Results with the Original Irwin and Klenow Dataset

Next, we took the original Irwin and Klenow dataset and eliminated all firms other than the 12 "broad" IC producers. The results are shown in Table 2.

The point estimates of reductions in R&D-to-sales ratios due to SEMATECH membership (S88) are generally larger in this subsample, though standard errors are quite large (as might be expected, given the small sample size), and the estimated coefficients are not statistically significant. The differenced model with the lagged R&D-to-sales ratio instrumented produced even larger point and long-run multiplier estimates, though standard errors were even larger. Note that all SEMATECH members belong to this group of "broad" IC producers.

TABLE 2 Results of Irwin and Klenow Dataset with 12 Broad IC Producers

	IK Original Broad Only		
	Unweighted OLS	IKAssets-WLS	FW-Assets WLS
RSL	0.37(0.06)<0.15>	0.36(0.07)<0.09>	0.36(0.07)<0.09>
Age35	1.75(1.48)<2.37>	3.98(2.84)<3.78>	4.12(3.04)<3.99>
Age6	3.37(1.42)<2.55>	4.65(2.74)<3.63>	4.63(2.94)<3.84>
S88	-0.21(0.90)<0.89>	-1.29(1.14)<0.90>	-1.35(1.08)<0.91>
Std. Err. of Regression	0.026	0.028	0.028
R^2	0.860	0.840	0.839
LRMult	-0.34(1.45)<1.44>	-2.03(1.78)<1.43>	-2.10(1.66)<1.41>
	Unweighted 1st Diff	IK-weighted 1st Diff	FW-Weighted 1st Diff
RSLD2	0.59(0.63)<0.57>	0.71(0.40)<0.43>	0.64(0.36)<0.42>
Age6D	-1.28(1.31)<0.89>	-0.86(1.85)<1.24>	-0.69(1.87)<1.19>
S88D	-1.34(1.53)<1.21>	-1.16(2.00)<1.11>	-1.21(1.81)<1.01>
LRS88D	-3.26(5.08)<4.42>	-3.99(8.15)<6.25>	-3.40(5.72)<4.28>
Std. Err. of Regression	0.029	0.032	0.031

< >Robust SE's

[23]We are referring here to *Ibid.*, Table 3.

"Narrow" producers were added to this sample and a Wald test was calculated for the hypothesis that both broad and narrow producers had identical coefficients for age dummies and lagged R&D to sales (none in the N group joined SEMATECH). We could not reject the hypothesis of homogeneity.

The top row of Table 3 shows the chi-squared values for the Wald test where, as before < > denotes the use of robust covariance estimates, and p-values are shown in parentheses. The remainder of Table 3 shows coefficient estimates and the estimated value of the long-run impact of SEMATECH membership.

Standard errors are now substantially lower, and with weighted least squares the R&D-saving effects of SEMATECH membership are now statistically significant. The differenced model, as before, has substantially larger standard errors.

Table 4 also shows that using current nominal assets (FW weights) as weights in lieu of real assets (asset weights deflated by an investment price index for the semiconductor industry, the original Irwin and Klenow procedure) makes little difference in estimated coefficients. When we extend the sample, we will use current-dollar nominal asset values since the semiconductor investment deflator is not available for recent years.

When fabless firms (F) were added to the sample, their inclusion could not be rejected when the weighted estimators were used (though unweighted least squares with non-robust covariance matrix estimates leads to rejecting homogeneity with a Wald test). It is worth noting that the pattern visible in this one case—with Wald tests leading to rejection of the homogeneity hypothesis using a non-robust covariance matrix estimate, but not being able to reject when robust estimates were used—was a pattern that was often repeated. Weighted least squares estimates suggest a long-run impact of SEMATECH in reducing member

TABLE 3 Coefficient Estimates and the Estimated Value of the Long-run Impact of Sematech Membership

ChiSq(all)	27.5(0.28)<25.6(0.38)>	9.23(1.00)<22.1(0.57)>	8.16(1.00)<22.5(0.61)>
	IK Original B+N		
	Unweighted OLS	IKAssets-WLS	FW-Assets WLS
RSL	0.29(0.03)<0.07>	0.34(0.04)<0.08>	0.34(0.04)<0.08>
Age35	1.69(1.11)<1.44>	2.74(1.70)<2.93>	2.86(1.81)<3.07>
Age6	1.69(1.10)<1.30>	3.39(1.66)<2.84>	3.39(1.77)<2.98>
S88	-0.68(0.71)<0.62>	-1.52(0.55)<0.59>	-1.52(0.54)<0.60>
Std. Err. of Regression	0.027	0.029	0.029
R^2	0.774	0.755	0.756
LRS88	-0.95(1.00)<0.87>	-2.31(0.83)<0.87>	-2.30(0.80)<0.89>
	Unweighted 1st Diff	IK-weighted 1st Diff	FW-Weighted 1st Diff
RSLD2	0.62(0.26)<0.24>	0.76(0.27)<0.43>	0.70(0.25)<0.42>
Age6D	-0.53(0.83)<0.62>	-0.47(1.10)<0.98>	-0.32(1.11)<0.94>
S88D	-0.73(1.39)<1.04>	0.05(1.15)<1.09>	-0.13(1.07)<1.03>
Std. Err. of Regression	0.034	0.038	0.037
LRS88D	-1.92(3.79)<2.90>	0.19(4.64)<4.64>	-0.44(3.49)<3.22>

< > robust SE's

TABLE 4 Current Nominal Assets as Weights in Lieu of Real Assets

ChiSq(all)	44.5(0.00)<21.8(0.19)>	4.54(1.00)<12.1(0.80)>	5.05(1.00)<12.3(0.78)>
	IK Original B+N+F		
	Unweighted OLS	IKAssets-WLS	FW-Assets WLS
RSL	0.25(0.03)<0.08>	0.36(0.04)<0.07>	0.35(0.04)<0.07>
Age35	4.02(0.96)<1.29>	3.40(1.40)<2.14>	3.45(1.50)<2.23>
Age6	4.79(0.92)<1.26>	4.23(1.37)<2.08>	4.25(1.46)<2.18>
S88	0.07(0.80)<0.66>	-1.21(0.52)<0.58>	-1.20(0.51)<0.58>
Std.Err. of Regression	0.032	0.033	0.033
R^2	0.751	0.732	0.733
LRMult	0.09(1.06)<0.87>	-1.88(0.80)<0.89>	-1.87(0.77)<0.89>

< > robust SE's

R&D by around 2 percent of sales. While standard errors remain relatively large, we reject the hypothesis that there was no impact or a positive impact in our weighted regressions.

We note that we rejected the inclusion of discrete device producers and materials and equipment producers when appropriate Wald tests were run.

New Data

One strategy for improving these estimates is to extend and refine the original Irwin and Klenow sample. We do so by adding additional IC producers and extending the sample to cover the 1993-1998 period. Table 5 reproduces some of these estimates for samples that covered broad producers only and samples that included broad, narrow, and fabless IC producers.

Standard errors, unfortunately, are huge. Point estimates for weighted least squares regressions are somewhat larger than in the shorter, narrower Irwin and Klenow sample.

Time-varying SEMATECH Effects

The fact that a larger, longer, more populous sample actually raised standard errors highlights the question of whether it is really reasonable to assume that the impact of SEMATECH was constant over time. An extended sample twice as long as the original sample would most likely accentuate this problem. Because SEMATECH got off to a slow start in 1988-1989 and shifted its focus in response to changing events over the period 1989-1992, it seemed reasonable to expect that even over the 1988-1992 period there would have been a ramp-up of sorts in SEMATECH activity. We therefore decided to experiment by letting the impact of SEMATECH vary from year to year.

By running a series of Wald tests for homogeneity of distinct annual SEMATECH dummies over the 1988-1998 period we narrowed down to a plausible set of periods over which it seemed reasonable to propose that SEMATECH's activities were relatively homogeneous. The Wald tests confirmed that the hypotheses

TABLE 5 Estimates of Samples Covering, Broad, Narrow, and Fabless IC Producers

	FW 7198 Broad Only		FW 7198 B+N		FW 7198 B+N+F	
	Unweighted OLS	FW-Assets WLS	Unweighted OLS	FW-Assets WLS	Unweighted OLS	FW-Assets WLS
RSL	0.47(0.05)<0.17>	0.74(0.06)<0.14>	0.32(0.04)<0.43>	0.51(0.04)<0.13>	-0.04(0.01)<0.07>	0.08(0.02)<0.15>
Age35	1.47(1.31)<2.42>	-0.96(0.65)<1.11>	9.14(7.26)<5.06>	-0.49(0.56)<0.97>	-2.17(9.25)<10.8>	-0.06(1.13)<0.99>
Age6	2.36(1.29)<2.76>	0.31(1.72)<2.17>	-4.61(7.28)<5.50>	-0.20(1.27)<1.55>	-17.9(9.35)<9.89>	-2.86(2.33)<2.42>
S88	1.24(0.74)<0.77>	-2.28(0.92)<1.13>	-3.96(5.04)<3.04>	-2.25(0.69)<0.98>	-6.53(8.47)<3.65>	-3.70(1.37)<1.24>
Std. Err. of Regression	0.026	0.032	0.236	0.025	0.408	0.463
R^2	0.845	0.789	0.441	0.780	0.625	0.526
LRMult	2.34(1.43)<1.63>	-8.74(3.68)<4.92>	-5.84(7.43)<4.95>	-4.58(1.41)<1.93>	-6.28(8.16)<3.54>	-4.00(1.48)<1.29>
	Unweighted 1st Diff	FW-WLS 1st Diff	Unweighted 1st Diff	FW-WLS 1st Diff	Unweighted 1st Diff	FW-WLS 1st Diff
RSLD2	0.82(0.47)<0.42>	0.74(0.22)<0.28>	0.88(0.08)<1.07>	0.77(0.14)<0.39>	0.75(0.08)<0.41>	0.72(0.07)<0.30>
Age6D	1.53(1.31)<0.88>	-0.75(2.43)<1.25>	-15.1(0.05)<12.8>	0.35(1.68)<1.17>	-1.13(7.86)<10.7>	4.47(3.28)<2.35>
S88D	1.02(1.47)<1.19>	-1.13(1.58)<0.58>	-0.55(9.43)<2.00>	-0.71(1.23)<0.73>	-6.55(17.1)<7.30>	-1.02(2.60)<0.63>
Std. Err. of Regression	0.031	0.031	0.254	0.247	0.473	0.470
LRMs88D	5.79(14.9)<12.6>	-4.38(6.68)<4.24>	-4.67(80.4)<45.8>	-3.12(5.49)<4.55>	-25.9(68.3)<61.9>	-3.68(9.46)<4.21>
		ChiSq(all)	13.11(1.00)<13.93(0.99)>	65.9(0.00)<30.5(0.39)>	60.9(0.00)<8.37(1.00)>	57.3(0.00)<26.8(0.26)>

< >Robust SE's

that these subperiods had a constant SEMATECH impact effect could not be rejected. The subperiods we ended up with using these procedures were 1988, 1989-1993, 1993-1997, and 1998. Estimates using these subperiods are shown in Table 6.

Once again, standard errors are still quite large. We cannot reject the hypothesis of a significant negative impact on member R&D over the 1989-1992 period in the weighted OLS estimates, but none of the SEMATECH variables is significant when we use the differenced model and an IV estimator, sadly, because our standard errors are so great. Interestingly, in all of the specifications and estimators 1998 stood out as significantly different from the 1993-1997 period. In industry R&D figures it is clear that, for the semiconductor industry overall, there was a major jump in R&D as a share of sales in 1998. It is interesting to speculate that the cumulative impact of investments in technology acceleration over the 1993-1997 period permitted SEMATECH members to "hang back" as other chip companies found themselves forced to invest major sums to catch up. The addition of some major global companies as international members in 1998 and the ongoing conversion of SEMATECH into International SEMATECH during this period may have brought in additional resources that allowed the incumbent U.S. members (the only ones in our sample prior to 1998) to not raise their R&D share of sales as much as the rest of the industry.

The Materials and Equipment Industry

Finally, we might ask if cooperation with SEMATECH had a discernable impact on companies supplying specialized materials and equipment to the semiconductor industry that were cooperating actively with SEMATECH. Given the discussion above and an evolving focus within SEMATECH on upgrading supplier industry technology, we might expect SEMATECH to have had some impact on supplier industry R&D.

We sampled U.S. semiconductor materials and equipment suppliers by adding selected firms from SIC code 3559 (a grab-bag category for semiconductor equipment suppliers) to the semiconductor equipment and materials suppliers included in the original Irwin and Klenow sample, and then coding a dummy variable to reflect participation in the SEMI/SEMATECH supplier consortium for a company in a given year. Results from fitting the model to this data are shown in Table 7.

As might perhaps be expected with such a highly diverse group of suppliers, the portion of the variance explained by an ordinary least square regression model is substantially less than for the more narrowly focused semiconductor producers analyzed earlier. The estimated impact of SEMI/SEMATECH membership on company R&D ratios is generally positive, but estimated standard errors are high, and we cannot reject the hypothesis of no (or even a negative) impact on R&D ratios within cooperating suppliers.

TABLE 6 Estimates using Subperiod for a Series of Wald Test for Homogeneity

	FW7198 B+N+F SY-Estimated		FW7198 B+N SY-Estimated	
	Unweighted OLS	FW-Assets WLS	Unweighted OLS	FW-Assets WLS
RSL	-0.04(0.01)<0.07>	0.08(0.02)<0.15>	0.32(0.04)<0.43>	0.52(0.04)<0.13>
Age35	-2.00(9.26)<10.7>	-0.14(1.13)<1.00>	9.34(7.28)<5.20>	-0.58(0.55)<0.98>
Age6	-17.5(9.38)<9.75>	-3.19(2.35)<2.46>	-4.35(7.31)<5.33>	-0.24(1.25)<1.57>
SY88	0.17(19.0)<1.90>	-2.13(3.47)<1.36>	0.65(11.4)<1.76>	-2.08(1.82)<1.16>
SY8992	-14.1(10.8)<6.86>	-4.75(1.90)<1.76>	-8.49(6.49)<6.35>	-2.29(0.97)<1.12>
SY9397	-0.05(12.9)<2.30>	-3.04(1.46)<1.29>	0.13(7.25)<1.54>	-1.57(0.72)<1.05>
SY98	5.45(25.5)<6.59>	-6.28(1.92)<2.92>	-2.18(15.0)<4.68>	-5.51(0.94)<1.86>
Std. Err. of Regression	0.408	0.464	0.236	0.251
R^2	0.626	0.526	0.443	0.403
LRMultSY88	0.17(18.3)<18.3>	-2.30(3.76)<1.52>	0.95(16.8)<2.42>	-4.32(3.79)<2.58>
ChiSq(all)	59.7(0.00)<8.64<1.00>	58.6(0.00)<27.7(0.23)>	11.9(1.00)<14.3(0.99)>	75.6(0.00)<32.0(0.32)>
	Unweighted 1st Diff	FW-Weighted 1st Diff	Unweighted 1st Diff	FW-Weighted 1st Diff
RSLD2	0.75(0.08)<0.41>	0.71(0.07)<0.29>	0.88(0.07)<1.06>	0.57(0.12)<0.35>
Age6D	-1.24(7.90)<10.8>	4.28(3.28)<2.32>	-15.3(5.21)<12.9>	0.14(1.52)<1.16>
SY88D	-5.55(18.5)<5.83>	-1.02(3.40)<0.99>	-0.42(10.3)<2.72>	-0.46(1.53)<0.96>
SY8992D	-9.38(21.0)<9.77>	-0.86(3.12)<0.99>	-4.71(11.5)<3.72>	-1.40(1.32)<0.85>
SY9397D	-5.88(27.6)<10.5>	-1.11(3.10)<0.90>	6.90(15.0)<8.18>	-0.97(1.30)<0.91>
SY98D	-0.18(39.5)<12.9>	-5.14(3.53)<1.73>	2.65(21.5)<7.96>	-5.10(1.48)<1.73>
Std. Err. of Regression	0.474	0.467	0.254	0.227
LRSY88D	-21.9(73.5)<51.9>	-3.53(11.7)<5.02>	-3.47(85.3)<41.0>	-1.07(3.58)<2.22>

< >Robust SE's

TABLE 7 Results of Using Sample of SIC Code 3559 Firms with the Irwin and Klenow Model

	EM3559	
	UnWeighted OLS	FW-Assets WLS
RSL	0.26(0.04)<0.08>	0.08(0.04)<0.08>
Age2	23.0(6.31)<12.3>	26.2(9.13)<7.57>
Age35	29.8(5.49)<12.0>	26.6(8.93)<7.43>
Age6	28.6(5.50)<12.0>	28.9(8.96)<7.49>
SS88	0.44(2.29)<1.08>	1.07(1.30)<1.16>
Std.Err. of Regression	0.103	0.119
R^2	0.545	0.393
LRMult	0.60(3.11)<1.46>	1.16(1.41)<1.23>
	UnWeighted 1st Diff	FW-Weighted 1st Diff
RSLD2	0.48(0.08)	0.27(0.09)
Age6D	4.04(1.82)	5.03(2.43)
SS88D	0.10(3.36)	-2.17(3.64)
Std.Err. of Regression	0.090	0.088
LRSS88D	0.20(6.42)	-2.97(5.02)

< > robust SE's

CONCLUSION

This paper has attempted to reexamine the impact of SEMATECH on semiconductor industry R&D by updating and improving the only published work on this issue. Unfortunately, the numbers in the original Irwin and Klenow study are marred by substantial errors in reporting results and a grab bag of econometric and data issues. When these problems were addressed and substantial volumes of additional data added to their sample, results of estimating the basic model being considered did not improve greatly. Estimated standard errors continued to be large, and parameter estimates were not particularly stable when different variants of the same basic estimators were run over different subsets of companies and periods.

Our conclusion is that the basic model specification used here is not capturing important aspects of the data well. Though we attempted as best we could to provide an analytical framework that justifies the model specification, we were forced to simply ignore the issues of the degree of substitutability or complementarity between consortium R&D and internal-company R&D (as does most of the theoretical literature), and to include in any analytical way the impacts of spillovers to internal-company R&D efforts. A model in which companies are assumed to have a constant R&D share relative to some industry-wide index that changes from year to year simply does not seem to work well. This is perhaps not surprising, since a casual perusal of National Science Foundation R&D data suggests that the Dasgupta-Stiglitz model—the only model we are aware of that can be manipulated to make a clear prediction of this sort—does not seem to fit real-world industry cross section data particularly well. There just seem to be too many other things going on that are not addressed in these stylized stories.

To the extent that these exercises can be believed (and we already have expressed our reservations), the data seem to suggest that SEMATECH reduced the R&D expenditures of its membership somewhat, as might have been predicted. However, we have taken pains to point out this cannot be interpreted as showing anything about the nature of the effort going on within SEMATECH, or as proving whether the effort justified a government subsidy. Finally, the underlying models of R&D cooperation that ultimately must be the basis of a scientific effort to untangle the chains of causality are simply too simplified at this point to capture the complexity of the real world of SEMATECH: a real world in which companies committed to R&D carried out within a joint venture while at the same time competing through internal R&D efforts that also may have spilled over to competitors, a real world in which the menu of consortium activities changes over time with experimentation and learning. At the end of the day the only absolutely certain thing about SEMATECH is that a substantial portion of its member companies must have found it to be of net value, having actually run the experiment of ending public subsidy and finding that they—its consumers—continued to buy its output.

APPENDIX A

A SIMPLE EXTENSION OF THE DASGUPTA-STIGLITZ MODEL OF R&D TO AN R&D CONSORTIUM

Assume companies invest in R&D to lower their unit costs of production, c. We assume that company attempts to protect proprietary information are successful and there are *no* spillovers between internal company R&D efforts. All companies belong to an organization called SEMATECH, with an exogenously set R&D budget of S. Company i invests x_i in order to lower its unit costs to $c(x_i + \beta S)$, with c a decreasing function of its argument—effective R&D—with $c' < 0$, $c'' > 0$. S is SEMATECH R&D, assumed a perfect substitute for internal company R&D, and β is the contribution of a dollar of shared, privately funded SEMATECH R&D to member-company effective R&D. If SEMATECH is less efficient or effective than a private company, $\beta < 1$. If it is equally effective, $\beta = 1$, and if there are matching federal funds, it is possible that $\beta > 1$. We assume that the SEMATECH budget is shared equally among members, so if there are N members, each pays a fixed sum of S/N in dues to join SEMATECH. For firms to voluntarily join SEMATECH, therefore, it is necessary that member-company benefit exceeds membership cost (i.e., that $\beta > 1/N$, which we shall assume).

Because there are no spillovers between internal company R&D efforts (and one company's internal R&D does not affect another's production costs), there is no need to distinguish between a first-stage R&D decision and a second-stage production decision, and we can treat the two decisions as simultaneous in analyzing a Cournot-Nash equilibrium, with each firm choosing its R&D and output levels taking those of its competitors as given. The profit of company *i* is given by

$$\Pi_i = \left[P(Q) - c\left(X_i + \beta S\right)\right] q_i - X_i - \frac{S}{N} \tag{1}$$

where the *X* and *q* variables are company *i*'s chosen R&D and output levels and *Q* is homogeneous industry output ($=\Sigma q_i$). We assume a constant price elasticity for industry demand, so optimal *q* for company *i* is determined by the standard markup condition,

$$\frac{P - c}{P} = -\frac{1}{\varepsilon} \cdot \frac{P}{Q} \cdot q_i \tag{2}$$

We also will assume a symmetric industry equilibrium, so $q_i = Q / N$ in equilibrium and (2) becomes

(3)
$$\frac{P(Q)-c}{P}=-\frac{1}{\varepsilon N}$$

We define α, the elasticity of unit cost with respect to R&D, as the percentage change in unit cost per percentage increase in effective R&D

(4)
$$\alpha = c' \cdot \frac{X_i + \beta S}{c} < 0$$

and assume it to be constant.

In addition to maximizing profit by choosing output a firm also chooses its optimal R&D X_i. The first-order maximization condition is

$$\frac{\partial \Pi_i}{X_i} = 0 = -c' q_i - 1$$

or taking into account (4) and the symmetric industry equilibrium,

(5)
$$N = \frac{-\alpha c Q}{X_i + \beta S}.$$

Note that for given N (number of firms in industry) symmetric industry equilibrium conditions (3) and (5) give us

(6)
$$PQ = \frac{N(X_i + \beta S)}{-\alpha\left(1 + \frac{1}{N\varepsilon}\right)}.$$

What happens then, if—given some N—we go from a no (non-trivial) SEMATECH ($S = 0$) equilibrium to an equilibrium with SEMATECH spending money on R&D projects ($S > 0$)? Inspection of first-order conditions (3) and (5) reveals that if company R&D spending X_i is cut by βS, the first-order conditions continue to hold. Thus, as Irwin and Klenow have argued, with pure information sharing (which is what is going on in this model) and SEMATECH effort a perfect substitute for internal R&D, the impact of a SEMATECH effort of size S will be to cut company effort by βS.

But do things stop here, where the Irwin and Klenow argument stops? Inspection of profit function (1) reveals that firms are now earning an additional profit if $\beta > 1/N$, which will always be true if membership is voluntary. Thus, firms are now earning an additional rent due to the benefits of information sharing, stimulating entry by new firms if entry is possible.

Next, consider the implications of free entry and a zero-profit, long-run-symmetric equilibrium. Then we have an additional condition in the long run,

$$NX_i + S = (P - c)Q \tag{7}$$

(i.e., variable profits should just cover fixed R&D costs).

Substituting (3) into (7) we have an expression for total industry R&D expense as a percent of sales in the long run,

$$\frac{NX_i + S}{PQ} = -\frac{1}{N\varepsilon} \tag{8}$$

which is a familiar result from the Dasgupta-Stiglitz model (with $S = 0$). Overall industry R&D (including SEMATECH projects) as a share of sales declines with number of firms in the industry (N) and price elasticity of demand. Internal industry R&D as a share of sales (i.e., excluding S) is smaller than it would be without SEMATECH.

Combining equation (8) with equation (6) we have

$$N(X_i + \beta S) = (NX_i + S)\alpha(1 + N\varepsilon). \tag{9}$$

With no SEMATECH ($S = 0$), this gives us the familiar Dasgupta and Stiglitz result that $N = \frac{1-\alpha}{\varepsilon\alpha}$, or equivalently (using [8]), that the R&D-to-sales ratio is $\frac{-\alpha}{1-\alpha}$.

APPENDIX B

ORIGINAL IRWIN AND KLENOW SAMPLE MEMBERSHIP

COMPANY	Broad	Narrow	Fabless	Discrete	Materials & Equipment	Other Electronics	Others	SEMATECH
DENSE-PAC MICROSYSTEMS INC			√					
LINEAR TECHNOLOGY CORP		√						
CYPRESS SEMICONDUCTOR CORP	√							
CHIPS&TECHNOLOGIES INC			√					
IMP INC		√						
OPTEK TECHNOLOGY INC				√				
DALLAS SEMICONDUCTOR CORP		√						
MAXIM INTEGRATED PRODUCTS		√						
LOGIC DEVICES INC		√						
AMERICAN MICROSYSTEMS							√	
APPLIED SOLAR ENERGY CORP							√	
TEXAS INSTRUMENTS INC	√							√
AVANTEK INC			√					
BURR-BROWN CORP		√						
CODI CORP				√				
INTL RECTIFIER CORP				√				
DIONICS INC		√						
ELECTRONIC ARRAYS INC		√						
JETRONIC INDUSTRIES INC				√				
GENERAL SEMICONDUCTOR INDS				√				
SOLITRON DEVICES INC				√				
HYTEK MICROSYSTEMS INC		√						
BOURNS INC						√		
INTECH INC						√		
INTEGRATED DEVICE TECH INC	√							
INTERSIL INC		√						
INTERSIL INC NEW		√						
ALPHA INDUSTRIES INC							√	
UNITRODE CORP		√						
LSI LOGIC CORP	√							√
MSI ELECTRONICS INC							√	
SEMTECH CORP	√							
NATIONAL SEMICONDUCTOR CORP	√							√
MICRON TECHNOLOGY INC	√							√
MICROSEMI CORP				√				
MICROWAVE SEMICONDUCTOR CORP						√		
MONOLITHIC MEMORIES INC		√						
MOSTEK CORP	√							
INTEL CORP	√							√
ANALOG DEVICES		√						
ADVANCED MICRO DEVICES	√							√
KYOCERA CORP -ADR					√			
SOLID STATE SCIENTIFIC						√		
RIPLEY CO INC						√		
SEEQ TE		√						
SEMICON INC				√				
SILICON SYSTEMS INC		√						
SILICONIX INC		√						
SILTEC CORP					√			
SUPERTEX INC		√						
TECCOR ELECTRONICS				√				
ALTERA CORP		√						
LATTICE SEMICONDUCTOR CORP			√					
XILINX INC		√						
ZILOG INC		√						
ATMEL CORP		√						
BROOKTREE CORP			√					
PMC-SIERRA INC		√						
INTEGRATED CIRCUIT SYSTEMS		√						
ADVANCED PHOTONIX INC -CLA			√					
TRANSTECTOR SYSTEMS INC		√						
VITESSE SEMICONDUCTOR CORP			√					
BKC SEMICONDUCTORS INC					√			
MICROCHIP TECHNOLOGY INC		√						
VLSI TECHNOLOGY INC	√							
XICOR INC			√					
APPIAN TECHNOLOGY INC			√					
ZITEL CORP							√	
EXAR CORP			√					
ELECTRONIC DEVICES INC				√				
PRECISION MONOLITHICS INC		√						

APPENDIX C

ADDITIONAL FIRMS ADDED TO SAMPLE IN THIS STUDY

COMPANY	Broad	Narrow	Fabless	Discrete	Materials & Equipment	Other Electronics	Others	SEMATECH
ACTEL CORP		√						
AEROFLEX INC			√					
ALLIANCE SEMICONDUCTOR CORP		√						
ANADIGICS INC		√						
AUREAL INC			√					
BENCHMARQ MICROELECTRONICS		√						
CATALYST SEMICONDUCTOR INC			√					
CREE RESEARCH INC				√				
CYRIX CORP			√					
ELANTEC SEMICONDUCTOR INC		√						
ELECTRONIC DESIGNS INC			√					
ESS TECHNOLOGY INC			√					
GALILEO TECHNOLOGY LTD			√					
GATEFIELD CORP		√						
HEI INC			√					
IBIS TECHNOLOGY INC					√			
INFORMATION STORAGE DEVICES				√				
INTEGRATED SILICON SOLUTION				√				
LEVEL ONE COMMUNICATIONS INC				√				
MEMC ELECTRONIC MATRIALS INC					√			
MICREL INC	√							
MICRO LINEAR CORP		√						
MITSUBISHI ELEC CORP -ADR	√							
MRV COMMUNICATIONS INC			√					
OAK TECHNOLOGY INC			√					
OPTI INC			√					
QUALITY SEMICONDUCTOR INC			√					
RAMTRON INTERNATIONAL CORP			√					
REMEC INC			√					
SDL INC					√			
SIGMA DESIGNS INC			√					
SILICON STORAGE TECHNOLOGY		√						
SIMTEK CORP		√						
SIPEX CORP		√						
SMART MODULAR TECHNOLGS INC				√				
SPECTRUM SIGNAL PROCESSING		√						
STANDARD MICROSYSTEMS CORP		√						
STMICROELECTRONICS N V	√							√
TELCOM SEMICONDUCTOR INC		√						
TOWER SEMICONDUCTOR LTD		√						
TRANSWITCH CORP					√			
TRIDENT MICROSYSTEMS INC			√					
UNIVERSAL DISPLAY CORP			√					
WHITE ELECTRIC DESIGNS CORP			√					
ZING TECHNOLOGIES INC			√					
ZORAN CORP		√						

APPENDIX D

EQUIPMENT COMPANIES

Company Name	cusip	SEMI/SEMATECH
AG ASSOCIATES INC	107310	√
AMISTAR CORP	3153510	
AMTECH SYSTEMS INC	3233250	
APPLIED MATERIALS INC	3822210	√
APPLIED SCI & TECH	3823610	
ASM INTERNATIONAL N V	N0704510	
ASM LITHOGRAPHY HOLDING NV	N0705911	
ASYST TECHNOLOGIES INC	04648X10	√
BE SEMICONDUCTOR INDUSTRIES	7332010	
BROOKS AUTOMATION INC	11434A10	√
BTU INTERNATIONAL INC	5603210	√
CFM TECHNOLOGIES INC	12525K10	√
CHEMINEER INC	16381810	
CRYO-CELL INTERNATIONAL INC	22889510	
CVD EQUIPMENT CORP	12660110	
CYMER INC	23257210	√
DT INDUSTRIES INC	23333J10	
DTM CORP	23333L10	
ELECTROGLAS INC	28532410	√
EMCORE CORP	29084610	√
ENGINEERED SYS & DEV CORP	29286810	
ETEC SYSTEMS INC	26922C10	√
FSI INTL INC	30263310	√
GASONICS INTERNATIONAL CORP	36727810	√
GCA CORP	36155620	
GENERAL SCANNING INC	37073710	
GENUS INC	37246110	√
GERBER SCIENTIFIC INC	37373010	
HELISYS INC	42328210	
HI-RISE RECYCLING SYS INC	42839610	
IBIS TECHNOLOGY INC	45090910	
ICOS VISION SYSTEMS CORP NV	B4923310	
INNOTECH INC	45766M10	
INTEGRATED PROCESS EQ	45812K10	
INTEVAC INC	46114810	
IONICS INC	46221810	
KULICKE & SOFFA INDUSTRIES	50124210	√
LAM RESEARCH CORP	51280710	√
MATERIALS RESEARCH	57668010	
MATTSON TECHNOLOGY INC	57722310	
MICRION CORP	59479P10	
MRS TECHNOLOGY INC	55347610	
NOVELLUS SYSTEMS INC	67000810	√
ONTRAK SYSTEMS INC	68337410	
OPAL INC	68347410	
PHOTRONICS INC	71940510	√
PLASMA-THERM INC	72790010	√
PRI AUTOMATION INC	69357H10	√
QC OPTICS INC	74693410	√
QUAD SYSTEMS CORP	74730Q10	
RIMAGE CORP	76672110	
SEMITOOL INC	81690910	√
SILICON VALLEY GROUP INC	82706610	√
SONO-TEK CORP	83548310	
SPEEDFAM-IPEC INC	84770510	√
SUBMICRON SYSTEMS CORP	86431310	
TEGAL CORP	87900810	√
TOOLEX INTERNATIONAL NV	N8715N10	
TRIKON TECHNOLOGIES INC	89618710	
TURN TECHNOLOGY INC	90021310	
VARIAN SEMICONDUCTOR EQUIPMT	92220710	√
VEECO INSTRUMENTS INC	92241710	√
VITRONICS CORP	92850310	
WAVEMAT INC	94356110	
YIELDUP INTL CORP	98583710	

V

APPENDIXES

Appendix A

Description of Focus Center Research Program Centers

As noted in the Introduction, there are currently four focus centers under The Focus Center Research Program (FCRP). They are:

The **Design and Test Focus Center,** commonly referred to as the Gigascale Silicon Research Center (GSRC), was founded in 1998 to explore all the aspects of semiconductor design and test issues. Design and Test refers primarily to the software programs used by the people who create microchips and the people who test them to see if they work. The University of California at Berkeley is the lead campus for this effort. The participating universities in the Design and Test Focus Center are Carnegie Mellon; MIT; Pennsylvania State; Princeton; Purdue; Stanford; University of California at Berkeley, Los Angeles, San Diego, Santa Barbara, and Santa Cruz; University of Michigan; UT Austin; and University of Wisconsin–Madison.

The **Interconnect Focus Center** was founded in 1998 to research all aspects of the wiring that connects the millions of transistors on a microchip—from process to system-level architecture. As the circuits on a semiconductor chip become ever faster, the pacing item increasingly becomes the time required to have each circuit communicate with the other circuits on the chip. The Interconnection Focus Center, led by the Georgia Institute of Technology, is working on novel ways to address this problem, including having the circuits communicate through optical or radio wave connects instead of today's copper connections. The six participating universities are Georgia Tech; MIT; Stanford; Rensselaer Polytechnic Institute; SUNY at Albany; and Cornell.

The **Materials, Structures, and Devices Focus Center** was formed in 2001 to scale the current CMOS process to its ultimate limit using novel transistor structures, and to explore hybrid chips where silicon CMOS devices are combined with new-frontier devices such as carbon nanotubes, organic semiconductors, or quantum-effect devices. The Focus Center is led by the Massachusetts Institute of Technology. The other participating universities are Cornell; Princeton; Purdue; Stanford; SUNY Albany; University of California at Berkeley and Los Angeles; UT Austin; and University of Virginia

The **Center for Circuits, Systems & Software (C2S2)** Focus Center was formed in 2001 to develop a wholly new generation of design techniques to convert semiconductor circuits into ultra-performance electronic products. Tomorrow's circuits must routinely move billions of bits per second through the air; perform billions of operations per milliwatt; access billions of bits of on-chip storage; and interact with a rich environment of communicating electrical, mechanical, optical, and biological systems. To convert tomorrow's transistors into this range of required performance requires a revolutionary rethinking of today's design strategies. The C2S2 Focus Center, led by Carnegie Mellon University, is exploring a variety of approaches to these problems. The other universities in this consortium are Cornell, Columbia, MIT, Princeton, Stanford, University of California at Berkeley, University of Illinois at Urbana-Champaign, University of Michigan, and University of Washington.

Appendix B

Biographies of Speakers*

MICHAEL BORRUS

Michael G. Borrus is a Managing Director of The Petkevich Group, LLC, a new financial services company offering capital and service for long-term value creation in high-growth industries. Most recently, Mr. Borrus was a Co-Director of the Berkeley Roundtable on the International Economy (BRIE) at the University of California, Berkeley; adjunct professor in UC Berkeley's College of Engineering; and a partner in Industry and Trade Strategies, a business consultancy.

As consultant, Mr. Borrus has worked with a wide variety of firms and governments in the U.S., Asia, and Europe. His consulting clients have included both large multinationals like Applied Materials, Motorola, Kawasaki Steel, and Teledenmark, and technology startups like Tachyon Networks and Solo Energy. Much of his consulting has focused on how business models need to adjust to successfully exploit new market opportunities or to adapt to new technologies and new competitors.

He is the author or editor of three books and over 70 chapters, articles, and monographs on a variety of topics including management of technology, high-technology competition, international trade and investment, and business strategies for information technology industries. Mr. Borrus is an honors graduate of Harvard Law School and a member of the California State Bar.

*As of October 2000. Includes biographies submitted by speakers.

PAPKEN DER TOROSSIAN

Papken Der Torossian is Chairman of the Board and Chief Executive Officer of Silicon Valley Group, Inc. Der Torossian joined SVG as president in 1984, became CEO in 1986, and was appointed Chairman of the Board in 1991. A respected industry spokesperson, he has served on the Executive and International committees of the American Electronics Association and as Chairman of the SEMI/SEMATECH Board of Directors. His over-30-year career is highlighted by three years with ECS Microsystems Inc., as President and CEO, and five years with Plantronics—as president of the Santa Cruz Division and, earlier, vice president of its telephone products group. In addition, he spent four years at Spectra-Physics and held a variety of management positions during a 12-year tenure at Hewlett-Packard. He received a bachelor's degree in mechanical engineering from the Massachusetts Institute of Technology and a master's degree from Stanford University.

PETER DRAHEIM

Since 1995, Peter Draheim has been Managing Director of Philips GmbH, Hamburg, and Chief Executive Officer of Philips Semiconductors SMST, Böblingen.

Previously, he was the Chief Executive Officer of Philips Medical Systems, Hamburg; the Director of Philips Semiconductors, Eindhoven; the Business Unit Manager for Microelectronics VALVO, Philips GmbH, Hamburg; the Sales and Marketing Manager for Consumer Integrated Circuits, VALVO, Philips GmbH, Hamburg; a member of the development management at the VALVO Röhren- und Halbleiterwerke Philips GmbH, Hamburg; and an assistant at the Institute for Electrophysics, Technical University Braunschweig.

Mr. Draheim studied physics at the Technical University Braunschweig.

KENNETH FLAMM

Kenneth Flamm, who joined the LBJ School of the University of Texas at Austin in fall 1998, is a 1973 honors graduate of Stanford University and received a Ph.D. in economics from M.I.T. in 1979.

From 1993 to 1995, Flamm served as Principal Deputy Assistant Secretary of Defense for Economic Security and as Special Assistant to the Deputy Secretary of Defense for Dual-Use Technology Policy. He was awarded the department's Distinguished Public Service Medal in 1995 by Defense Secretary William J. Perry. Prior to his service at the Defense Department, he spent 11 years as a Senior Fellow in the Foreign Policy Studies Program at the Brookings Institution.

Flamm has been a professor of economics at the Instituto Tecnológico A. de México in Mexico City, the University of Massachusetts, and the George Washington University. He has also been an adviser to the Director General of Income Policy in the Mexican Ministry of Finance and a consultant to the Organization for Economic Cooperation and Development, the World Bank, the National Academy of Sciences, the Latin American Economic System, the U.S. Department of Defense, the U.S. Department of Justice, the U.S Agency for International Development, and the Office of Technology Assessment of the U.S. Congress.

Among Dr. Flamm's publications are *Mismanaged Trade? Strategic Policy and the Semiconductor Industry* (1996), *Changing the Rules: Technological Change, International Competition, and Regulation in Communications* (ed., with Robert Crandell, 1989), *Creating the Computer (1988), and Targeting the Computer* (1987). He is currently working on an analytical study of the post-Cold War defense industrial base.

Flamm, an expert on international trade and the high-technology industry, teaches classes in microeconomic theory, international trade, and defense economics.

EDWARD D. GRAHAM, JR.

Ed Graham joined SISA (Semiconductor Equipment Suppliers Association) as its fourth President in December 1999. He came to SISA from Sandia National Laboratories, where his career spanned more than 32 years. At Sandia, he was most recently the Director of Operations and Engineering. During a portion of his career at Sandia, Graham had responsibility for all semiconductors in all nuclear weapons systems.

Ed Graham has received three degrees in electrical engineering: a B.S. degree from Mississippi State University, an M.S. degree from the University of New Mexico, and a Ph.D. from North Carolina State University. In addition, he is an Extra Class Radio Operator [N5HH].

Currently, Ed is commuting between Albuquerque, New Mexico, and Austin, Texas. When he finds some free time, his other interests include skiing, hiking (with wife Sandra), reading, collecting slide rules, and observing the stock market.

MASATAKA HIROSE

Masataka Hirose was born in Gifu, Japan, on September 30, 1939. He received the B.S. and M.S. degrees in electronic engineering from Nagoya University, Nagoya, Japan, in 1963 and 1967, respectively, and the Ph. D. degree in electronic engineering from Tohoku University, Sendai, Japan, in 1975.

From 1963 to 1964, Dr. Hirose worked in the Central Research Laboratory, Fuji Electric Co., Ltd. Since 1970, he had been with Department of Electrical Engineering, Hiroshima University, Higashi-Hiroshima, Japan, and appointed as Professor from 1982 to 2001. From 1986 to 1996, he was Director of the Research Center for Integrated Systems, Hiroshima University. From 1996 to 2001, he was Director of the Research Center for Nanodevices and Systems. Since April in 2001, he has been Director General, Advanced Semiconductor Research Center of the National Institute of Advanced Industrial Science and Technology in Tsukuba, Japan. His research interests include ULSI devices and processes, and silicon quantum nanodevices. Since 1996, he has been Chairman of the 165 Research Committee for Ultra Large Scale Integrated Devices and Systems, Japan Society for the Promotion of Science.

Dr. Hirose is a member of IEEE, serving as an editor of *Transactions on Electron Devices* from 1999 to 2002; the Japan Society of Applied Physics; and the Institute of Electronics, Information and Communication Engineers, Japan.

GENDA J. HU

Genda Hu received a B.S.E.E. from National Cheng Kung University, Taiwan, R.O.C., and a Ph.D. from Princeton University in 1973 and 1979, respectively. After graduation he worked at IBM's T. J. Watson Research Center as a Research Staff Member, where he pioneered the company's CMOS technology development. Later Dr. Hu joined Xerox Palo Alto Research Center and demonstrated the industry's first latch-up free CMOS operation. He also demonstrated the role of n+/p+ poly gate for future CMOS scaling. In 1984, he participated in the startup of Sierra Semiconductor. As a Director of Device Engineering he contributed in developing 4 generations (3um, 2um, 1.5um & 1um) of CMOS technologies and more than 100 products in analog, digital, and EEPROM. In 1990, Dr. Hu joined Cypress Semiconductor as Director of Non-volatile Technology, where he completed development of a 0.8-um BiCMOS and a 0.65-um CMOS UV EPROM/PLD technology. He then served as Vice President of Technology at ISD, San Jose, California, from 1995 to 1996. At ISD, he was in charge of bringing up multiple wafer foundries and implementing multi-level analog storage in Flash technology.

Since July 1996 Dr. Hu has been the General Director of the Electronics Research and Service Organization (ERSO) of the Industrial Technology Research Institute (ITRI) in Taiwan. He was credited as the key player in achieving the landmark victory over Micron's anti-dumping charge against Taiwan's DRAM industry in 1999 and SRAM industry in 2000. In May 2000, Dr. Hu joined TSMC as Vice President of Advanced Technology development.

Dr. Hu is an IEEE Fellow and is also currently President of the Taiwan Semiconductor Industry Association (TSIA).

KALMAN KAUFMAN

Mr. Kaufman was named Applied Materials' Corporate Vice President of Strategic Planning and New Business Development in May 1997. He joined

Applied Materials as General Manager of the Implant Division in April 1994. In December 1995, he was appointed President of the Thermal Process and Implant group and Corporate Vice President of Applied Materials.

Prior to joining Applied Materials, Mr. Kaufman was Divisional Manager and General Manager of Kulicke & Soffa. Before that, he was President of KLA Instruments, Israel.

Mr. Kaufman earned his Master of Science degree in physics and his Bachelor of Science degree in mechanical engineering from the Technion University in Israel.

CHIEN-YUAN LIN

Chien-Yuan Lin is a Professor in the Graduate Institute of Building and Planning at the National Taiwan University in Taipei.

Dr. Lin's major areas of research include industrial/business park planning & development; technology-park and local development; and land-use planning and control systems. Within these areas he has published more than 35 journal papers, 70 conference papers, and 44 research reports since 1987. Most of his publications are related to national industrial park development policy and planning.

Dr. Lin received a B.A. in land economics from the National Chengchi University, Taiwan; an M.E. in urban planning from the National Taiwan University; and a Ph.D in transportation planning from the University of Washington.

MICHAEL LUGER

Michael Luger is Professor of Public Policy Analysis, Management, and Planning, and Director of the Office of Economic Development, at the University of North Carolina at Chapel Hill. He formerly served as Chairman of the Department of Public Policy Analysis and as the Carl E. Pegg Professor of City and Regional Planning.

Dr. Luger's Ph.D. is in economics (Berkeley, 1981). He also holds an M.P.A. (Princeton, 1976) and M.C.P. (Berkeley, 1978). His A.B. (architecture) is from Princeton. His scholarly work is in the areas of regional economic development, public finance, infrastructure development, and science and technology policy. He is the author of many books and articles on those topics.

Professor Luger has advised many organizations about economic development, including the European Commission, OECD, UNDP, the World Bank, and numerous foreign governments.

Prior to moving to UNC-CH, Luger taught at Duke University (economics and public policy) and the University of Maryland (economics). He recently has served as chair or member of the Durham Board of Adjustment, Area Transportation Authority, and Merger Commission.

TOSHIAKI MASUHARA

Toshiaki Masuhara obtained B.S. and M.S. degrees in electrical engineering from the University of Kyoto, Japan, in 1967 and 1969. He obtained a Ph.D. in electrical engineering from the University of Kyoto in 1977. In 1969, he became a member of the technical staff at the Hitachi Central Research Laboratory, where he initially worked on depletion-load NMOS integrated circuits and later on modeling and analysis of MOS transistors. In 1974-75, he was a special student in the Electrical Engineering and Computer Science Department of the University of California, Berkeley, where he worked on double-diffused MOS transistors and a new CMOS process. In the ensuing years, he initiated a project to develop new high-speed CMOS static memories with NMOS cells; supervised research groups working on high-speed GaAs and bipolar integrated circuits, solar cells, and imager tubes; and was responsible for the design of VLSIs, components, and PCBs. In 1993 Dr. Masuhara became General Manager of Technology Development Operation, and in 1977 he became General Manager of the Semiconductor Manufacturing Technology Center. He became Senior Chief Engineer in 1998.

Dr. Masuhara is a Fellow of IEEE (since 1994, with the citation "For Contribution in the invention and development of NMOS circuits and high-speed CMOS static memories") and a member of the Institute of Electrical, Information and Communication Engineers of Japan. He has been an administrative committee member of the Solid-State Circuits Society of IEEE since 1998. He was the program co-chair and chairman of the 1992 and 1993 Symposium on VLSI Circuits and was the co-chair and the chairman of the 1996 and 1997 Symposium on VLSI Circuits. He received the IEEE Solid-State Circuits Award in 1990 for his contributions to NMOS depletion-load circuits and the development of high-speed CMOS memories. Since 2000, he has been a chairman of the Semiconductor Technology Roadmap Committee of Japan. He has received a Significant Invention Award, Japan, in 1994; four Significant Invention Awards, Tokyo, Japan, in 1984, 1985, 1988, and 1992; and Significant Invention Awards, Yamanashi, Japan, in 1995, and Gunma, Japan, in 1996.

W. CLARK MCFADDEN

Mr. McFadden specializes in international corporate transactions, especially the formulation of joint ventures, consortia, and international investigations, and enforcement proceedings. Mr. McFadden has a broad background in foreign affairs and international trade, including experience with Congressional committees, the U.S. Department of Defense, and the National Security Council.

In 1986, he was appointed General Counsel, President's Special Review Board, to investigate the National Security Council system ("Tower Commission"). In 1979, Mr. McFadden served as Special Counsel to the Senate Foreign

Relations Committee on the Strategic Arms Limitation Treaty (SALT II). Previously, from 1973-1976, he worked as General Counsel, Senate Armed Services Committee and was responsible to the Committee for all legislative, investigatory, and oversight activities. Mr. McFadden has a B.A. from Williams College (1968), M.B.A. from Harvard University (1972), and J.D. from Harvard Law School (1972).

GORDON E. MOORE

Gordon E. Moore is currently Chairman Emeritus of Intel Corporation. Moore co-founded Intel in 1968, serving initially as Executive Vice President. He became President and Chief Executive Officer in 1975 and held that post until elected Chairman and Chief Executive Officer in 1979.

He remained CEO until 1987 and was named Chairman Emeritus in 1997.

Gordon Moore is widely known for "Moore's Law," in which he predicted that the number of transistors that the industry would be able to place on a computer chip would double every year. In 1995, he updated his prediction to once every two years. While originally intended as a rule of thumb in 1965, it has become the guiding principle for the industry to deliver ever-more-powerful semiconductor chips at proportionate decreases in cost.

Dr. Moore earned a B.S. in chemistry from the University of California at Berkeley and a Ph.D. in chemistry and physics from the California Institute of Technology. He was born in San Francisco, California, on January 3, 1929.

Gordon Moore is a director of Varian Associates, Gilead Sciences, Inc., and Transamerica Corporation. He is a member of the National Academy of Engineering, a Fellow of the IEEE, and Chairman of the Board of Trustees of the California Institute of Technology. He received the National Medal of Technology in 1990 from then-President George H.W. Bush.

DAVID MOWERY

David Mowery is Milton W. Terrill Professor of Business at the Walter A. Haas School of Business at the University of California, Berkeley, and Director of the Haas School's Ph.D. program. He received his undergraduate and Ph.D. degrees in economics from Stanford University and was a postdoctoral fellow at the Harvard Business School. Dr. Mowery has taught at Carnegie-Mellon University, served as the Study Director for the Panel on Technology and Employment of the National Academy of Sciences, and served in the Office of the United States Trade Representative as a Council on Foreign Relations International Affairs Fellow. He has been a member of a number of National Research Council panels, including those on the Competitive Status of the U.S. Civil Aviation Industry, on the Causes and Consequences of the Internationalization of U.S. Manufacturing, on the Federal Role in Civilian Technology Development, on U.S. Strategies for the Children's Vaccine Initiative, and on Applications of Biotechnology to Contraceptive Research and Development. His research deals with the economics of technological innovation and with the effects of public policies on innovation. He has testified before Congressional committees and served as an adviser for the Organization for Economic Cooperation and Development, various federal agencies, and industrial firms.

MICHAEL R. POLCARI

Michael Polcari received his Bachelor of Science degree in physics from the University of Notre Dame and his Master of Science and Doctor of Philosophy degrees in physics from Stevens Institute of Technology. From 1977-1978, he was employed at National Micronetics in Kingston, New York, as a process development engineer in thin films. In 1978, he joined IBM at Kingston, New York, as a process development engineer on a silicon pilot line, moving to manager of process development in 1980.

In 1982, Dr. Polcari joined the IBM Research Division at the T. J. Watson Research Center as a research staff member in the Silicon Technology department. He held various management positions in the Yorktown silicon facility until October 1992, including management of this silicon fabricator. From 1992 until 1994, he was responsible for the Advanced Lithography Systems Department of IBM's Semiconductor Research and Development Center (SRDC) in East Fishkill, New York.

In January 1994, he was appointed Research Director, Silicon Technology, and Director, Advanced Semiconductor Technology Laboratory, in the SRDC. In this assignment, he was responsible for silicon process technology in the Research Division and Advanced Semiconductor Process Development in the SRDC.

In August 1999, Dr. Polcari was appointed Vice President, Procurement Engineering, IBM Global Procurement in Somers, New York. In his present position, he is responsible for driving technology convergence and qualification processes across a number of high-usage commodities within IBM. He is also responsible for re-engineering all procurement technical resources within the company to achieve maximum effectiveness in executing critical procurement engineering tasks across the varied IBM Divisions.

He has served as IBM's representative on the Board of Directors of the Semiconductor Research Corporation and in 1999 was the SRC Board Chairman. He has also served on the advisory boards of Stanford's Center for Integrated Systems and MIT's Microsystems Technology Laboratory.

Dr. Polcari is a member of IEEE, SPIE, ECS, and APS.

GEORGE SCALISE

George Scalise is President of the Semiconductor Industry Association (SIA), the premier trade association representing the microchip industry. As President, Scalise directs and oversees SIA programs focused on public policy, technology, workforce, international trade and government affairs, environment safety and health, and communications.

Mr. Scalise has had a long career in the semiconductor and related industries, bringing with him over 30 years of industry experience. Prior to joining the SIA in June 1997, Scalise served as the Executive Vice President of Operations and Chief Administrative Officer at Apple Computer. Preceding Apple, he worked in numerous executive positions at National Semiconductor Corporation, Maxtor Corporation, Advanced Micro Devices, Fairchild Semiconductor, and Motorola Semiconductor.

Mr. Scalise is a highly respected technology and public policy spokesperson for the industry. He has a special interest and expertise in international trade and competition issues. For over eight years, Scalise was the chairman of SIA's Public Policy Committee, shaping and implementing the semiconductor industry's agenda on major policy issues. Additionally, he was a founder, member, and the Chairman of the Board of the Semiconductor Research Corporation (SRC), an industry-funded organization that provides resources for pre-competitive semiconductor research at American universities. For three years, he also served on the Board of Directors of SEMATECH, a research consortium created to gain manufacturing advantage in semiconductor technology.

Mr. Scalise is active on many boards and advisory committees. In December 1999, he was elected to the Board of Directors of the Federal Reserve Bank of San Francisco, Twelfth Federal Reserve District, to represent non-banking interests in the District's nine states.

He also serves on the boards of Cadence Design Systems, Network Equipment Technologies, and the Foreign Policy Association.

Mr. Scalise has served on a number of university and government boards, including the University of Southern California School of Engineering Board of Councilors, the Santa Clara University Leavey School of Business Advisory Board, the University of Texas at Austin Engineering Foundation Advisory Committee, Purdue University Engineering Visiting Committee, the Secretary of Energy Advisory Board for the U.S. Department of Energy (as chairman), and the Joint High Level Advisory Panel of the US-Israel Science and Technology Committee.

George Scalise graduated from Purdue University with a Bachelor of Science in mechanical engineering.

HIDEO SETOYA

On April 1, 1996, Mr. Hideo Setoya was appointed as Executive Director of the ASET. Before joining ASET, Mr. Setoya had been working for the Ministry of International Trade and Industry since 1971. His experience in the government includes planning and management of technological research and development programs and administration of electronics and information industries. He served at MITI as the Director of the Information Systems Development Division, the Information Technology Standard Division, the Industry Department of Tohoku Regional Bureau, and the Security Export Control Office. He also was assigned to MITI-related organizations as the Director of Geothermal Research Department of NEDO (New Energy and Industrial Technology Development Organization), the Technology Research Department of MMAJ (Metal Mining Agency of Japan), and the Research Division of Chicago Center of JETRO (Japan External Trade Organization).

Mr. Setoya received a bachelor,s degree from the Department of Precision Machinery Engineering, Faculty of Engineering, the University of Tokyo in 1971.

BILL SPENCER

Bill Spencer recently retired as Chairman of SEMATECH, a research and development consortium consisting of fourteen international corporations involved in semiconductor manufacturing. From 1990-1997, he served as President and Chief Executive Officer of SEMATECH. Prior to 1990, he was Group Vice President and Senior Technical Officer at Xerox Corporation in Stamford, Connecticut, as well as Vice President and Manager of the Xerox Palo Alto Research Center (PARC). He was Director of Systems Development and also Director of Microelectronics at Sandia National Laboratories from 1973 to 1981, prior to joining Xerox. He began his career at Bell Telephone Laboratories in 1959. He received his Ph.D. and M.S. from Kansas State University, and an A.B. from William Jewell College in Missouri.

Dr. Spencer is also a Research Professor of Medicine at the University of New Mexico, where the first implantable electronic drug delivery systems were developed jointly with Sandia National Labs. For this work, he received the Regents Meritorious Service Medal and, later, a Doctor of Science degree from William Jewell College. Until recently he served as a Director of Adobe Systems and a member of the Board of Trustees of the Computer Museum and the Austin Symphony. Currently Dr. Spencer is a Director of the Investment Corporation of America and SRI International. He is also a member of the Board of Trustees of William Jewell College.

Dr. Spencer has served on several National Research Council studies in the areas of technology, trade, cooperation, and competition. In 1998, he co-chaired, with former Attorney General Richard Thornburgh, an NRC workshop on "Harnessing Technology for America's Future Economic Growth." Also, in 1998-1999 he served as a visiting professor at the Haas School of Business and the College of Engineering at the University of California at Berkeley.

YOICHI UNNO

Yoichi Unno was born in Tokyo in 1936. He received a B.S. in electrical engineering from Keio University in 1960. He joined Tokyo Shibaura Electric Company (now Toshiba) in 1960 where he worked as research engineer in the field of electron devices. In 1980 he became Director of the Electron Devices Laboratory at the R&D Center of Toshiba.

In 1985 he joined the Semiconductor group of Toshiba, where he was responsible for general management of the Semiconductor Engineering Laboratory. He became the technology executive of the Semiconductor Division in 1990 and general manager of the Microelectronics Engineering Laboratory in 1993.

He joined Semiconductor Industry Research Institute Japan (SIRIJ) as acting executive director in 2000. Since 1990 he has been project leader of the New Semiconductor Century Committee of SIRIJ.

CHARLES W. WESSNER

Dr. Wessner is the Director of the Program on Technology and Competitiveness for the National Research Council's Board on Science, Technology, and Economic Policy. Dr. Wessner began his federal career with the U.S. Treasury, served overseas as an international civil servant with the OECD and as a senior officer with the U.S. Diplomatic Corps, and directed the Office of International Technology Policy in the Technology Administration of the Department of Commerce. Since joining the National Research Council, he has led several major studies working closely with the senior levels of the U.S. government, leading industrialists, and prominent academics. Recent work includes a White House-initiated study on "The Impact of Offsets on the U.S. Aerospace Industry" and a major international study on "Competition and Cooperation in National Competition for High Technology Industry" in cooperation with the HWWA in Hamburg and the IFW in Kiel, Germany.

Currently, he is directing a portfolio of activities centered around "Government-Industry Partnerships for the Development of New Technologies" and initiating work on "Measuring and Sustaining the New Economy." The Partnerships program constitutes one of the first program-based efforts to assess U.S. policy on government-industry partnerships. Recent publications include *Conflict and Cooperation in National Competition for High Technology Industry*, *Policy Issues in Aerospace Offsets*, *International Friction and Cooperation in High-Technology Development and Trade*, *Trends and Challenges in Aerospace Offsets*, *New Vistas in Transatlantic Science and Technology Cooperation*, *Industry-Labo-*

ratory Partnerships: A Review of the Sandia Science and Technology Park Initiative, *The Advanced Technology Program: Challenges and Opportunities*, and *The Small Business Innovation Research Program: Challenges and Opportunities*. Dr. Wessner holds degrees in international affairs from Lafayette College (Phi Beta Kappa) and the Fletcher School of Law and Diplomacy, where he obtained an M.A., an M.A.L.D., and a Ph.D. as a Shell Fellow.

PATRICK WINDHAM

Until April 1998, Mr. Windham served as Senior Professional Staff Member for the Subcommittee on Science, Technology, and Space of the U.S. Senate's Committee on Commerce, Science, and Transportation. He helped the Senators oversee and draft legislation for several major civilian R&D agencies with responsibility for science, technology, and U.S. competitiveness; industry-government-university R&D partnerships; state economic development; federal laboratory technology transfer; high-performance computing; and computer encryption. From 1982 to 1984, he served as a legislative aide in the personal office of Senator Ernest Hollings. From 1976 to 1978, he worked as a Congressional fellow with the Senate Commerce Committee, then returned to California from 1978 to 1982 to complete doctoral course work and exams in political science at the University of California at Berkeley. Mr. Windham holds a Masters of Public Policy from the University of California at Berkeley and a B.A. from Stanford University. He is currently an independent, California-based consultant on science and technology policy issues and an adjunct professor at Stanford University.

Appendix C

Participants List

Arthur Alexander
Japan Economic Institute

Jeff Alexander
Washington CORE

Vincent Alexandre
Embassy of France

Kevin Allbaugh
Praxair Semiconductor Materials

Pablo Amor
European Commission

Alan Anderson
National Research Council

David Anderson
SEMATECH

Adrienne Andrews
Korea Economic Institute

Maggie Angell
Dewey Ballantine

Jose Araujo
George Washington University

Tom Arrison
National Research Council

Philip Aspden
National Research Council

Luis Baco
McDermott, Will & Emery

Wilhelm Beinvogl*
Infineon

Arpad Bergh
Optoelectronics Industry Development Association

*Speaker

Richard Bissell
National Research Council

Carolina Blake
NASA

Robert Blankenbaker
Department of Commerce

Dorothea Blouin
Department of Commerce

Michael Borrus*
The Pekevich Group, LLC

Brad Botwin
Department of Commerce

Tony Brunello
Pew Center
on Global Climate Change

Yvan Capouet
European Commission

Elias Carayannis
George Washington University

Bill Chang
National Science Foundation

McAlister Clabaugh
National Research Council

Vicki Cochran
Pew Center
on Global Climate Change

Eliot Cohen
Palisades Institute
for Research Services

Camille Collett
National Research Council

Reid Cramer
Office of Management and Budget

Mark Crawford
Department of Commerce

Reginald Cunningham
National Research Council

K.C. Das
Virginia Office of Innovative
Technology

Laurent de Mercey
Embassy of France

Brian Delroy
Embassy of Australia

Papken Der Torossian*
Silicon Valley Group

David Dierksheide
National Research Council

Bertha Dong
General Accounting Office

Margaret Donnelly
Department of Commerce

Peter Draheim*
Submicron Semiconductor
Technologies

*Speaker

George Elliott
National Research Council

Karin Ernst

Maryann Feldman
Johns Hopkins University

Kevin Finneran
National Research Council

Kenneth Flamm*
University of Texas at Austin

John Flannery
Office of Congresswoman Lofgren

Glenn Fong
Thunderbird, The American Graduate School of International Management

Ray Fornes
National Research Council

Bill Frymoyer
House Democratic Policy Committee

Tatsuya Fujiwara
Japan Electronics Bureau

Toshitaki Fukushima
SIRIJ

Gurmukh Gill
Department of Commerce

Edward Graham*
SISA

Denis Gray
North Carolina State University

Jeffrey Gren
Department of Commerce

Victoria Hadfield
SEMI

Gerald Hane
White House Office of Science and Technology Policy

Gilbert Herrera
Sandia National Laboratories

David Heyman
Department of Energy

Masataka Hirose*
Hiroshima University

Genda Hu*
Taiwan Semiconductor Manufacturing Company

Kent Hughes
The Woodrow Wilson Center

John Jankowski
National Science Foundation

Dale Jorgenson
Harvard University

William Joyner
Semiconductor Research Corporation

Erik Kamerbeek*
Semiconductor and Materials European Association

*Speaker

Steve Kang
University of California at Santa Cruz

Kalman Kaufman*
Applied Materials

John Kelly*
Novellus Systems

Kathleen Kingscott
IBM

Jüergen Knorr*
MEDEA

Jeffrey Kueter
NACFAM

Eduardo Lachida
Asian Wall Street Journal

Conway Lackman
National Institute of Standards and Technology

Rolf Lehming
National Science Foundation

Chien-Yuan Lin*
National Taiwan University

Greg Linden
University of California at Berkeley

Zoe Lofgren
U.S. House of Representatives

William Long
Business Performance Research Associates

Michael Luger*
University of North Carolina at Chapel Hill

Jeffrey Macher
Georgetown University

Jack Martinez
National Institute of Standards and Technology

Toshiaki Masuhara*
Hitachi

Juri Matisoo
Semiconductor Industry Association

Jeff Mayer
Department of Commerce

Sonny Maynard
Microelectronics Advanced Research Corporation

Purabi Mazumdar
NIST

Clark McFadden*
Dewey Ballantine

Egils Milbergs
NACFAM

James Mink
National Science Foundation

Sabrina Montes
Department of Commerce

Gordon Moore*
Intel Corporation

*Speaker

William Morin
R. Wayne Sayer and Associates

Akihiko Morino*
Selete

David Mowery*
University of California at Berkeley

Dong-Kyu Na
Wharton School

Barbara Norton
Office of the U.S. Trade Representative

Robert Norwood
NASA

Kees Planqué
Embassy of the Netherlands

Michael Polcari*
IBM

Barry Press
Washington CORE

Lee Price
Department of Commerce

Dan Radack
DARPA

Samuel Rankin
American Mathematical Society

Alan Rapoport
National Science Foundation

Lawrence Rausch
National Science Foundation

Hector Rivera
Department of Commerce

Robin Roark
Department of Commerce

Andras Roboz
Embassy of Hungary

Dorothy Robyn
National Economic Council

Sally Rood
Federal Laboratory Consortium

Gabrielle Sarrano
George Mason University

R. Wayne Sayer
R. Wayne Sayer and Associates

George Scalise*
Semiconductor Industry Association

Craig Schultz
National Research Council

Hideo Setoya*
ASET

Vashisht Sharma
Institute for Defense Analysis

Minjae Song
Harvard University

Bill Spencer*
International SEMATECH

*Speaker

Lester Su
Office of Congressman Ehlers

Toyoki Takemoto
STARC

Richard Thayer
Telecommunications and Technologies International

Mamoru Tsukamoto
Japan Electronics Bureau

Yoichi Unno*
SIRIJ

Nicholas Vonortas
George Washington University

Qifei Wang
University of Texas at Austin

Pamela Whitney
National Research Council

Charles Wessner*
National Research Council

Joel Widder
National Science Foundation

Pat Windham*
Windham Consulting

Jack Yadvish
NASA

Rosemarie Ziedonis
Wharton School

*Speaker

Appendix D

Bibliography

American Institute of Physics. 1997. "Stark Warning Issued on Advanced Technology Program." *Bulletin of Science Policy News*. 45(March 21).

Anthes, Gary H. 1997. "DARPA Looks Ahead." *Computerworld*. July 28.

Arrow, K.J. 1962. "Economic Welfare and the Allocation of Resources for Invention," in National Bureau of Economic Research, *The Rate and Direction of Inventive Activity: Economic and Social Factors*. Princeton: Princeton University Press.

Asian Technology Information Program. 1994. *Korean Engineering Professors Form a National Group*. Cambridge, MA: Harvard University Press.

Audretsch, David B. and Roy Thurik. 1999. *Innovation, Industry, Evolution, and Employment*. Cambridge, UK: Cambridge University Press.

Baily, M.N., and A.K. Chakrabarti. 1988. *Innovation and the Productivity Crisis*. Washington, D.C.: The Brookings Institution.

Beijing Morning Post. 2000. "Beijing Issued Implementation Rules on Recruiting Talents from Outside City." August 14.

Bloch, Erich and Jerry Sheehan. 1999. "Federal IT Funding: Creating an Infrastructure, Growing an Infrastructure, Growing an Industry." *IMP Magazine*. March.

Bodway, George, Sunlin Chou, Mark Melliar-Smith, and Peter Verhofstadt. 1995. *SRC White Paper on Research Investment Gap Analysis*. Research Triangle Park, NC: Semiconductor Research Corporation.

Brander, J.A. and B.J. Spencer. 1983. "International R&D Rivalry and Industrial Strategy." *Review of Economic Studies*. 50(4): 720-722.

Branstetter, L.G. and M. Sakakibara. 1998. "Japanese Research Consortia: A Microeconometric Analysis of Industrial Policy." *Journal of Industrial Economics*. 46(2): 207-233.

Branstetter, L.G. and M. Sakakibara. 2000. *When Do Research Consortia Work Well and Why? Evidence from Japanese Panel Data*. NBER Working Paper No. W7972. October.

Branstetter, L.G. and M. Sakakibara. 2002. "When Do Research Consortia Work Well and Why? Evidence from Japanese Panel Data." *American Economic Review*. 92(1): 143-159.

Braun, E. and S. MacDonald. 1978. *Revolution in Miniature*. Second Edition. Cambridge, UK: Cambridge University Press.

Browning, Larry D., and Judy C. Shetler. 2000. *SEMATECH: Saving the U.S. Semiconductor Industry*. College Station: Texas A&M University Press.

Bureau of Economic Analysis. 1999. *Statistical Abstract of the United States: 2001*. Washington, D.C: U.S. Government Printing Office.

Burggraaf, Peter. 2000. "2000 Begins with a Revised Industry Roadmap." *Solid State Technology*. January.

Burggraaf, Peter. 2000. "A Closer Look at Some of the Most Difficult Processing Challenges." *Solid State Technology*. September.

Cahners Semiconductor International. 2000. "Top Fabs of 2000." May.

Cavin, Ralph K. III, Daniel J.C. Herr and Victor Zhirnov. Undated. *Semiconductor Research Needs in the Physical Sciences*. SRC Working Paper. Research Triangle Park, NC: Semiconductor Research Corporation.

Chinaonline. 2000. "State Council Releases Text of Software, Integrated Circuit Policies." August 2.

Choson Ilbo. 2001. "U.S. Questions Bailout of Hyundai Electronics." January 31.

Chugan Choson. 2001. "The Government Above the Market, the Market Above Hyundai?" March 29.

Clark, Gordon L., Maryann P. Feldman, and Meric S. Gertler, eds. 2000. *The Oxford Handbook of Economic Geography*. Oxford: Oxford University Press.

Clarke, Peter and Yoshiko Hara. 2000. "Analysis: Japan, Europe Ride 300-mm Wave." *Electronic Engineering Times*. May 1.

CMP Media Inc. "Taiwan's IC Makers Eye China—May Contract Low Cost Plants There, but the Island's Government is Balking." <http://www.ebonline.com>.

Coburn, C. and D. Berglund. 1995. *Partnerships: A Compendium of State and Federal Cooperative Technology Programs*. Columbus, OH: Battelle Press.

Cohen , Linda R. and Roger G. Noll. 1991. *The Technology Pork Barrel*. Washington, D.C.: The Brookings Institution.

Council of Economic Advisers. 1995. *Supporting Research and Development to Promote Economic Growth: The Federal Government's Role*. Washington, D.C.: U.S. Government Printing Office.

Council of Economic Advisers. 2001. *Economic Report of the President*. Washington D.C.: U.S. Government Printing Office.

D'Aspremont, C., and A. Jacquemin. 1988. "Cooperative and Noncooperative R&D in Duopoly with Spillovers." *American Economic Review*. Vol. 78.

Dasgupta, P. and J. Stiglitz, 1980. "Industrial Structure and the Nature of Innovative Activity." *Economic Journal*. 90(June): 266-293.

David, Paul. 2000. "Understanding Digital Technology's Evolution and the Path of Measured Productivity Growth: Present and Future in the Mirror of the Past," in E. Brynjolfsson and B. Kahin, eds., *Understanding the Digital Economy: Data, Tools, and Research*. Cambridge, MA: MIT Press.

Dieffendorff, Keith. 2000. "Extreme Lithography." *Microprocessor Report*. June.

Dixit, A.K. and A.S. Kyle. 1985. "The Use of Protection and Subsidies for Entry." *American Economic Review*. 75(1): 139-52.

The Economist. 2001. "The Great Chip Glut." August 11.

Electronic Design. 2000. "DNA Motor Drives Nanotechnology Toward the Post-CMOS Era." September 18.

Electronic Engineering Times. 1998. "Joint European Electronics Consortium–Cadence Targets RF Design." March 16.

Electronic Engineering Times. 2000. "Chip Giants Prep. Fabs as Wafer Suppliers Gear Up–Europe, Japan Tool 3mm." May 1. (Nexis reprint).

Electronic Engineering Times. 2000. "Lab Looks to Wring Two More Generations from Today's Techniques, Materials—Intel, CMOS up to 70-nm Task." December 11.

Electronique International Hebdo. 1996. "LETI Joins Forces with Equipment Manufacturers in 0.25-Micron Technologies." June 27.

Elektronik. 1997. "Laser 2000:1KW from a Matchbox." July 22.
European Commission. 1994. *Official Journal of the European Communities.* L.126/1. May 18.
FBIS. 1994. Taro Okabe, "Semiconductor Industry Research Institute Resolves Policy Framework." *Nikkei Microdevices* (FTS19941201000063). December.
FBIS. 1995. Translation of "For Research and Industry." *Markt und Technik* (FTS19951208000472). December 8.
FBIS. 1995. Translation of "JESSI Has Developed Single-Chip Solutions for GSM Telephones." *Mikroelektronik und Mikrosystemtechnik* (FTS19951005000666). October 5.
FBIS. 1995. Translation of "Largest Eureka Project a Resounding Success." *Economische Zaken* (FTS19951006000613). October 6.
FBIS. 1995. Translation of Emi Yokata, "Japanese Manufacturers to Launch Joint Research on New Chip Technology to Counterattack U.S. Manufacturers." *Ekonomisuto* (FTS19951031000245). October 31.
FBIS. 1996. Translation of Andreas Beuthner, "Fraunhofer Tech Transfer to Industry Viewed." *Computer Zeitung* (FTS19960815000685). August 15.
FBIS. 1996. Translation of Casper Busse, "Billions in Support Smooth Tough Road." *Handelsblatt* (FTS19960226000662). January 2.
FBIS. 1996. Translation of "Dresden Becomes Capital of Microelectronics–Biederkopf's Dreams." *Markt and Technik.* January 12.
FBIS. 1996. Translation of Elisabeth Feder, "GaAs Technology is Becoming an Individual Priority in Germany." *Electronique International Hebdo.* May 16.
FBIS. 1996. Translation of Philippe Le Coeur, "European Semiconductor Industry in Economic Turnaround." *Le Monde* (FTS19960130001915). January 30.
FBIS. 1996. Translation of "LETI Joins Forces with Equipment Manufacturers in 0.25-Micron Technologies." *Electronique International Hebdo* (FTS19960919001047). June 27.
FBIS. 1996. Translation of Leo Ploner, "Compute Electronically, Compete Optically. Photonics: Focus on Research Attention Not the Optical Computer but on the Synergy of Microelectronics and Optics." *VDI Nachrichten* (FTS1996100400963). October 6.
FBIS. 1996. Translation of "Remarkable Results of JESSI 'Phase Shift' Project." *Computable* (FTS19960524001698). May 24.
FBIS. 1996. Translation of Richard Sietmann, "With Research and Development in First Place; Belgian Microelectronics Institute IMEC in Demand as Partner for Cooperation in International Research and Development." *VDI Nachrichten* (FTS1996012600087). January 26.
FBIS. 1997. "Media Report ROK, Japanese Strategy to Control DRAM Prices." Summarizing articles in *Digital Choson Ilbo*, January 29, 1997; *Choson Ilbo,* April 17, 1997; *Chugang Maegyong*, April 30, 1997; and *Maeil Kjongje Sinmun*, April 8, 1997, (FTS1997062Y002280). June 14.
FBIS. 1997. Reprint of Nir Nathan, "We've Taken Silicon Valley." *Tel Aviv Globes.* May 6.
FBIS. 1997. Reprint of Shamsul Akmar and Lee Yuk Peng, "U.S. Firms Investment Arm Tie Up for Wafer Fab Project." *The Star* (FTS19971015000152). October 15.
FBIS. 1997. Reprint of Jennifer Lien, "Germany to Put up First Silicon Wafer Manufacturing Plant." *Singapore Business Times* (FTS19970729000242). July 29.
FBIS. 1997. "Research for the Next Millenium." *Electronik* (FTS19970801001209). January 21.
FBIS. 1997. Translation of Frederic Fassot, "A European Plan for Passive Components." *Electronique International Hebdo* (FTS19980122000510). December 11.
FBIS. 1997. Translation of *Photonik–Foerderschwerpunkt–Photonik in Rahmen des Foerderkonzepts Informationstechnik.* Cologne. (FTS1997073100045497G16107). August 5.
FBIS. 1997. Translation of "Playing to Our Strengths." *Bild der Wissenschaft* (FTS19970730000530). July 1.
FBIS. 1997. Translation of "State Encourages Foreign-Founded Micorelectronics, Basic Electronic Product Items." *Zhongguo Dianzi Bao* (FTS19970728001781). January 14.

FBIS. 1997. Translation of Stephen Stroeytkens, "Flemish Technology Will Make Video Cameras Very Cheap." *Groot-Bijgaarden De Standaard* (FTS19970916000718). September 15.

FBIS. 1997. Translation of Guenter Weimann, "Compound Semiconductors for Mobile Communications." *Fraunhofer Magazin* (FTS19971119000552). November 19.

FBIS. 1998. "Philips Semiconductor Plant in Singapore." *Algemeen Dagblad* (BR3009120698). September 30.

FBIS. 1998. Translation of Anne Chatel-Demerge, "The SGS-Thomson Group Invests 3 Billion Francs in Grenoble." *Les Echos* (FTS19980512000875). May 12.

FBIS. 1998. Translation of Chong Ch'ang-hun, "Taedok to Become Mecca for Venture Firms." *Chonja Sinmun* (FTS19980705000165). April 10.

FBIS. 1998. Translation of "Europe's Largest Electronics Project Starts in Dresden." *Frankfurter Zeitung/Blick Durch Die Wirtschaft.* February 4.

FBIS. 1998. Translation of "Gilles Musi, "Strategy–SGS-Thomson Investing Almost 3 Billion Francs in Grenoble." *La Tribune* (Internet Version) (FTS19980513000697). May 13.

FBIS. 1999. "How Does China Attract Foreign Direct Investment." *Guangzhou Gangao Jingji* (FTS19990407001093). February 15.

FBIS. 1999. Translation of Motohiko Kitahara, "Interview with Hsuan Ming-chih of UMC Regarding Taiwan's Semiconductor Foundry Strategy." *Nikkei Sangyo Shimbun* (FTS19991003000139). August 25.

FBIS. 1999. Translation of "NGC and Hitachi Join Other Three Big Makers in Commissioning Taiwanese Companies with their LSI Production." *Kagaku Kogyo Nippo* (FTS19990427000356). March 30.

FBIS. 1999. Translation of "No End to Global Competition." *Elektronik* (FTS19991013001537). October 13.

FBIS. 2000. "Japan Ministry Contracts 5 Firms for Development of High Speed Optical Network." *Nikkan Kogyo Shimbun* (JPP20011150000004). November 15.

FBIS. 2000. Reprint of "Jiang Zemin's Son, Taiwan Company Team Up on Chip Venture." *Taipei Times* (CPP20000511000120). May 12.

FBIS. 2000. Reprint of Wen-Hung Fung,"Taiwan's IT Production Continues Moving Into China." Taipei *Central News Agency*. (CPP20000511000146). May 11.

FBIS. 2000. "Taiwan Company to Setup up 6-Inch Chip Foundry in China." *Taipei Times* (CPP20001003000135). October 3.

FBIS. 2000. "Taiwan's High-Tech Advantage Over China Eroding." *Taipei Times* (CPP20001130000168). November 30.

FBIS. 2000. "Taiwan Semiconductor Chair Going to China for 'Bridge Tournament.'" *Taipei Times.* (CPP20001211000132). December 11.

FBIS. 2000. Translation of "Asuka Joint Industry Project for Micro-Lithography System LSI Outlined." *Kagaku Kogyo Nippo* (JPP20001004000002). October 3.

FBIS. 2000. Translation of "Billions for Chip Research." *VDI Nachrichten* (EUP200006226). June 9.

FBIS. 2000. Translation of Yi Che, "Reportage: Taedok Valley–Over 14,000 Researchers with a Master's Degree or Better." *Chugan Choson* (KPP20000510000022). May 4.

FBIS. 2000. Translation of Chu Dechao, "Overview of the Semiconductor Industry in China." *Tokyo Semiconductor FDP World* (JPP20001127000111). November 2000.

FBIS. 2000. Translation of "Development of 0.07 to 0.05 micron LSI Technology." *Nikkei Microdevices* (JPP20001018000105). October.

FBIS. 2000. Translation of Kazunori Ishiguro, "IT War: US Strategy to Contain the NTT Optical Empire." *Shokun* (JPP20000821000008). August 1.

FBIS. 2000. Translation of Osamu Katayama, "With the Start of Digital TV Broadcasts, Japanese Home Appliance Makers Now Engaged in a Fierce Struggle for Dominance." *Chuo Koron* (JPP2000127000051). December 1.

FBIS. 2000. Reprint of Matthew Miller,"Beijing Approves Motorola Base, to Work with Taiwan Tycoon, Jiang Zemin's Son." *South China Morning Post.* (CPP20008220000287). August 22.
FBIS. 2000. Translation of A. Minamikawa, "Investment in 300 mm Plants Heating Up, 32 Heating Up, 32 New Lines to be Built." *Nikkei Microdevices* (JPP20000706000024). June 1.
FBIS. 2000. Translation of Y. Nagahiro, "Hitachi-UMC Tie-Up to Generate Test Cases that Will Forecast the Future of LSI Technology and Business." *Nikkei Microdevices* (JPP200002080000029). February 8.
FBIS. 2000. Translation of Y. Nagahiro, "World's First 300mm Production Line–Establishing World Standard 300 mm Technology." *Nikkei Microdevices* (JPP20000516000091). May 1.
FBIS. 2000. Translation of Michio Mizogami, "Japanese LSI, LCD Makers Take Note, Part 2: Management Holding Down Production Technology is Reason for Japan's Decline." *Nikkei Microdevices* (JPP20000911000014). September.
FBIS. 2000. Translation of Michio Mizogami, "Prescription for Japan's Revival." *Nikkei Microdevices.* August.
FBIS. 2000. Translation of "Sakakibara on NTT Connection Fee Cuts." *Mainichi Shimbun* (JPP20000519000006). May 19.
FBIS. 2000. Translation of "The Strengths of Research Europe." *La Tribune* (EUP20000524000474). May 24.
FBIS. 2000. Translation of Hajime Susaki, chairman of NEC Corporation "Japanese Semiconductor Industry's Competitiveness: LSI Industry in Jeopardy." *Nikkei Microdevices.* December.
FBIS. 2000. Translation of Hitoishi Tanaka, "New Development in Japan's Economic Diplomacy: Free Trade Agreements." *Chuo Koron* (JPP20001019000039). November 1.
FBIS. 2000. Translation of Pai Te-hua, Wang Chung-ning and Wang Shih-Chi. "Construction of Taiwan-Funded Microchip Plant Begins in Shanghai." *Chung-Kuo Shih-Puo* (CPP2000112000035). November 20.
FBIS. 2000. Translation of interview by Zhou Wei of Zhu Lilan,"PRC Official Discusses Nurturing High-Tech Enterprises." *Jingji Ribao* (CPP20000926000052). September 22.
FBIS. 2000. Translation of Qu Weizhi, Vice-Minister of Information Industry. "How to Develop Integrated Circuits Industry." R*enmin Ribao.* (CPP20000515000066). May 15.
FBIS. 2000. Translation of D. Wendeln, "Europe on the Way to World Class in Microtechnology and Nanotechnology." *VDI Nachrichten* (EUP20001011000369). October 6.
FBIS. 2000. Translation of Xia Wensi, "The Eldest Young Master Jiang Sailing Smoothly Through Business World–China's Telecom King Jiang Mianheng." *Kai Fang* (CPP20001211300004O). December 5.
FBIS. 2000. Translation of Suo Yan,"PRC's State Council Drafts New Policies to Develop IT Industry." *Renmin Ribao* (CPP20000712000106). July 12.
FBIS. 2000. Translation of Kazuhiro Yoshihura, "World Expanding With Nano Technology." *Kagaku Gijutsu Janaru* (JPP20010118000005). December 1.
FBIS. 2001. "China-Taiwan Princelings Joint Venture Stalled." *Taipei Times* (CPP20010116000160). January 13.
FBIS. 2001. "LSI Makers Eye China for Developing New Business Strategies." *Nikkei Microdevices* (JPP20010307000003). March.
FBIS. 2001. "PRC: Taiwan Computer Firm CEO Cited On Importance of Mainland Expansion." *China Daily* (Hong Kong Edition, Internet Version) (CAP20010118000038). January 18.
FBIS. 2001. "Taiwan Business Leader's Advice on High Tech Investment in China: Don't." *Taipei Times* (CPP20010116000158). January 13.
FBIS. 2001. Translation of "3 major European LSI Makers Show Stable Growth Through Large Investments." *Nikkei Microdevices* (JPP20010131000003). January 1.
FBIS. 2001. Translation of K. Amagasa, "Nano-Intelligent Materials That Utilize Self-Organization." *Tokyo Trigger* (JPP20010327000012). March 1.

FBIS. 2001. Translation of "Can Next-Generation Semiconductor Project Save Japan?" *Nikkei Microdevices* (JPP20001103000001). November 1.

FBIS. 2001. Translation of Peter Chang, "Expanding Si Foundry Business with 300mm and 0.13 Micron Technology." *Nikkei Microdevices* (JPP20010126000005). January 1.

FBIS. 2001. Translation of "From Stagnation to Growth, the Push to Strengthen Design." *Nikkei Microdevices* (JPP20010125000012). January 2.

FBIS. 2001. Translation of "The Full Story on Nanotechnology; the U.S. Aims for it after IT! Japan Bets Recovery on it!" *Tokyo Trigger* (JPP20001031600001l). March 1.

FBIS. 2001. Translation of "IT Strategy Head Interviewed on Obstacles to IT Revolution." *ChuoKoron* (JPP20001019000040). November 1.

FBIS. 2001. Translation of "Japan's NEC to Expand Semiconductor Production in China." *Kagaku Kogyo Nippo* (JPP20010202000013). January 12.

FBIS. 2001. Translation of "Japan Pioneers Ultra-Low-Energy, High-Density Ion Implantation." *Tokyo Trigger* (JPP20010327000004). March 1

FBIS. 2001. Translation of "Japan's NEC to Expand Semiconductor Production in China." *Kagaku Kogyo Nippo* (Nikkei Telecom Database Version) (JPP20010202000013). February 2.

FBIS. 2001. Translation of M. Kimura, "Industry, Government and Universities United in Enthusiasm and Talent for LSIs." *Nikkei Microdevices* (JPP20010307000001). March 7.

FBIS. 2001. Translation of "Nanontechnology Challenge: Hidden Power in Nanometer Size, Key to Japan's Revival." *Nikkei Sangyo Shimbun* (JPP20010223000047). February 12.

FBIS. 2001. Translation of "Preferential Treatment for Hyundai." *Choson Ilho* (KPP20010202000082). February 2.

FBIS. 2001. Translation of Hajime Sasaki, "Lack of Awareness of LSIs as a Leading Industry, Foundation of Industry Weakening." *Nikkei Microdevices* (JPP20001108000046). November.

FBIS. 2001. Translation of Ulrich Schumacher, "2001 to be a Good Year, Expand DRAM in Areas other than PC." *Nikkei Microdevices* (JPP20010131000003). January.

FBIS. 2001. Translation of "Special Projects–Part One–Industry, Government and Universities United in Enthusiasm and Talent for LSIs and LCDs." *Nikkei Microdevices* (JPP20010307000001). March.

FBIS. 2001. Translation of "United States, Japan Compete for Nanotechnology Research." *Nikkon Kogyo Shimbun* (Nikkei Telecom Database Version) (JPP20010125000031). January 1.

FBIS. 2001. Translation of Yatsuka Yoshida, "Asuka Project Targets Semiconductor Industry Revival." *Tokyo Trigger* (JPP20010316000013). March 16.

FBIS. 2002. "MOEA Announces Regulations on Wafer Fab Transfers to Mainland China," *Taipei Central News Agency* (08:55 GMT, August 9, 2002) (CPP20020809000092).

FBIS. 2002. Translation of Hsu Yu-chun, "Tsai Ying-wen says Punishment will be Meted Out to Enterprises Which have Moved to the Mainland," *Ching-Chi Jih-Pao* (March 2, 2002), p. 1 (CPP20020307000018). March 7.

Flamm, K. 1988. *Creating the Computer: Government, Industry, and High-technology.* Washington, D.C.: The Brookings Institution Press.

Flamm, K. 1996. "Japan's New Semiconductor Technology Programs." Asia Technology Information Program Report No. ATIP 96.091. Tokyo. November.

Flamm, K. 1996. *Mismanaged Trade? Strategic Policy and the Semiconductor Industry.* Washington, D.C.: Brookings Institution Press.

Flat Panel Display Task Force. 1994. *Building U.S. Capabilities in Flat Panel Displays: Final Report.* Washington D.C.: U.S. Department of Defense. October.

Fransman, M. 1992. *The Market and Beyond: Cooperation and Competition in Information Technology Development in the Japanese System.* Cambridge: Cambridge University Press.

Future Horizons. 1999. "An Executive Report on the MEDEA Programme: Mid-Term Assessment." Online report, <HtmlResAnchor http://www.minez.nl/kamerbrieven1999/execut.pdf>, *accessed on* July 19.

Gibson, D.V. and E. M. Rogers. 1994. *R&D Collaboration on Trial.* Boston: Harvard Business School Press.

Good, Mary L. 2001. "Scientist's Call to Action: S.F. conference opens with plea for Cabinet position." *San Francisco Chronicle.* February 16.

Greenspan, Alan. 2000. *Technological Innovation and the Economy,* Remarks Before the White House Conference on the New Economy. Washington, D.C.: Federal Reserve Board. April 5.

Griliches, Zvi. 1990. *The Search for R&D Spillovers.* Cambridge, MA: Harvard University Press.

Ham, R.M., G. Linden, and M. Appleyard. 1998. "The Evolving Role of Semiconductor Consortia in the United States and Japan." *California Management Review.* 41: 137-163.

Hawley, Ellis W. 1966. *The New Deal and the Problem of Monopoly.* Princeton: Princeton University Press.

Helpman, Elhanan and Manuel Trajtenberg: 1998. "Diffusion of General Purpose Technologies," in E. Helpman, ed. *General Purpose Technologies and Economic Growth.* Cambridge: MIT Press.

Helpman, Elhanan and Manuel Trajtenberg. 1998. "A Time to Sow and a Time to Reap: Growth Based on General Purpose Technologies," in E. Helpman, ed. *General Purpose Technologies and Economic Growth.* Cambridge: MIT Press.

Hill, Christopher T. 1998. "The Advanced Technology Program: Opportunities for Enhancement," in Lewis M. Branscomb and James H. Keller, eds. *Investing in Innovation: Creating a Research and Innovation Policy That Works.* Cambridge: MIT Press.

The Hindu. 2000. "Further Boost to Ties with Singapore." June 30.

Hofstadter, Richard. 1948. *The American Political Tradition.* New York: A.A. Knopf.

Holton, Conrad W. 1996. "Federal Funds for Chip Research Dwindle Under Congressional Pressure." *Solid State Technology.* July.

Horrigan, John B. 1999. "Cooperating Competitors: A Comparison of MCC and SEMATECH." Monograph. Washington D.C.: National Research Council.

Howell, Thomas R. 1997. "Dumping: Still a Problem in International Trade," in National Research Council, *International Friction and Cooperation in High-Technology Development and Trade.* Washington, D.C.: National Academy Press.

Hsiao, C. 1986. *Analysis of Panel Data.* Cambridge: Cambridge University Press.

Hutcheson, G. Dan and Jerry D. Hutcheson. 1996. "Technology and Economics in the Semiconductor Industry." *Scientific American.* January. pp. 66-73.

IMEC. 2000. *IMEC Annual Report. Edition of Nihon Handotai Nenkan.* Tokyo: Press Journal. March 25.

Irwin, Douglas A. and Peter J. Klenow. 1994. "Learning-by-Doing Spillovers in the Semiconductor Industry." *Journal of Political Economy.* 102(6): 1200-1227.

Irwin, Douglas A. and Peter J. Klenow. 1996. "High-Tech R&D Subsidies: Estimating the Effects of SEMATECH." *Journal of International Economics.* 40(3/4): 323-344.

IST Advisory Group. 1999. *IST Workprogramme.* 2000. <http://www.cordis.lv/ist/present.htm>.

ITRI. 1999. *Annual Report* ITRI.

Japan Information Processing Development Corporation. 2000. *Information Whitepaper.* Tokyo: Computer Age. June 16.

Johnson, Grant. 1999. "Is China's Semiconductor Market Worth the Risk for Multinationals? Definitely!" *Electronic News.* March. <http://www.instat.com/insights/semi/1999/china42999.htm>.

Jorgenson, Dale W. 2001. "Information Technology and the U.S. Economy." *The American Economic Review.* 91(1): 1-32.

Jorgenson, Dale W. and Kevin J. Stiroh. 2000. "Raising the Speed Limit: U.S. Economic Growth in the Information Age." *Brookings Papers on Economic Activity.* Washington, D.C.: The Brookings Institution.

JTEC. 1988. *The Japanese Exploratory Research for Advanced Technology (ERATO).* JTECH Panel Report. December.

Kamien, M.I., E. Muller and I. Zang. 1992. "Research Joint Ventures and R&D Cartels." *American Economic Review*. 82(5): 1293-1306.

Katz, M.L. 1986. "An Analysis of Cooperative Research and Development." *Rand Journal of Economics*. 17(4): 527-543.

Katz, M.L. and J.A. Ordover. 1990. "R&D Cooperation and Competition." *Brookings Papers on Economic Activity: Microeconomics*. Washington, D.C.: The Brookings Institution.

Kenney, Martin. 2000. *Understanding Silicon Valley: The Anatomy of an Entrepreneurial Region*. Stanford: Stanford University Press.

Kleinman, Daniel Lee. 1995. *Politics on the Endless Frontier: Postwar Research Policy in the United States*. Durham, N.C.: Duke University Press.

Koizumi, Kei, Albert H. Teich, Stephen D. Nelson, and Joanne Padrón Carney. 2000. *Congressional Action on Research and Development in the FY 2001 Budget*. Washington, D.C.: American Association for the Advancement of Science.

Kondratieff, Nikolai D. 1935. "The Long Waves in Economic Life." *Review of Economic Statistics*. 17(6). November.

Koopmann, Georg and Hans-Eckart Scharer, 1996. *The Economics of High-Technology Competition and Cooperation in Global Markets*. Baden-Baden, Germany: HWWA Institute for Economic Research.

Krempl, Stefan. 2000. "EU Commissioner: we need a sense of urgency." *Munich Telepolis*. April 17.

Krugman, Paul. 1990. *Rethinking International Trade*. Cambridge, MA: MIT Press.

Krugman, Paul. 1991. *Geography and Trade*. Cambridge, MA: MIT Press.

Krugman, Paul. 1994. *Peddling Prosperity*. New York: W.W. Norton Press.

Krugman, P. and M. Obsfeldt. 1994. *International Economics: Theory and Policy,* 3d Edition, New York: Addison-Wesley Publishing Company.

Kubota, Akira. 2000. "Semiconductors Support the IT Revolution: Silicon Technology the Key in the Next Decade." *Nikkei Microdevices*. October.

Levin, R.C. 1982. "The Semiconductor Industry," in Richard R. Nelson, ed.. *Government and Technical Progress: A Cross-Industry Analysis*. New York: Pergamon Press.

Liesman, Steve. 2002. "Productivity Growth May Be Here to Stay." *Wall Street Journal*. January 7.

Linden, Greg, David C. Mowery, and Rosemarie Ham Ziedonis. 2001. "National Technology Policy in Global Markets," in Maryann P. Feldman and Albert N. Link eds., *Innovation Policy in the Knowledge-based Economy*. Boston: Kluwer Academic Publishers.

Link, Albert. 1998. *Public/Private Partnerships as a Tool to Support Industrial R&D: Experiences in the United States*. Paper prepared for the working group on Innovation Policy, Paris.

Link, A.N., D.J. Teece, and W.F. Finan. 1996. "Estimating the Benefits from Collaboration: The Case of SEMATECH." *Review of Industrial Organization*. 11: 737-751.

Lipsey, Richard G., Cliff Bekar, and Kenneth Carlaw. 1998. "General Purpose Technologies: What Requires Explanation," in E. Helpman, ed., *General Purpose Technologies and Economic Growth*. Cambridge, MA: MIT Press.

Macher, Jeffrey T., David C. Mowery and David C. Hodges. 1998. "Reversal of Fortune? The Recovery of the U.S. Semiconductor Industry." *California Management Review*. 41(1): 107-136.

Macher, Jeffrey T., David C. Mowery, and David A. Hodges. 1999. "Semiconductors," In National Research Council, *U.S. Industry in 2000: Studies in Competitive Performance*. Washington, D.C.: National Academy Press.

Mann, Charles C. 2000. "The End of Moore's Law? Technology Information." *Technology Review* . May/June.

Mansfield, Edwin. 1991. "Academic Research and Industrial Innovation." *Research Policy*. 20(1):1-12.

Matthews, John A., and Dong-Sung Cho. 2000. *Tiger Technology: The Creation of a Semiconductor Industry in East Asia*. Cambridge: Cambridge University Press.

Merrill, Stephen A., and Michael McGeary. 1999. "Who's Balancing the Federal Research Portfolio and How?" *Science*. 285: 1679-1680.

Mizogami, Michio. 2000. "Japanese LSI, LCD Makers Take Note, Part 2: Management Holding Down Production Technology is the Reason for Japan's Decline." *Nikkei Microdevices*. September.

Moore, Gordon E. 1965. "Cramming more components onto integrated circuits." *Electronics*. 38(8). April 19.

Moore, Gordon E. 1997. "The Continuing Silicon Technology Evolution Inside the PC Platform." *Intel Developer Update*. Issue 2. October 15.

Mowery, David. 1998. "Collaborative R&D: how effective is it?" *Issues in Science and Technology*. 15(1): 37-44.

Mowery, David. 2001. "Using Cooperative Research and Development Arrangements as S&T Indicators: What Do We Have and What Would We Like?" in National Science Foundation, Division of Science Resources Studies. *Strategic Research Partnerships: Proceedings from an NSF Workshop*. NSF 01-336. Project Officers: John E. Jankowski, Albert N. Link, Nicholas S. Vonortas. National Science Foundation: Arlington, VA.

Nadiri, M. Ishaq. 1993. *Innovations and Technological Spillovers*. NBER Working Paper 4423. August.

Nance, Scott. 2000. "Broad Federal Research Required to Keep Semiconductors on Track." *New Technology Week*. October 30.

National Research Council. 1986. *The Positive Sum Strategy: Harnessing Technology for Economic Growth*. Washington, D.C.: National Academy Press.

National Research Council. 1987. *Semiconductor Industry and the National Laboratories: Part of a National Strategy*. Washington, D.C.: National Academy Press.

National Research Council. 1991. *Mathematical Sciences, Technology, and Economic Competitiveness*. James, G. Glimm, ed. Washington, D.C.: National Academy Press.

National Research Council. 1995. *Allocating Federal Funds for R&D*. Washington, D.C.: National Academy Press.

National Research Council. 1996. *Conflict and Cooperation in National Competition for High-Technology Industry*. Washington, D.C.: National Academy Press.

National Research Council. 1999. *The Advanced Technology Program: Challenges and Opportunities*. Charles W. Wessner, ed. Washington, D.C.: National Academy Press.

National Research Council. 1999. *Funding a Revolution: Government Support for Computing Research*. Washington, D.C.: National Academy Press.

National Research Council. 1999. *Industry-Laboratory Partnerships: A Review of the Sandia Science and Technology Park Initiative*. Charles W. Wessner, ed. Washington, D.C.: National Academy Press.

National Research Council. 1999. *New Vistas in Transatlantic Science and Technology Cooperation*. Charles W. Wessner, ed. Washington, D.C.: National Academy Press.

National Research Council. 1999. *The Small Business Innovation Research Program: Challenges and Opportunities*. Charles W. Wessner, ed. Washington, D.C.: National Academy Press.

National Research Council. 2000. *The Small Business Innovation Research Program: A Review of the Department of Defense Fast Track Initiative*. Charles W. Wessner, ed. Washington, D.C.: National Academy Press.

National Research Council. 2000. *U.S. Industry in 2000: Studies in Competitive Performance*. Washington, D.C.: National Academy Press.

National Research Council. 2001. *The Advanced Technology Program: Assessing Outcomes*. Charles W. Wessner, ed. Washington D.C.: National Academy Press.

National Research Council. 2001. *Attracting Science and Mathematics Ph.D.'s to Secondary School Education*. Washington, D.C.: National Academy Press.

National Research Council. 2001. *Building a Workforce for the Information Economy.* Washington, D.C.: National Academy Press.

National Research Council. 2001. *Capitalizing on New Needs and New Opportunities: Government-Industry Partnerships in Biotechnology and Information Technologies.* Charles W. Wessner, ed. Washington, D.C.: National Academy Press.

National Research Council. 2001. *A Review of the New Initiatives at the NASA Ames Research Center.* Charles W. Wessner, ed. Washington, D.C.: National Academy Press.

National Research Council. 2001. *Trends in Federal Support of Research and Graduate Education.* Washington, D.C.: National Academy Press.

National Research Council. 2002. *Government-Industry Partnerships for the Development of New Technologies: Summary Report.* Charles W. Wessner, ed. Washington, D.C.: National Academies Press.

National Research Council. 2002. *Measuring and Sustaining the New Economy.* Dale W. Jorgenson and Charles W. Wessner, eds. Washington, D.C.: National Academy Press.

National Science Foundation. 2000. "National expenditures for R&D, from funding sectors to performing sectors: 1993-2000." *National Patterns of R&D Resources: 2000 Data Update.* National Science Foundation: Arlington, VA.

National Science and Technology Council. 2000. Committee on Technology: Subcommittee on Nanoscale Science, Engineering and Technology. *National Nanotechnology Initiative: The Initiative and Its Implementation Plan.* National Science and Technology Council.

New York Times. 2002. "China Finds Way to Beat Chip Limits," May 6. p. 4.

Nikkei Microdevices. 2000. "Investment in 300mm Plants Heating Up; 32 New Lines to be Built." June

Nikkei Microdevices. 2001. "The Game–System LSIs for Cell Phones; Making the Most of Equipment Know-How in Design." January.

Nikkei Microdevices. 2001. "Industry, Government, and Universities United in Enthusiasm and Talent for LSIs and then LCDs." March.

Nikkei Microdevices. 2001. "LSI Makers Eye China for Developing New Business Strategies." March.

Office of Science and Technology Policy. 2000. *Fact Sheet on How Federal R&D Investments Drive the U.S. Economy.* Washington, D.C.: Executive Office of the President.

Okimoto, Daniel I. 1989. *Between MITI and the Market: Japanese Industrial Policy for High Technology.* Stanford, California: Stanford University Press.

Ophuls, William. 1977. *Ecology and the Politics of Scarcity: Prologue to a Political Theory of the Steady State.* San Francisco: Freeman Press.

Packan, Paul A. 1999. "Pushing the Limits: Integrated Circuits Run Into Limits Due to Transistors." *Science.* September 24.

Pecht, Michael, Weifeng Liu and David Hodges. 2000. *China's Semiconductor Industry.* Washington, D.C.: Office of Naval Research.

Poyago-Theotoky, Joanna. 1995 "Equilibrium and Optimal Size of a Research Joint Venture in an Oligopoly with Spillovers." *Journal of Industrial Economics.* 43(2): 209-226.

Press Journal. 1998. *1998 Edition of Nihon Handotai Nenkan.* Tokyo: Press Journal. March 25.

Procassini, Andrew A. 1995. *Competitors in Alliance: Industrial Associations, Global Rivalries, and Business-Government Relations.* Westport, CT: Quorum Books.

Rausch, Lawrence M. 1995. *Asia's New High-Tech Competitors.* NSF 95-309. Arlington VA: National Science Foundation.

Rodgers, T.J. 1995. "A CEO Against Corporate Welfare," Testimony before the House Subcommittee on Government Management, Information, and Technology of the Government Reform and Oversight Committee, September 6, 1995.

Romer, Paul. 1990. "Endogenous Technological Change." *Journal of Political Economy.* 98(5).

Rosenbloom , Richard and Bill Spencer. 1996. *Engines of Innovation: U.S. Industrial Research at the End of an Era.* Boston: Harvard Business School Press.

Sakakibara, M. 1997. "Evaluating Government-sponsored R&D Consortia in Japan: Who Benefits and How?" *Research Policy*. 26(4-5): 447-473.

Sasaki, Hajime. 2000. "Lack of Awareness of LSIs as a Leading Industry, Foundation of the Industry Weakening." *Nikkei Microdevices*. November.

Saxenian, Annalee. 1994. *Regional Advantage: Culture and Competition in Silicon Valley and Route 128*. Cambridge MA: Harvard University Press.

Semiconductor Industry Association. 1999. *Annual Databook*. San Jose, CA: Semiconductor Industry Association.

Semiconductor Industry Association. 1999. *International Technology Roadmap for Semiconductors*. San Jose, CA: Semiconductor Industry Association.

Semiconductor Industry Association. 2000. *International Technology Roadmap for Semiconductors: Executive Summary*. San Jose, CA: Semiconductor Industry Association.

Semiconductor Industry Association. 2001. *Annual Databook: Review of Global and U.S. Semiconductor Competitive Trends, 1978-2000*. San Jose, CA: Semiconductor Industry Association.

Semiconductor Industry Association. 2001. *Turbocharging the U.S. Economy*. San Jose, CA: Semiconductor Industry Association. August.

Sigurdson, J. *Industry and State Partnership in Japan: The Very Large Scale Integrated Circuits (VLSI) Project*. Lund, Sweden: Research Policy Institute.

Simpson, R.D. and N.S. Vonortas. 1994. "Cournot Equilibrium with Imperfectly Appropriable R&D." *Journal of Industrial Economics*. 42(1): 79-92.

SinaNet. 2000. "China Actively Participates in Fight for Overseas Talent." August 14.

Singapore Investment News. 2000. "Singapore's Transition into a Knowledge-Based Economy—Investing in the Future, Today." April 1.

Solid State Technology. 1998. "Upbeat Europe Aims to Upgrade its Chip Technology." December.

Solid State Technology. 1999. "Technology Networking Will Speed European Research." March 1.

Solid State Technology. 2000. "1999 Roadmap: Solutions and Caveats." 43(5): 35-46.

Solid State Technology. 2000. "European Market Roundup: Continued Growth Predicted Through 2001." March 1.

Solid State Technology. 2000. "Fab Trends: Are Europe and Asia Leading the Way?" 2000. March 1.

Solow, Robert. 1987. "We'd Better Watch Out." *New York Times Book Review*. July 12.

Solow, Robert M., Michael Dertouzos, and Richard Lester. 1989. *Made in America*. Cambridge, MA: MIT Press.

South China Morning Post. 2000. "Taiwan Legislator Warns of High Tech Exodus." Online report. July 6. <http://www.nikkeibp.asiabiztech.com/archive/onnet/200007onnet.html>.

Spencer, W.J. and P. Grindley. 1993. "SEMATECH After Five Years: High Technology Consortia and U.S. Competitiveness." *California Management Review*. 35(4): 9-32.

Stokes, Donald. 1997. *Pasteur's Quadrant*. Washington D.C.: Brookings Institution Press.

The Straits Times. 1994. "Government Makes First Major Investment from Cluster Development Fund in Wafer Plant." February 26.

Taipei Times. 2001. "China-Taiwan Princelings' Joint Venture 'Stalled.'" January 13.

Taiwan Central News Agency. 2000. "Taiwan's IT Production Continues Moving Into China." May 11.

Taiwan Central News Agency. 2000. "Taiwan's Electronics Industry Migrates to Mainland China." August 24.

"Taiwan's Semiconductor Industry's Sales Jump in 1999; After 2000 Taiwan to Take the Offensive in Assembly, Testing and Design Fields." 1999. *Semiconductor FDP World*. Industrial Technology and Research Institute. *Annual Report 1999*.

Technecon Analytic Research. Inc. 1999. *1999 Annual Databook: Review of Global and U.S. Semiconductor Competitive Trends 1978-1998*. Technecon Analytic Research, Inc.

Tilton, J.E. 1971. *International Diffusion of Technology: The Case of Semiconductors.* Washington, D.C.: Brookings Institution.

Tokyo Semiconductor FDP World. 2000. "Overview of the Semiconductor Market in China." November.

Tokyo Trigger. 1998. "Toward the Creation of New Industries." December.

La Tribune. 2000. "The Strengths of Research Europe." May 24.

Tsuchiyama, Ray. 2000. "Deconstructing Phone Culture: How Japan Became a Leader in Mobile Internet." *ACCJ Journal.* July.

Tyson, Laura. 1992. *Who's Bashing Whom? Trade Conflict in High Technology Industries.* Washington D.C.: Institute for International Economics.

U.S. General Accounting Office. 2002. *Export Controls: Rapid Advances in China's Semiconductor Industry Underscore Need for Fundamental U.S. Policy Review.* Washington, D.C.: General Accounting Office. GAO-02-620

U.S.-Japan Technology Management Center. 1999. *Technology Standards & Standardization Processes–Their Impact on Innovation and International Competitiveness.* June.

Utterback, James and Albert Murray. 1977. *The Influence of Defense Procurement and Sponsorship of Research and Development on the Development of the Civilian Electronics Industry.* Final Report of U.S. National Bureau of Standards, Experimental Technology Incentives Program. Cambridge, MA: The M.I.T. Center for Policy Alternatives, 1977.

Varmus, Harold. 1996. "The Impact of Physics on Biology and Medicine." Plenary talk, Centennial Meeting of the American Physical Society. Atlanta.

Varmus, Harold. 2000. "Squeeze on Science." *Washington Post.* October 4.

VDI Nachrichten. 2000. "Europe on the Way to World Class Microtechnology and Nanotechnology." October 6.

Vogel, David. 1996. *Kindred Strangers: The Uneasy Relationship Between Politics and Business in America.* Princeton: Princeton University Press.

Westphal, Larry E. 1975. "Planning with Economies of Scale," in Charles R. Blitzer, Peter B. Clark, and Lance Taylor, eds., *Economy-Wide Models and Development Planning.* Oxford: Oxford University Press.

Wolff , Alan Wm., Thomas R. Howell, Brent L. Bartlett, and R. Michael Gadbaw. eds. 1995. *Conflict Among Nations: Trade Policies in the 1990s.* San Francisco: Westview Press.

World Trade Organization. 2001. *Accession of the People's Republic of China* (Decision of November 2001), WT/L/432 (November 23, 2001), Parts I.5, I.7.

Zachary, G. Paschal. 1997. *Endless Frontier: Vannevar Bush, Engineer of the American Century.* New York: The Free Press.